Creeping Environmental Problems
and **Sustainable Development**
in the **Aral Sea Basin**

Environmental degradation in the Aral Sea basin has been a touchstone for increasing public awareness of environmental issues. The Aral crisis has been touted as a 'quiet Chernobyl' and as one of the worst human-made environmental catastrophes of the twentieth century. Just a few decades ago, it was the fourth-largest inland body of water in the world. Today, it has fallen to sixth place . . . and it continues to shrink.

This multidisciplinary book is the first to comprehensively describe the slow onset of low grade but incremental changes (i.e., creeping environmental change) which affected the region. Over a dozen researchers explore every facet of this environmental disaster: changes in landscape, water level and salinity, river flow changes, fish population dynamics, desertification, public health, and political decision-making. The demise of the sea cannot be blamed on natural factors. Its sorry state is clearly the result of decisions made to irrigate the fertile but dry sands of Central Asian deserts for the sake of cotton production. This involved a hidden cost to the inhabitants of the region which far outweighed the benefits derived. In addition to the sharp reduction in the size of the sea and in the quality of its water, environmental degradation has had a drastic negative effect on human health in the region. The book is an attempt to 'set the record straight' on how decision-makers allowed small incremental changes to grow into an environmental and societal nightmare.

This book presents a set of case studies on a region of worldwide environmental interest, and outlines many lessons to be learned for other areas undergoing detrimental creeping environmental change. It therefore provides an important multidisciplinary example of how to approach such environmental disasters for students and researchers of environmental studies, global change, political science and history.

MICHAEL H. GLANTZ is a Senior Scientist in the Environmental and Societal Impacts Group, a program at the US National Center for Atmospheric Research (NCAR). He is interested in how climate affects society and how society affects climate, especially how the interaction between climate anomalies and human activities affect quality of life issues. He is a member of numerous national and international committees and advisory bodies related to environmental issues. In 1987 his *Scientific American* article on drought in Africa was given an award by World Hunger Year. In March 1990 he received the prestigious 'Global 500' award from the United Nations Environment Programme. He has written and edited several books and is the author of numerous articles on issues related to climate, environment, and policy. His most recent books are: *Scientific, Environmental, and Political Issues in the Circum-Caspian Region* (Kluwer Academic Publishers, 1997), *Currents of Change: El Niño's Impact on Climate and Society* (Cambridge University Press, 1996), *Drought Follows the Plow: Cultivating Marginal Areas* (Cambridge University Press, 1994), *Climate Variability, Climate Change and Fisheries* (Cambridge University Press, 1992), *Teleconnections Linking Worldwide Climate Anomalies: Scientific Basis and Societal Impact* (Cambridge University Press, 1991), *Societal Responses to Regional Climate Change: Forecasting by Analogy* (Westview Press, 1988), and *Drought and Hunger in Africa* (Cambridge University Press, 1987 and 1988).

one of the worst human-made environmental catastrophes of the twentieth century

Creeping Environmental Problems and **Sustainable Development** in the **Aral Sea Basin**

Edited by
MICHAEL H. GLANTZ

CAMBRIDGE
UNIVERSITY PRESS

Produced with technical and financial support from the United Nations Environment Programme

CAMBRIDGE UNIVERSITY PRESS
Cambridge, New York, Melbourne, Madrid, Cape Town, Singapore, São Paulo

Cambridge University Press
The Edinburgh Building, Cambridge CB2 8RU, UK

Published in the United States of America by Cambridge University Press, New York

www.cambridge.org
Information on this title: www.cambridge.org/9780521620864

© Cambridge University Press 1999

First published 1999
This digitally printed version 2008

A catalogue record for this publication is available from the British Library

Library of Congress Cataloguing in Publication data

Glantz, Michael H.
    Creeping environmental problems and sustainable development in the
Aral Sea basin / edited by Michael Glantz.
        p. cm.
    Includes index.
    ISBN 0 521 62086 4 (hb)
    1. Environmental degradation – Aral Sea Watershed (Uzbekistan and Kazakhstan)
    2. Sustainable development – Aral Sea Watershed (Uzbekistan and Kazakhstan)
    I. Title.
    GE160.U6G55   1999
    363.7'009587 – dc21   98-24788 CIP

ISBN 978-0-521-62086-4 hardback
ISBN 978-0-521-08138-2 paperback

CONTENTS

CONTRIBUTORS

NIKOLAI V. ALADIN, Zoological Institute, Russian Academy of Sciences, 1, University quay, St. Petersburg 199034, Russia

VITALYI N. BORTNIK, State Oceanograpic Institute, Federal Service of Hydrometeorology and Nature Protection, Kropotkinskyi per 6, Moscow 119838, Russia

LEONID I. ELPINER, Water Problems Institute, Russian Academy of Sciences, 3, Gubkina Str., Moscow 117971, Russia

MICHAEL H. GLANTZ, National Center for Atmospheric Research, PO Box 3000, Boulder, Colorado, USA 80307

ANATOLY N. KRUTOV, The World Bank, Resident Mission in Uzbekistan, 43 Academician Suleimanova Str., 700017 Tashkent, Uzbekistan

NINA M. NOVIKOVA, Water Problems Institute, Russian Academy of Sciences, 3, Gubkina Str., Moscow 117971, Russia

NIKOLAI S. ORLOVSKY, Ben-Gurion University, Desert Institute, Sde Boker, Israel

ASOMITDIN A. RAFIKOV, Geography Department, Presidium of the Academy of Sciences, Republic of Uzbekistan, Gogol St. 70, Tashkent, Uzbekistan

V.V. SUMAROKOVA, State Hydrological Institute, 2nd Line of Vasiljevskyi Ostrov 23, St. Petersburg 119053, Russia

K.V. TSYTSENKO, State Hydrological Institute, 2nd Line of Vasiljevskyi Ostrov 23, St. Petersburg 119053, Russia

ELIZABETH A. VOSTOKOVA, Institute of Ecology and Evolution, Russian Academy of Sciences, Leninsky Prospect 33, Moscow 117071, Russia

ILIYA ZHOLDASOVA, Institute of Ecology and Biology, Karakalpak Department, Uzbek Academy of Sciences, Berdach St. 41, Nukus 742000, Karakalpakstan, Uzbekistan

ALEXANDER N. ZOLOTOKRYLIN, Institute of Geography, Russian Academy of Sciences, Staromonetny 29, Moscow 109107, Russia

IGOR S. ZONN, Russian National Committee for UNEP, Baumanskaya St. 43/1, Moscow 107005, Russia

INTRODUCTION

This book has proven to be a labor of love. It began in 1994 with support from the United Nations Environment Programme (UNEP) Water Unit's director, Walter Rast. The idea was to document the incremental changes that have taken place in the Aral Sea basin in the past several decades. As is now well known, the Aral Sea has dropped in level about 17 meters in the short time span of three-and-a-half decades, and has dropped in volume by two-thirds. The Aral Sea's commercial fishing industry has collapsed. And as a result of chemical fertilizers and pesticides in the runoff from the fields to the rivers and the sea, human health in the region surrounding the Aral coastline (called the Priaralye) has been greatly affected.

The approach taken was to identify researchers who have spent years, if not decades, monitoring some aspects of environmental change in the Aral Sea basin. It therefore involved researchers from a variety of disciplines and countries who dedicated, and continue to dedicate, their professional lives to improving our understanding of environmental changes at the regional level. The environmental aspects presented include the following: landscape changes, changes in sea water quality and quantity, desertification processes, regional climate change, changes in the deltas, human health, political ideo-logical changes related to the environment, streamflow variations, fisheries, and environmental impacts of the Karakum Canal.

The framework suggested as a guideline to these researchers in the prep-aration of their assessments was to enable them to view the changes that they were to write about as creeping environmental problems (or CEP). CEP are long-term, low-grade, incremental but cumulative environmental problems. Each researcher was asked to try to identify with hindsight predetermined thresholds of change. The thresholds included the following: awareness of a change in the environment (not necessarily seen as a problem but only as a change); awareness that the change had become an environmental problem; awareness that the problem had become a crisis; awareness of the need to act to address the CEP; and actions actually taken to address the crisis. Each author recorded the progression of change through the thresholds in his/her own way, as no rigid outline was imposed. The idea was to get the researcher's perceptions of change in the particular location in the Aral basin and with the particular environmental factor on which he or she had focused. Several authors put their findings with regard to thresholds in the form of charts. Others chose to discuss these threshold changes in their text.

ELISABETH VOSTOKOVA discussed Aral basin landscape changes that she had witnessed over a period of more than thirty years. Landscape refers to large areas containing different types of ecosystems and vegetation (plant) communities. At first, her observations were on the ground and, later as satellite imagery became available, she continued her monitoring of changes from space imagery as well.

VITALYI BORTNIK is an authority on the status of the Aral Sea's water quantity and quality. His work has been instrumental in monitoring the changes in the Aral Sea level, surface area, water volume, and salinity levels of the sea. His assessment suggests that as obvious as the human impacts of water diversions from the region's two major rivers may be, there is also an impact of the natural variability of the regional climate on Aral Sea level.

Arid lands are known to be quite fragile and therefore vulnerable to the activities of human settlements. The Aral Sea is sandwiched between two major deserts, the Karakum and the Kyzylkum. As the sea dries up, the newly exposed seabed becomes vulnerable to wind erosion. Plants will not grow on this salt-laden soil. As there is nothing to stop the soils from blowing away, the region becomes a source of salt and dust storms. The water that is diverted from the rivers, the Amudarya and the Syrdarya, is used for irrigation of desert sands, primarily for the production of cotton and, to a lesser extent, rice. As the water runs off from the fields, carrying with it chemical fertilizers and pesticides, it is later reapplied to fields further downstream. The soils become increasingly saline and eventually crop yields and total production drop, and the land has to be abandoned. These are some of the desertification problems discussed by ASOMITDIN RAFIKOV in his chapter.

ALEXANDER ZOLOTOKRYLIN presents data in support of the view that the climate in the region of the Aral Sea basins has changed over the past several decades. While some of those changes are natural in origin (e.g., climate varies on a variety of time scales from months to millennia), other climate changes may have been induced by the shrinkage of the sea. It is generally suggested that the winters have become colder and the summers hotter in the past few decades. In other words, the regional climate has become more continental.

NINA NOVIKOVA has spent much of her professional life working in the delta of the Amudarya. She provides the reader with detailed description of vegetative changes over time in the deltaic area. She discusses the impact of reduced river flow into the delta and the loss of lakes and a degradation in the types of vegetation in the area as a result of increasing desiccation in the delta and its surrounding area.

One of the major concerns of groups around the world is the poor health status of much of the population of the Priaralye. LEONID ELPINER notes that the degradation of health in the region had been registered for some decades, but it was not officially permitted to be discussed or presented to the public. Only with *glasnost* and *perestroika* in the USSR in the mid-1980s were such data allowed to see the light of day, so to speak. Elpiner shows through statistics the poor state of health of inhabitants closest to the sea, compared with those in Uzbekistan or in the former Soviet Union as a whole. He lists numerous diseases and other health problems plaguing people in the Dashowuz part of Turkmenistan, the Kyzyl-Orda region in Kazakstan and in Karakalpakstan, an autonomous political unit in Uzbekistan.

IGOR ZONN traces the political context in which creeping environmental changes in Central Asian states have taken place. He begins with Lenin's plans to transform nature in Central Asia and follow up with Stalin's grandiose schemes to make Central Asia the source of 'white gold' – cotton – for the textile factories and military activities in Russia and for export to foreign markets. This chapter answers some of the questions people often raise when learning of the demise of the Aral Sea: how could such an environmental catastrophe occur in such a short period of time?

K.V. TSYTSENKO and V.V. SUMAROKOVA focused their research on the two major rivers feeding the Aral Sea, the Amudarya and the Syrdarya. They discuss interannual variability in river flow, as well as interdecadal changes and what those variations have meant for the condition of the sea. They discussed changes in the quantity and quality of river water, as these rivers were recipients of return flow and contaminated water runoff from the fields. The rivers are the lifeline of the sea, and they are the lifeline of the irrigated activities along their courses.

One of the first and most visible physical and socioeconomic impacts of the contamination of sea water was on the sea's fish population and its commercial fishery. ILIYA ZHOLDASOVA has studied fish populations in the Aral Sea and its deltas for several decades. She has observed considerable change in both fish spawning habitats and in the fish populations themselves. She provides fairly detailed accounts of the fate of Aral fish populations that were endemic to the sea, as well as those that have been introduced. Most popular articles on the Aral region note that the fishery had failed by the late 1970s, and that fish had to be imported from the Pacific Ocean and the Baltic Sea for processing in Muynak (Karakalpakstan) factories in order to provide employment to a large part of the local population (on the order of tens of thousands of fish industry workers).

NIKOLAI ORLOVSKY, former Deputy Director of the Institute of Deserts of the Turkmen Academy of Sciences, reports on the environmental impacts of the Karakum Canal. This constructed canal is the longest in the world, registering a length of about 1400 km. It draws a considerable portion of water from the Amudarya. The canal passes by several oases in Turkmenistan (around which major population centers have developed). Aside from the adverse environmental impacts associated with this unlined canal cut out of barren desert sands, the Karakum Canal is an apparent irritant to other Central Asian Republics, as it deprives the Aral Sea of about 15 km³ each year; it deprives the Uzbek Republic from using that volume of water further downstream for watering its own fertile but dry desert sands; and it takes the water out of the Aral basin and puts it into the Caspian basin.

ANATOLY KRUTOV supplies an overview to the environmental problems in one of the key Central Asia Republics implicated in the Aral crisis: Uzbekistan. He discusses the plethora of environmental problems, as well as the numerous legislative attempts to address those problems, mostly failed attempts. He notes that the recent increase in flow in the rivers and into the sea may be only a temporary respite from the environmental catastrophe that awaits the sea, in the absence of effective governmental responses to the identified creeping environmental problems.

The final chapter was prepared by NIKOLAI ALADIN. Aladin is well known in the former Soviet Union for his repeated field trips to the shores of the Small Aral Sea. He studied changes in fish populations, among other aquatic organisms in the Aral Sea and in the Small Aral for almost two decades. His studies have been labor-intensive and represent a considerable monitoring effort. He notes that the sea's characteristics have varied throughout time, with evidence that the sea level had been much lower and the sea had even disappeared, only to return. He suggests that the sea has in fact been influenced by human activities for a few thousand years, that the recent level of human impacts is much greater and, therefore, much more damaging to aquatic ecosystems.

This set of studies is intended to provide a baseline assessment of some of the creeping environmental problems in the Aral Sea basin. When first proposed to some potential funding sources, the editor was advised that there was little interest in how the Aral Sea environmental crisis had developed and that the current interest was in preparing the Central Asian Republics for the future and in 'saving the sea'. But UNEP supported the view that it was important to attempt to reconstruct the history of how the Aral crisis developed over time, in the hope that lessons could be learned on how to proceed into the future.

Environmental groups around the world have developed a strong interest in the Aral Sea, following the exposure to the world of the state of the sea's degradation in the mid-1980s. A considerable amount of lip service had been paid to 'saving the sea' in the early 1990s. UNEP produced a diagnostic study of the problems of the Aral Sea, which served to spark renewed interest in the region. The World Bank then reluctantly got involved in the Aral region, drawing up numerous plans for multilateral cooperation to save the sea and to develop the economies of the Central Asian Republics.

As noted in some of the chapters in this volume, there have been some positive changes in the region, in terms of agricultural activities and water use. More water has been getting to the sea (a series of wet years in the early 1990s), and there was a reduced use of chemicals on the land because of the high cost of these agricultural inputs. However, there is some evidence that 'saving the sea' *per se* has been given a much lower priority than was the case in the early 1990s. While governments talk about it, it appears that little can or will be done about it by policy-makers in the region. However, one must wonder if the interest of the global community in the plight of Central Asian Republics would remain high if the sea were allowed to disappear. The sea may be more important as a symbol of human misuse of the environment and as a symbol of how much damage humans can do in a short period of time in the absence of concern for the state of the environment. Saving the sea would not be just a symbolic act, however, but it may prove to be an action that serves to sustain interest in and support for the economic development fate of the Central Asian Republics.

Note on Russian names

It is important to note that an attempt was made to achieve consistency in the transliteration of Russian terms and location names. However, this proved to be an almost impossible task. Compounding the problem of transliteration is the fact that the spelling of locations in Central Asia has changed since the breakup of the Soviet Union in December 1991, as each of the newly independent republics sought to nationalize their country's names. It is also important to alert the reader to the differences among references at the end of each chapter. They are not necessarily filled with the same level of completeness of reference information. This is partly the result of different styles of reference between the United States, the Soviet Union and the republics of the former Soviet Union. Nevertheless, the information provided in the references will enable the reader to locate the source of that information. I hope that this does not detract from the importance of the information provided by the contributors in their chapters.

Acknowledgments

This book has involved the dedicated work of several people whom I would like to acknowledge with my sincere appreciation. First and foremost, I must thank D. JAN STEWART for her tireless effort in producing numerous drafts of this manuscript. Sincere thanks also go to JAN HOPPER, who worked diligently in making the first effort to input the entire manuscript into the computer. JUSTIN KITSUTAKA provided excellent graphic support for most of the figures in the book.

Scientific coordination and support was supplied by IGOR ZONN. It goes without saying that the manuscript and the logistics of organizing the contributors and the translation of their papers, as well as the endless queries to the authors in the former Soviet Union could not have been done without the friendship and dedication to this project of IGOR ZONN. NINA NOVIKOVA was instrumental in identifying and seeking answers to problems generated by translation from Russian to English. We met on several occasions in Moscow (trips of opportunity) to iron out technical problems, including those introduced through translation and differences in the way scientific concepts are defined in different cultures.

I would also like to thank the contributors for their continued interest in this project and in their desire to assist in identifying thresholds of change for their specific creeping environmental change and problem. They have been and continue to be dedicated researchers and dedicated practitioners who hold onto the possibility that a concerned effort could 'save the Aral Sea' and the ecosystems and populations dependent upon it. They have more than a century of combined experience in the Aral region. This book provides them with a chance yet again to share their knowledge and expertise with the broad community of people interested in the future of the Aral Sea and its inhabitants.

Finally, not the least important is the support (moral and financial) that the Water Unit of the United Nations Environment Programme (Nairobi, Kenya) provided for the initiation of this project. Their moral support was a crucial factor in seeing this manuscript through to completion. GERHART SCHNEIDER, TAKAHIRO NAKAMURA, and WALTER RAST of the Water Unit provided a useful critique of a draft of the manuscript.

MICHAEL H. GLANTZ
Boulder, Colorado
Summer 1998

1 Sustainable development and creeping environmental problems in the Aral Sea region

MICHAEL H. GLANTZ

The Aral Sea region (Figure 1.1) has been characterized in the popular press and in the scientific literature as a region deep in crisis: an environmental crisis, a health crisis, a development crisis, and most of all a water crisis. Clearly, the rapid shrinking of the Aral Sea in Central Asia has captured the attention, and to some extent the interest, of governments, environment and development organizations, the public, and the media around the globe. Once considered a quiet catastrophe, one that has evolved slowly, almost imperceptibly, over the past few decades, the demise of the Aral Sea is now acknowledged as one of the major human-induced environmental disasters of the twentieth century. In the late 1980s, the Soviet Union issued a set of disaster stamps, one of which related to the demise of the Aral Sea (Figure 1.2).

The blame for this situation has been put on such factors as the domination of the region by Soviet authorities who ruled from Moscow, over-dependence on the cultivation of cotton, the rapid expansion of irrigated agriculture, totalitarian regimes, a controlled news media, inappropriate use of cost-benefit analyses, and the Cold War.

Figure 1.1 The Aral Sea region.

Figure 1.2 Russian postage stamp depicting the Aral Sea. Note ship trapped by receding sea level.

Those harmed by the crisis include, but are not limited to, the following: human populations (especially women and children) in the regions adjacent to the sea and in the lower reaches of the Aral basin's two major rivers (the Amudarya and the Syrdarya), regional vegetation and animals, fish and other living organisms in the aquatic environment, soil quality, air quality, ground and surface water quality, environmental sustainability and societal resilience, and some Central Asian administrative districts.

We now know about most of the environment-related problems in the Aral Sea region and we are now learning through anecdotes that various people in the former Soviet Union (and likely in other countries as well) have known about them for a very long time, almost from their inception (e.g., Goldman, 1972). In fact, signs of change were appearing everywhere throughout the first twenty years of the Aral Sea problem (1960–80): wind erosion, salt-laden dust storms, destruction of vital fish spawning grounds and the subsequent collapse of fisheries, increased salinity of sea water, waterlogging and secondary salinization of soils, disruption of navigation, the division of the sea into separate parts as a result of sea level decline, the need for extra-basin water resources to stabilize the sea level, the loss of wildlife in the littoral areas, the large reduction of streamflow from the region's two major rivers, a dramatic change in regional climate, the disappearance of pasturelands, and so forth. In fact, there were several scientists in the Soviet Union and outside of it who made projections about the fate of the sea and the territory surrounding it. For example, Davis (1956) noted:

> Some of the inland seas and lakes have recently been the scene of extensive human activity which has had notable effects upon coastlines ... Among these are the changes in the offshore areas and coasts of the Caspian and Aral seas owing to large-scale development of dams for power and irrigation on the rivers supplying water to these seas. [An] extensive lowering of water level is beginning in the Aral Sea basin with the development of irrigation projects on the Amu Darya and Syr Darya, which supply most of the water to this sea. It is the aim of these projects eventually to divert for irrigation most or all of the waters of the rivers from entering the sea. It has been calculated that within twenty-five years the water area of this sea will shrink to half the size that it was in 1940, when the irrigation projects began. This would bring about an increase of nearly 13 000 square miles of land area. (DAVIS, 1956, p. 517)

Clearly, a considerable amount of information already exists in disparate sources about the Aral basin and the various physical processes of environmental change and environmental degradation.

But political leaders, among others with decision-making power, have not acted on many of these changes in the past. Why? Is it that there have been no financial resources available to do so? Is it that there has been no desire on the part of national, regional, or local leaders to do so? Has it been because there is no *perceived* reason among policy-makers at any level to take immediate action (e.g., did they happen to believe that the sea was not worth saving because its waters could be used more cost effectively elsewhere? Were they led to believe that water would likely be diverted from Siberian rivers to the arid lands of Central Asia)? In fact, at least as early as 1927, Soviet scientists exposed the ultimate fate of the Aral Sea if water diversions from the Amudarya and Syrdarya were not limited in the future. Tsinzerling (1927) constructed scenarios of impacts based on increased amounts of water diversions from these rivers. His scenarios were mimicked in the region by the decades of events that followed.

I would argue that a major part of the environmental and health problem in the Aral Sea basin relates to the nature of these adverse environmental changes and to the nature of human society, especially in the way people look at slow-onset, low-grade, long-term and cumulative environmental changes (e.g., creeping environmental problems or CEPs).

---

creeping \krē-piŋ*adj* : developing or advancing by slow imperceptible degrees < a period of ~ inflation > –

from Webster's Ninth New Collegiate Dictionary, 1991.

A major feature shared by various CEPs is that a change in this type of environmental problem is not much worse today than it was yesterday; nor is the rate or degree of change tomorrow likely to be much different than it is today.

So, for the most part, societies (individuals as well as government bureaucrats) frequently do not recognize changes that would prompt them to treat their environments any differently than they had on previous days. Yet, incremental changes in environmental conditions often accumulate over time with the eventual result that, after some perceived threshold of change has been crossed, those previously imperceptible increments of change 'suddenly' appear as serious crises. If no action is taken, as is often the case, those incremental changes will likely continue to build until they emerge as full-blown disaster(s). In the Aral Sea region, the traditional indicators of these crises relate primarily to the declining levels of the sea; they include changes in water quantity and quality, water diversions, water use, and water-related diseases.

It is important to recognize many of the environmental changes in the Aral Sea region as CEPs with likely adverse consequences at some time 'down the road'. It is also important to realize that, although technologies might exist somewhere in the world 'to save us' from the worst consequences of local or regional environmental changes, governments affected by the CEPs might not be able to afford them. Therefore, ways must be devised to deal more effectively with CEPs than we apparently do at present. We must learn to deal better either with their underlying causes, their consequences, or their characteristics (such as rates of change).

Introduction to the notion of creeping environmental problems

Just about anywhere one lives, people are constantly bombarded with bad news about the environment. Some of that news is about environmental problems of a global nature (e.g., global warming, ozone depletion) and some of it is about problems at the local level. Some of these problems have long lead times before their adverse consequences become apparent, while for others adverse consequences can develop over relatively shorter time frames (e.g., tropical deforestation). The list of environment-related problems around the globe is quite long and, unfortunately, is still growing: air pollution, acid rain, global warming, ozone depletion, deforestation, desertification, droughts, famines, water quality, and the accumulation of nuclear, toxic, and solid waste. Each is the result of long-term, low-grade, and slow-onset cumulative processes. Each is a creeping environmental problem.[1]

---

1. In a letter critical of the US National Research Council report *Confronting Natural Disasters* (NRC, 1987), the writer (Smith, 1988) noted that the report chose to focus solely on a particular set of 'natural hazards' that happen to be initiated by events that are 'sudden and of short duration'. To do so excludes other hazards that cause orders of magnitude more human damage. The report identifies one class of these other hazards: long-term problems such as desertification, deforestation, and drought. It goes on, however, to reject them because 'mitigating these hazards requires a greater ecological or social emphasis, and civil engineering approaches are less critical.'

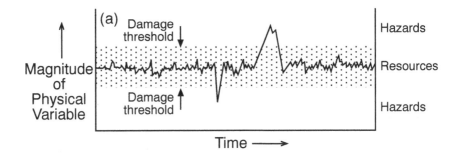

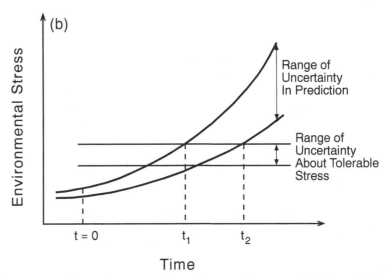

Figure 1.3 (a) Schematic of a rapid-onset natural hazard (Burton and Hewett, 1974). (b) Schematic of slow-onset (creeping) environmental problems (Döös, 1994).

Decision-makers worldwide have had considerable difficulty in addressing ways to slow down, arrest, or reverse these gradually occurring adverse changes. While societies respond relatively quickly to step-like adverse changes in the environment or to problems perceived by experts or the public as crises – for instance 'rapid-onset hazards', such as earthquakes and flash floods (Palm, 1990) – they have much more difficulty in developing an awareness of the risks associated with slow-onset, long-term, low-grade, cumulative change (Figure 1.3).

Thresholds

For each of the creeping environmental changes there may be identifiable thresholds beyond which continued degradation of the environment will increase the likelihood of major, even irreversible, changes in the environment. While our concern should be focused on *thresholds* of environmental change, thresholds are usually easier to talk about than to detect.

For CEPs such as desertification and water quality degradation, at first changes may be noted by individuals at the local level, but may not be considered an immediate or even a potential threat. Such changes in their earliest stages will likely go unreported to local or regional authorities or to national researchers.

Once a creeping environmental change is perceived to have intensified in time, space, or impact, it may be brought to the attention of authorities by local inhabitants, officials, or by environmental researchers who happen to be working in that particular locale. A further deepening or broadening of the adverse consequences associated with environmental change could generate concern at the national policy-making level. At this point the international media can also get involved, generating international awareness of the local or regional problem. Who it is that might be the first to generate awareness of a creeping environmental change and of subsequent thresholds of awareness can vary from one region to the next and from one type of creeping environmental problem to another: it could be a farmer, an hydrologist, a scientist, a policy-maker, or a news reporter.

Because these full-blown problems derive from slow-onset, low-grade, long-term and cumulative environmental changes, it is not easy to identify universally accepted, objective, quantitative indicators for thresholds. Nevertheless, several generic thresholds could be subjectively identified for the evolution of CEPs: a first threshold relates to awareness of a change in the environment that has not yet been perceived as a problem; a second threshold could relate to the awareness that a previously undetected environmental change has become a problem; a third threshold relates to the realization that the problem has reached a crisis stage; a fourth threshold relates to the realization that there is a need to take action to cope with the problem; a fifth threshold is one beyond which direct and specific actions (not just the convening of conferences or workshops) are taken to resolve the CEP.

Why do CEPs continue?

Creeping environmental problems change the environment in a negative, cumulative and, at least for some period of time, an invisible way. As a result of these minor insults to the environment over time, during which no obvious step-like changes occur, both governments and individuals tend to assume 'business as usual' attitudes. People fear change (e.g., Hoffer, 1952) and, unless a crisis situation is perceived, they are not likely to change their behavior in the absence of any incentive to do so.

Most environmental changes are surrounded by scientific uncertainties. For example, are they primarily natural or human-induced changes? Lack of scientific certainty is often cited as another reason for political inaction on

CEPs. Yet, policy-makers are constantly forced to make policy decisions in the midst of uncertainty. For most CEPs, there is often a minority voice, often quite loud, which insists on highlighting the scientific uncertainties, as opposed to emphasizing what is known. Such conflicting interpretations of the science among factions within the scientific community tend to weaken the resolve of those who are expected to act (the public, policy-makers, the media). Thus, the selective use of information on creeping environmental issues drawn from the scientific literature allows policy-makers to pursue any decisions they wish, regardless of the true validity of the scientific information used. Whenever scientific uncertainty is perceived to have been used as an excuse for avoiding political risks associated with decision-making, it should be explicitly challenged as simply an excuse (a tactical measure) to delay meaningful action. Scientific uncertainties will always surround CEPs, and decision-makers must learn to cope with them.

Another reason why CEPs continue is that many changes to the environment are not considered detrimental in their early stages. Such changes would likely be viewed as environmental transformation, not degradation. For example, the cutting down of a small part of a mangrove forest to create a shrimp pond would not necessarily signal a stage in the destruction of a mangrove forest ecosystem (transformation). If, however, numerous ponds were to be constructed in the same location, then the mangrove forest ecosystem and its interactions with other ecosystems would eventually cease.

The willingness of some people to take slightly higher risks also explains inaction on CEPs. Considerable discussion exists in the scientific literature and the popular media about people who are risk-takers and about those who are risk-averse. The former are gamblers, while the latter tend to be more conservative in their approaches (and responses) to environmental change. Yet another risk-related category is that of the risk-maker.

Risk-makers are those decision-makers whose decisions make risks for others, but not necessarily for themselves. For example, reluctance to take action either to slow down or stop desertification processes threatening a village situated far from the capital city where the politicians live will likely have little, if any, direct or immediate adverse political fallout on decision-makers at the national level. Their inaction generates increased risks for the inhabitants of the threatened village, but not necessarily for themselves. With regard to the declining level of the Aral Sea, in reality there were no direct adverse impacts on those policy-makers in the Kremlin, or even in Tashkent, who made decisions about agricultural development in Central Asia, decisions that ultimately led to the degradation of the Aral Sea environment. This can be viewed as a variation of the NIMBY syndrome related to environmental pollution (i.e., 'you can pollute anywhere you want, but not in my back yard'; hence, Not In My Back Yard). Often, environmental change is of little concern unless it directly affects someone's home or workplace.

Yet another constraint on timely action to address a CEP involves the fact that what appears to be an environmental crisis to one person may be considered an opportunity by someone else. While some people may be concerned about environmental degradation, others might believe such degradation is a necessary – and acceptable – tradeoff for improving regional economic development prospects.

Creeping environmental problems in the Aral Sea basin

In the late 1950s, the Aral Sea was the fourth largest inland body of water on the planet, with a surface area of 66 000 km². In 1960 the mean level of the Aral Sea was measured at 53.4 m, and it contained about 1090 km³ of water.

The perennial flows of the basin's two major river systems, the Amudarya and Syrdarya, had until recently sustained a stable Aral Sea level. Over the centuries, about half of the flow of the two rivers reached the Aral. A flourishing fishing industry existed, based on the exploitation of around 20 commercially valuable species. The forests and wetlands surrounding the sea, especially in the Syrdarya and Amudarya deltas, were biologically productive, containing unique species of flora and fauna that had adapted to the natural saline characteristics of the sea. Historically, the levels of the Aral Sea were rather stable, fluctuating less than a meter in the first half of the twentieth century, and by no more than four meters during the preceding 200-year period.

In the span of just four decades, the Aral Sea basin was transformed into a major world-class ecological and socio-economic disaster (Micklin and Williams, 1996). Since the beginning of the 1960s, when the leaders of the Soviet Union embarked on a program to increase river diversions in order to expand irrigated cotton production in this arid region, the Aral Sea level dropped continually and dramatically. In fact, the annual average rates of sea level decline had actually accelerated: from 0.21 meters/year in the 1960s, to 0.6 m/yr in the 1970s, and reaching 0.8 m/yr in the early 1980s (1981–86) (Mnatsakanian, 1992). In all, the sea's level has declined by about 17 meters, and its surface area has been reduced by half. Today it has fallen to sixth place, with respect to its size, as an inland body of water. The initial and primary focus of attention has been on the declining level of the sea, in part because that change has been highly visible (especially from space). However, it is but one of several creeping changes in the Aral basin to have occurred during the past half-century.

Other creeping environmental problems in the basin include reduced inflow to the sea from the Amudarya and Syrdarya, monocropping of cotton and of rice, declining water quality, salt and dust storms, salinization of water and soils, vegetation changes, and escalating health effects. Because of the low-grade nature of these and other problems, high-level policy-makers, as well as decision-makers at other levels, have apparently had difficulties in identifying them as problems and then, once identified as such, in coping with them. As with CEPs elsewhere, it has been difficult to identify in advance thresholds of environmental change in the Aral basin that could serve to catalyze action to arrest environmental degradation. The following list of examples of CEPs in the Aral basin is meant to be suggestive and not exhaustive.

EXPANSION OF COTTON ACREAGE

The desire of Soviet leaders to expand cotton production onto desert lands increased the dependence of Central Asian Republics on irrigation and monocropping. Monocropping has adverse impacts on soil conditions, which prompts increasing dependence on mechanization, pesticides, herbicides, and fertilizers. Socio-economically, these policies are also risky in the sense that a regional economy based on production of a single agricultural crop is highly vulnerable to the variability of climate from year to year and from decade to decade, as well as to the 'whims' of demands, and therefore price, of the marketplace. The chart in Figure 1.4 depicts agricultural water use in the Amudarya and Syrdarya basins as of the late 1980s.

A sizeable portion of Central Asia's agricultural production is dependent on irrigation. Irrigated agriculture in the region predates by millennia the era of Tsarist conquests in the eighteenth and nineteenth centuries. What is 'new' about irrigation, however, is the huge amount of water diverted from the region's two major rivers, the Amudarya and the Syrdarya. Table 1.1 shows the expansion of cotton acreage in Central Asia between 1940 and 1986. The demands of cotton production for irrigation water are high (Table 1.2). Each year increasing amounts of water had been required to irrigate new fields and

Figure 1.4 Agricultural water use in the Amudarya and Syrdarya basins (Tsutsui, 1991).

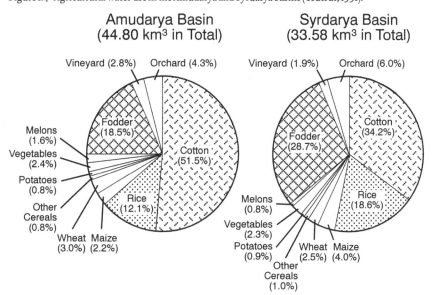

Table 1.1 Cotton sowings (× million hectares)

Unit	1940	1971–75[a]	1976–80[a]	1981–85[a]	1985	1986	Increase 1940–86 (%)
Uzbekistan	0.924	1.718	1.823	1.932	1.993	2.053	122
Tajikistan	0.106	0.264	0.295	0.308	0.312	0.314	196
Turkmenistan	0.151	0.438	0.504	0.534	0.560	0.650	330

Note: [a]Average per year for this period.

Source: Critchlow (1991).

Table 1.2 Land under irrigation (× 1000 hectares)

Country	1950	1960	1965	1970	1975	1980	1985	1986
Uzbekistan	2276	2570	2639	2750	2995	3527	3908	4171
Tajikistan	361	427	442	524	566	627	660	703
Turkmenistan	454	496	509	670	855	960	1160	1350

Source: Zonn (this volume).

for the flushing of salts from the old ones. In addition, starting in 1954 with the construction of the Karakum Canal in Turkmenistan, relatively large amounts of water had been diverted each year from the Amudarya to irrigate lands in that republic. The current withdrawals for the Karakum Canal are estimated to be about 15–20 km³ per year (or 23–30% of the Amudarya's total annual flow).

SEA LEVEL DECLINE

The decline in the level of the Aral Sea has received considerable political attention, both domestically and internationally. It became a highly visible problem in the mid-1980s. Increasing water diversions from the two main regional rivers robbed the sea and deltas of their annual fresh-water replenishment. The rate of decline of the sea can be seen in Figure 1.5. Note also that declining levels were accompanied by an even more rapid reduction in the volume of the sea and by an increase in sea-water salinity.

Another problem related to sea level decline and reduced sea surface area has been the increase in the number, frequency, and impacts of dust storms. In the mid-1970s, dust storms captured the attention of Soviet policy-makers when cosmonauts, during one of their space missions, identified major dust storms raging over the exposed seabed in the receding southeastern part of the Aral Sea. The exposed seabed enabled winds to pick up dust laden with a variety of chemicals and carry it hundreds of kilometers from the original site. Farms downwind of the storms were covered with these dry depositions,

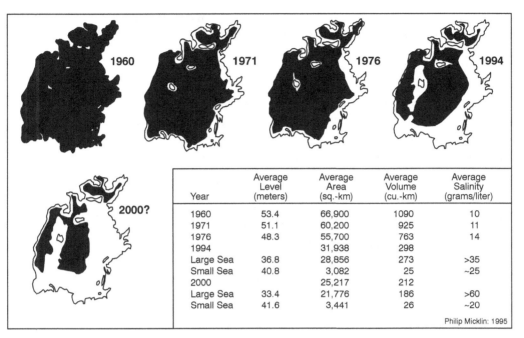

Year	Average Level (meters)	Average Area (sq.-km)	Average Volume (cu.-km)	Average Salinity (grams/liter)
1960	53.4	66,900	1090	10
1971	51.1	60,200	925	11
1976	48.3	55,700	763	14
1994		31,938	298	
Large Sea	36.8	28,856	273	>35
Small Sea	40.8	3,082	25	~25
2000		25,217	212	
Large Sea	33.4	21,776	186	>60
Small Sea	41.6	3,441	26	~20

Philip Micklin: 1995

Figure 1.5 Levels of the Aral Sea (Micklin and Williams, 1996).

prompting farmers to claim that the productivity of their land, as well as their health, were being adversely affected. Since then, the number and intensity of these dust storms along the exposed seabed had increased. The appearance of these storms exposed Soviet leaders, and belatedly the rest of the world, to yet another consequence of diverting water from Central Asia's two main rivers.

DECREASING FLOWS OF THE AMUDARYA AND SYRDARYA INTO THE SEA

Historically, the Amudarya had supplied about 70% of the water to the Aral Sea, more than twice the flow of the second major regional river, the Syrdarya. From the early 1960s, the decline in Syrdarya flow was noticed and, by the late 1970s, no water from this river reached the sea. As for the Amudarya, a sizeable amount of its water has been diverted into the Karakum Canal. In the mid-1990s, the last extension of the Karakum Canal into south-western Turkmenistan was completed, an event that will likely translate into additional diversions of water from the overdrawn Amudarya. There were a few years in the late 1980s when virtually no water from the Amudarya reached the sea. In the early 1990s, however, after several years of favorable snowpack in the Pamirs, water reached the sea and its deltas once again.

DECLINING WATER QUALITY IN THE RIVERS AND IN THE SEA

As fields in Central Asia were continually irrigated on a large scale, soil fertility declined rapidly. This prompted farmers to use increasing amounts of fertilizers, herbicides, and pesticides in order to maintain, and even expand, cotton production. Many of these chemicals found their way, through return flow, to the rivers, as well as into the region's groundwater. In order to avoid, or rather delay, the continuing salinization of soils, increasing amounts of water had to be used to flush the land in an attempt to make it free of salts and other compounds. Much of this drainage water ended up in the region's rivers and, eventually, the sea itself. Drainage canals were constructed to divert some of the contaminated water away from the sea, and much of it ended up in a regional desert depression, known as Lake Sarakamysh.

DEGRADATION OF DELTA ECOSYSTEMS

Another example of the ecological consequences of reduced streamflow into the sea has been the degradation of the highly productive Amudarya and Syrdarya deltas, a problem which has become increasingly pronounced during the past thirty years (Novikova, this volume; Smith, 1994). One of the consequences of the desiccation of the deltas has been a pronounced reduction in vegetative cover, a loss that destroyed habitats for wildlife and migratory birds. Wildlife had, to a large extent, disappeared from the delta regions. Worse yet, forest ecosystems, such as the unique tugai forest, have been devastated as the soils dried out or became salinized or waterlogged, depending on local soil conditions. Frederick (1991) highlighted the economic importance of the deltas in the recent past, noting that they provided a 'feeding base for livestock, a source of reeds for industry, spawning grounds for fish, and sites of commercial hunting and trapping.' Each of these delta-related ecological and societal processes has either been sharply curtailed or terminated altogether.

Today, Uzbek and Kazak leaders, supported by recent recommendations from World Bank consultants and staff, propose to rejuvenate deltaic ecosystems, apparently abandoning some of the more ambitious schemes designed to save the entire Aral Sea.

DECLINE OF FISH POPULATIONS IN THE ARAL SEA

Along with a decline in the quality of river water came a decline in the quality of Aral Sea water. At a 1977 Soviet conference on the environmental impact of a drop in the level of the Aral Sea, a paper prepared by two Uzbek scientists reported that there had been a sharp reduction in fish landings (Gorodetskaya and Kes, 1978). As a result of the desiccation of the sea's fish spawning grounds, they suggested that the demise of the region's commercial fisheries was imminent. Borovsky (1980) also suggested that the demise of the Aral Sea's fisheries would be one of the first consequences of declining sea levels. Reteyum (1991) wrote that 'in 1965, the Council of Ministers of the USSR passed a special resolution, On Measures to Preserve the Fishery Importance of the Aral Sea'. He cited this as one of the examples in support of

Table 1.3 Decline in fish catches over time

Year	Metric tons of fish
1960	43 430
1965	31 040
1970	17 460
1975	2940
1980	0
1985	0
1990	0

Source: From Létolle and Mainguet (1993), p. 182.

his belief that signs of deterioration in the Aral Basin were seen as early as the mid-1960s.

Table 1.3 shows the sharp decline over time in fish catches. This decline provided a visible threshold for decision-makers to see that their inaction with regard to declining sea level and water quality had adverse biological consequences. By the late 1970s, it was quite clear that the Aral Sea's fisheries were in irreversible decline. A once-thriving fishing industry was being slowly destroyed by increasing amounts of pollutants entering the sea from its two feeder rivers. In addition, the salinity of Aral Sea water increased to such an extent that several areas had salinity levels equivalent to that of the open ocean.

References to Aral Sea fishery problems were registered by Goldman (1972) in his book on environmental pollution in the Soviet Union.

> Although the quality of the fish native to the Aral Sea was not as high as it was in the Caspian, the impact of polluting and shrinking the Aral has been even greater. From a typical haul of 40 000 metric tons in 1962, the catch dropped to 20 000 metric tons in 1967 (*Soviet Geography*, 1969). Apparently by 1970 it had fallen to 16 000–18 000 metric tons (*Sotsialisticheskaia industriia*, 1970). And as the salt content of the sea rises, the expectation is that the remaining fish in the Aral will rapidly be annihilated. (*Kazakhstanskaia pravda*, 1969)

Today, no fish are caught commercially in the sea; the former sea ports of Muynak (to the south) and Aralsk (to the north) are now stranded several tens of kilometers from the receding shoreline. Up to the early 1990s, fish had been shipped in from distant locations (the Arctic, the Baltic, the Pacific) for processing in Muynak's fish cannery. But this expensive option came to an abrupt halt in 1994. The loss of the Aral Sea's fisheries sparked the collapse of the entire industry, causing unemployment and the decline of the economies of former coastal towns such as Muynak and Aralsk.

INCREASES IN HUMAN DISEASES

The dependence of several Central Asia Republics on cotton production has not only adversely affected the physical environment, by upsetting fragile ecological balances in many parts of the Aral Basin, but it has also had a major impact on human health (see Elpiner, this volume). Documented widespread regional health effects have only recently been reported to the public: high infant and maternal mortality and morbidity rates, a sharp increase in esophageal cancers directly attributable to 'poisoned' water resources, gastrointestinal problems, typhoid, high rates of congenital deformities, outbreaks of viral hepatitis, the contamination of mothers' milk, and life expectancies in some areas about 20 years less than for the Commonwealth of Independent States (CIS) in general. Groundwater supplies, too, have been contaminated as a result of the widespread and wanton use of chemicals on irrigated cotton fields. By all statistical measures, the region's human health profile fares poorly in comparison to the rest of the CIS (Feshbach and Friendly, 1992; Ellis, 1990). Adverse impacts of all-out cotton production on health have been compounded by the relative paucity of medical and health facilities in the Aral Basin. In addition, water treatment facilities in the region are wholly inadequate (and in many areas nonexistent), necessitating the use of untreated surface waters from rivers, irrigation canals, and drainage ditches for municipal purposes.

Systematic research on public health conditions in the Aral Sea basin began in the mid-1970s. From that time, the negative dynamics of deteriorating public health conditions in the region were observed. Had such research been undertaken and its results exposed earlier, adverse public health conditions would have been identified by the end of the 1960s, and would probably have been linked to the presence of pesticides (Elpiner, 1990). In addition, Kuznetsov (1992) noted that 'unfortunately, secrecy over an entire series of research results in the 1970s, especially medical-epidemiological data, precluded their publication at that time and the predictions associated with them did not become available to the public in time.'

One situation deserves special mention, namely, the deteriorating health conditions of the Turkic-speaking people living in the semi-autonomous republic of Karakalpakstan, situated in northwest Uzbekistan, along the southern shore of the Aral Sea. More than one million people have been affected:

> There is a shortage of clean water, and there is not enough even for drinking. In several parts of the region the consumption of water per person per day is about 5 liters, compared to an average of 200 to 300 liters. The mineralization (salt content) of this water stands at 2 to 4 grams per liter, and the bacteria content exceeds the maximum permissible concentration by 5 to 10 times. Through the dispensary system the Ministry of Health discovered a truly tragic picture: 60 percent of those examined – children and adults – have serious health problems; 80 percent of pregnant women suffer from anemia; intestinal infections are widespread; the infant mortality rate is much higher than national average figures and in several regions reaches 82 deaths for every 1000 live births. Diseases never before seen here are appearing, for example gallstones and kidney stones.   (Rudenko, 1989, p. 44)

Another human health tragedy that deserves mention is in Dashowuz (or Dashkhovuz) in Turkmenistan. A report on *The Health of Mothers and Children in the Aral Sea Region of Turkmenistan* (Radzinsky, 1994) noted the impact on health of the contamination of the Dashkhovuz water supply, attributing much of the problem to the agricultural sector. The report stated that:

> The middle and lower reaches of the Amu Darya run through
> Turkmenistan . . . the quality of the Amu Darya water is a cause for concern,
> because its mineralization and chemical pollution are increasing uncontrolla-
> bly . . . The increasing deterioration of drinking water quality and its contami-
> nation with toxic chemicals and pathogenic microorganisms are closely
> linked to the extensive use of large quantities of mineral fertilizers, pesti-
> cides, and defoliants in the cotton-growing regions of Turkmenistan, espe-
> cially in the fields of the Dashowuz Oasis, which is the largest cotton-growing
> region in the country . . . The application of such large amounts of mineral fer-
> tilizers, pesticides, and defoliants for so many years could not help but affect
> the environment of the Aral region and has naturally had a detrimental
> impact on the health of the local population.   (Radzinsky, 1994, pp. 5–6)

In the absence of any major improvement in regional health care or in detoxifying water and land resources in the Aral basin, the only way out for regional inhabitants, other than perpetuating the status quo, has been emigration. However, despite Soviet plans to encourage those most directly affected (e.g., the people of Karakalpakstan) to migrate to areas outside Central Asia, few have opted to leave their homelands. Thus, with few meaningful actions to improve the health of the people or the environment in the Aral basin, the total sum of misery can only increase, particularly since the region boasts an extremely high population growth rate, ranging from 2.6% to 3.2% per annum. If current growth rates continue, the population of the five Central Asian Republics could double to 60 million by the early decades of the twenty-first century.

A most recent attempt to address the health problems in the Aral Sea basin's disaster zone was an assessment of the well-known nongovernmental organization, *Medicins Sans Frontières* (MSF, Doctors Without Borders). In mid-1997, it sent an exploratory team to Karakalpakstan (in Uzbekistan) and Dashkhovuz (northern Turkmenistan) to assess the severity of the needs of the local inhabitants for the purpose of developing possible MSF programs in the region. This is not a typical activity of the MSF, as it usually responds to conflict and refugee situations in which there has been a breakdown in health services. Its recent assessment of the needs of Mother and Child Health underscores the entire situation in the region, and it will decide how it can best address it, given that it would be such a unique situation for MSF. MSF is no stranger to dealing, under very difficult conditions, with high rates of infant mortality, diarrhea, acute respiratory infections, and the spread of infectious diseases.

Exactly when the adverse environmental changes in the Aral Sea basin were awarded crisis status has been a source of controversy. All observers seem to agree that 1960 marked an important turning point for regional environmental quality. Whatever environmental conditions existed before 1960 were considered to have been more sustainable than those which followed that watershed year. The year 1986 can also be viewed as important, as Gorbachev's policy of *glasnost* (openness) began to take hold and discussions of the degradation within the Aral Sea basin became both public and vociferous. The year 1991, too, has been cited as an important marker for the Central Asian Republics; they began to view themselves as republics independent from the former Soviet Union.

One could also consider 1992 to have been an important year with regard to the recognition of a crisis in the Aral Sea region. Late that year the United Nations Environment Programme (UNEP) completed what some might call a pioneering effort to produce a 'Diagnostic Study of the Aral Sea' crisis. The study (UNEP, 1992) highlighted several problems facing leaders of the newly independent Central Asian Republics. As a result of this particular UNEP activity, the interest in and concern about the fate of the region was given higher visibility around the globe. This prompted other international organizations, especially nongovernmental organizations (NGOs), to deal with requests from Central Asian leaders to help resolve their environmental crises. 'Save the Aral Sea' then became the rallying cry of governmental and nongovernmental organizations alike.

'Saving the Sea,' however, is easier said than done. Any actions implemented to save the Sea carry with them major disruptions in the way things are usually done. As a result, support for 'business as usual' seems to be the order of the day. As noted earlier, a reliance on 'business as usual' can be blamed on the way that people and, therefore, societies tend to look at CEPs. They see little urgency associated with a particular creeping environmental change, so they tend to postpone dealing with it until it is almost too late.

Today, several of the incremental changes in the Aral basin are viewed as having developed to such an extent that they have turned into full-blown crises. Such belated responses by society to CEPs are not unusual; they occur in rich and poor countries, in capitalist and socialist countries, and in democratic and authoritarian regimes. They occur in response to local CEPs as well as to regional, international and the truly global ones. It appears to be a problem not of social organization but one of human nature. If Central Asian Republics in the Aral Sea basin, along with the international donor community, can be convinced to address cooperatively and in a timely way creeping environmental problems and the human activities that caused them, their adverse consequences could be mitigated and, perhaps, even arrested and reversed.

CEPs and sustainable development

Ever since the notion of 'sustainable development' was first raised in 1980 in an IUCN World Conservation Strategy document and, later, highlighted in the Brundtland Commission report in 1987 (WCED, 1987) and, again, in Agenda 21 (Sitarz, 1993) at the 1992 Earth Summit in Rio de Janeiro (Brazil), its usage has spawned what could be called a growth industry in academic research and economic development circles. Scores of definitions now exist, as do position statements on the notion and critical reviews of it. It seems that there are so many meanings associated with sustainable development, that policy-makers can find one in the scientific or economic literature to support any particular policy they wish to pursue in relation to development and the use of natural resources.

Discussions of sustainable development usually focus on one, or a combination of, the following processes: ecological, economic or social sustainability (e.g., Redclift, 1987). Saving ecosystems has been increasingly downplayed by those who favor economic growth and development over the prevention of environmental degradation. Natural resources are to be used, they might argue, and have little intrinsic value in themselves. The economic sustainability of a political system may overlook the need of people in general ('need' is subjectively defined and has many meanings); for example, one might argue that 'the poor will always be with us; let us try to keep their numbers relatively small.' Environment-oriented decision-makers favor the need to sustain ecosystems. Social sustainability suggests that governments have a responsibility to sustain a culture or way of life, which harks back to protecting the environmental setting on which it depends.

With regard to coping with environmental changes in the Aral Sea basin, going back to basics, perhaps, might not be a bad idea. These basics, according to my preferences, have been succinctly stated by Lélé (1994): (a) What is to be sustained? A particular resource or ecosystem in a particular form? The income it generates? Or the lifestyle it supports? (b) For how long? A few generations? Or forever? (c) How? That is, through what social process? Involving what tradeoffs against other social goals?

With regard to the Aral Sea region, more than a few questions need to be answered. For example, what is it that governments are seeking to sustain? Soil fertility? Human health? Fish populations? The economy? A ways of life? The deltas? The well-being of the region's leaders? How long do we wish to sustain it? The Aral Sea, as we know it, has been around for tens of thousands of years. Is that long enough? Can we now alter it in pursuit of human goals? Will the activities pursued in order to achieve sustainable development goals by, say, the year 2003 enable them to achieve the same goals by the year 2023? In other words, are short-term attempts at sustainability compatible with achieving sustainability in the long run?

Discussions of sustainable development generally raise questions concerning the relationship between present and future generations, often referred to as intergenerational equity issues (Partridge, 1981). Some people jokingly say that the future has done nothing for us, so we have little responsibility to future generations. Others contend that we have a moral responsibility to take into consideration the impacts of today's decisions on future generations – children, grandchildren and great-grandchildren. Still others argue that we do not know what future generations will either want or need, so there is little we can do for them from the vantage point of the present.

We frequently forget that there are five generations alive at any given time. Generational representatives can be brought together for intergenerational discussions about sustainable use of the environment. Great-grandchildren can ask of their great-grandparents why they let the environment deteriorate to the extent that it has. Great-grandparents, for their part, can ask their great-grandchildren what 'things' they are willing to give up so that environmental conditions can be improved. These generations can communicate with one another and can determine to some extent what future generations might want.

We are already aware of existing environmental sensitivities in the region. However, it would be useful to jump ahead a few decades into the future, in an attempt to identify new societal sensitivities. We need to 'leap-frog' well into the future in order to create a vision of sustainable economic development. For example, where do the Central Asian Republics want to be in 2003? And in 2023? What will it cost them to reach their goals? Assume that, as of today, there are few problems that have reached crisis stages, such as those related to water quantity, water quality, human health, ecological health, and population. Can we identify 'new sensitivities' that national and regional policy makers might have to face in the future if they were to pursue a 'business as usual' strategy?

The Aral Sea is not the only inland drainage basin that is facing severe pressures related to the issue of sustainable development, or of ecological or social sustainability. One might also look at other similar regions dependent on finite water resources to see how well (or poorly) they might have dealt with their situation. How did they approach sustainable development? What sensitivities have they come up against, with regard to population- environment-development interactions?

For example, in a recent book on the Great Lakes of North America, called *The Late, Great Lakes: An Environmental History*, Ashworth (1987) drew attention to the ever-present conflict between the *in situ* use of existing natural resources – in this case, seemingly abundant water supplies – and the need for that water in other distant locations to sustain economic growth and development.

The Great Lakes have several environmental problems: the lake beds and waters are suffering from varying degrees of pollution, several species of fish are no longer suitable for human consumption, lake flora and fauna have been adversely affected by more than a century of chemical and other toxic effluents. In addition, if global warming of the atmosphere were to occur, as suggested by various atmospheric scientists, computer models postulate that the water level in the Great Lakes would decline. A decline in water level would expose toxic lake-bed sediment to wind action and, therefore, its distribution throughout the basin and beyond.

In addition to all of these problems plaguing the Great Lakes sandwiched between two of the richest, most technologically advanced countries in the world, there is a growing demand on the basin States and Provinces to 'share' their lakes' waters with other nonriparian states. The following lengthy quote from Ashworth captures this dilemma for Great Lakes policy-makers.

> All of these problems, however, are pallid beside the threat looming on the western [US] horizon, where mining and agricultural interests are readying large-scale plans to lay pipelines to the Great Lakes, supplying by pump and pipe the water God doesn't supply by rain to the arid [parts of the west].

> The concept is simple. The need is *here*; the water is *there*; and the shortest distance between two points is a straight line, preferably a round, hollow one, made of concrete and filled with water ... The plans to pump water west amount to nothing short of a plan to drain the Great Lakes. Drain the Great Lakes? It sounds preposterous, and it is, but not because it cannot be done. It can. The technology exists; the need exists ... It comes from the assumption, basic to the idea of diversion, that the water can somehow be put to better use on the Plains than it can be in the Lakes. (Ashworth, 1987, p. 9)

Inland drainage basins such as these, and many of the rivers and streams that feed them, have become the repository for various kinds of waste. This is a problem faced by both industrialized and agricultural countries. There is no easy way to reverse the adverse impacts on the environment of such accumulated chemical and toxic waste. But there are ways, at least in theory, to prevent long-term, low-grade environmental changes from becoming major cumulative environmental disasters.

On the one hand we talk easily of sustainable development goals, while on the other hand we cannot deal effectively and in a timely way with creeping environmental problems. CEPs challenge our ability to achieve sustainable development because they are often not readily apparent. CEPs have plagued societies throughout history, right up to the present; and there appears to be little hope that this will change. Only by encouraging leaders from all levels to respond decisively and effectively to CEPs can one hope to improve the way in which societies interact with their physical and biological environments. Improved societal responses to CEPs can enable countries to achieve whatever 'sustainable development' goals they have set for themselves.

Timely responses to CEPs: What can be done?

US political scientist Anthony Downs, in his article called 'Up and down with ecology: the "issue-attention cycle"', identified five stages in the dynamics of the issue-attention cycle. These stages are, by analogy, similar to the way society, over time, tends to deal with creeping environmental problems: (1) the pre-problem stage, (2) the alarmed-discovery-and-euphoric-enthusiasm stage, (3) realizing-the-cost-of-significant-progress stage, (4) the gradual-decline-of-intense-public-interest stage, (5) and the post-problem stage (Downs, 1972).

With regard to the interest in environmental changes in the Aral Sea region, the pre-problem stage began in the mid-1960s and perhaps earlier, when the first signs of unusual environmental changes began to appear. In this period, some Soviet scientists did draw attention to potential severe and sometimes irreversible environmental changes in the sea and its deltas. They did so at risks to their careers.

Stage two was delayed until the mid-1980s because of the nature of the political system, the nature of the problem (i.e., a CEP), and because of the proposed technological fixes (e.g., the diversion of Siberian river water to Central Asia). Once the fate of the Aral region and its inhabitants became officially exposed to the public within and outside the Soviet Union, the international community, as well as national groups, began mobilizing to help national and regional authorities to address the Aral 'problem'.

Stage three began with the identification of the magnitude of the environmental problem faced by the region's authorities and inhabitants. However, once the high costs associated with correcting those problems were identified, enthusiasm about resolving the problems, at least in the short term, tended to wane. Another factor blunting enthusiasm was the realization that unlimited funding from international organizations such as the World Bank would not be available. Republics in the region quickly came to realize that the funding that would be made available would come with 'strings' attached.

This realization has been followed by stage four, a decline in interest in resolving the difficult environmental problems of the region, and a search for one grandiose costly plan to make all of the region's environmental ills go away. In this phase, realistic 'can-do' proposals can emerge, along with an improved appreciation of the problems faced and the limited funds available to address them. Environmental problems are then prioritized.

The fifth stage of the cycle (the post-problem stage) has not yet been realized in the Aral region. In this phase one can assume that 'environmental awareness' intensified and that the problems generated by human involvement in creeping environmental change had begun to be addressed. Capacity building and institutional development receive greater attention, given the demonstrated linkages between environmental quality and sustainable development, however one chooses to define it. Improved management of resources and improved interactions between human activities and the natural resource base are often generated by a political, as well as an economic, development goal.

Most creeping environmental problems confronting societies involve human activities. Delayed responses to such changes, until a crisis situation emerges, are costly to both society and the environment. Thus, it is necessary to improve societal understanding of the dynamics, as well as the implications, of CEPs in order to prompt more appropriate, effective, and timely responses to them by policy-makers.

Rich countries, despite their present-day claims to be so poor, may be able to get away with responding to their full-blown environmental crises by 'throwing large sums of money at them.' However, in countries with limited, scarce or dwindling resources, 'muddling through' is likely to lead to an environmental crisis or, even worse, the realization of a dreaded, irreversible situation for which they have few, if any, resources with which to respond.

Regional organizations in Central Asia: five heads are better than one

It is often difficult to bring together, into an effective regional organization, states that have been independent for a long time. It is even more difficult to do so with newly independent states, as each state seeks to develop its own national identity and policies.

States in the Aral Sea region have an opportunity to address a common set of problems; problems that stem from a key shared regional resource: water. Each state has a water-related 'bargaining chip', so to speak, with regard to other states in the region. The leaders of the five Central Asian Republics still have a unique opportunity to develop a truly effective (cooperative as opposed to competitive) regional organization centered on the management and use of regional water supplies. Given inherent limitations on the availability of national and international financial resources to resolve all present-day environmental problems in the Aral basin, an effective Central Asian regional organization in which no single country dominates could go a long way towards arresting regional CEPs and in achieving a regionally defined form of sustainable development. Such an organization could help the republics cope with new environmental sensitivities that will likely emerge in the next few decades.

HOW IMPORTANT IS THE ARAL SEA ANYWAY?

The Aral Sea may be more important to the region than many observers realize, especially if states in the region hope to achieve any degree of sustainability. In fact, one could argue that the sea itself is a key to the region's future well-being, both symbolically as well as realistically. Symbolically, it is much more than a useless body of water in a sandy desert depression. It has intrinsic value as a body of water sandwiched between two deserts. The relatively rapid demise of the sea has captured the attention of the international community. Realistically, it affects regional inhabitants. It affects regional climatic conditions. Its declining sea level has generated various proposals to institute heroic, high-tech schemes to preserve it (melt glaciers, pump Caspian water uphill, etc.). Its decline has also generated ill-will among people within and between the region's republics.

'I HAVE MET THE ENEMY AND HE IS US!'

Clearly, it is much easier to identify problems than it is to resolve them. Yet, societies everywhere have poor records in dealing with CEPs, let alone resolving them. We should correct this mismatch between the rates of environmental change and the rates at which decisions are taken to address them. Methods need to be devised to slow down, arrest, and wherever possible, reverse the CEPs which plague the environment and inhabitants in the Aral basin.

Human behavior issues must be addressed. To avoid dealing with behavioral factors that impinge on regional environmental quality would likely mean that similar behavioral processes will occur in those regions where governments and scientists hope to make gains (e.g., the deltas). The underlying causes of environmental degradation in the Aral Sea basin must be confronted.

A major goal must be to 'Save the Sea' in some form. This does not necessarily mean that it must be restored to its pre-1960 level; nor does it mean that the sea should be abandoned altogether, focusing instead on protecting only the deltas. Central Asian leaders must look to the future with a vision. They must identify where they want to be in the year 2003 and then again twenty years later, so that we can identify critical human, as well as other, resources that will be required to get us there. In other words, identify the level of environmental health leaders want to bring to the region by a designated point in time by which goals are to be met, and then try to work toward achieving it.

Conclusion: Steps decision-makers can take now to encourage sustainable resource management in the Aral Sea basin

Whatever the notion of sustainable development means, it does not mean that a nation or a region must live off its own resources. Autarchy clearly does not work well, especially in today's interconnected world. Nations have resources that they can either exploit or trade for other resources that they need. What sustainable development does mean, among other things, is that a nation must not overexploit its natural and human resources. The following discussion suggests some steps that can be taken immediately by political leaders of the Aral Sea basin states.

1. It is very important to improve awareness of the interdependence between the officially designated disaster zone around the Sea and upstream regions.
   Their fates are geographically entwined. That awareness should serve to reinforce the value of a regional political organization of equals; a regional organ-

ization, unlike most others around the globe where one or two countries tend to dominate the process.

2. It is also important to improve the awareness of the tradeoffs between environment and development. Today there are numerous examples of successes and failures in national and regional attempts at sustainable economic development. Examples relevant to the region should be collected and 'mined' for insights into what might or might not work in the Central Asian political, environmental and social context. For example, methods could be devised to enable upper basin states to share in the downstream profits derived from water used for economic development. This would provide an incentive to upper basin states to protect the water quantity and water quality that they pass on to lower basin states.

3. It is imperative that societies improve their awareness of the nature of creeping environmental problems and the ways in which societies have dealt with them. Most environment-related problems are partly or wholly human-induced. Early intervention in these creeping processes of change can improve the chances for sustainable development in the long term.

4. Numerous reports on development planning, in both industrialized and in developing countries around the world, have criticized the lack of involvement of local people in national planning efforts. This shortcoming, in fact, has been blamed for the failure of many development projects to live up to their stated goals. Capacity building within countries includes the involvement of the public in decision-making processes. There is a fallacy that must be dealt with here: namely that 'experts' are only those who come from another country. There is considerable expertise, actual and potential, within the Aral Sea basin countries. This reservoir of knowledge must be nurtured and brought into the development process. It is particularly vital that local people, who are directly affected by a given environmental problem, have a stake in the success of development planning activities.

5. There are no readily apparent quick-fixes – technological or otherwise – to resolve the environmental and, therefore, the sustainable development crises afflicting the Aral Sea basin. Problems in the region have been accumulating throughout the past century (not just since 1960). They may require 'creeping solutions' – incremental steps that can be taken to improve the health of the people, the economy and the environment. In this way, solutions will also work to achieve the region's vision and goals for sustainable development (see Agarwal, 1996). By addressing creeping environmental problems through planned, incremental steps, the nations of the region may improve their efforts at sustainable resource management.

As a final comment, the international community and the Central Asian Republics have defined a disaster zone that encompasses the southern region of the Aral Sea. It includes the Karakalpak Republic and the Khorezm Region of Uzbekistan, the Kyzyl-Orda Region of Kazakstan, and the Dashkovuz Region of Turkmenistan. This is unfortunate, because the river systems of the Amudarya and the Syrdarya, in fact, define the disaster zone. There will be no way to resolve the crises in the Aral region without recognizing explicitly the interconnectedness of the administrative units that share the resources provided by these river basins.

In ecology there is a saying: 'you can't do just one thing.' In other words, you cannot change one element of an ecosystem without inadvertently having an effect on other elements of that ecosystem. Similarly, anything that affects the flow of waters along their natural course has an impact elsewhere. It is important that the Aral Sea basin be viewed holistically as a 'meta-ecosystem': a system that cannot be separated into its many linked parts. Collective problems must be met with collective solutions.

References

Agarwal, A., 1996: *The Curse of the White Gold: The Aral Sea Crisis*. New Delhi, India: Centre for Science and Environment. March, 49-51.

Ashworth, W., 1987: *The Late, Great Lakes: An Environmental History*. Detroit, Michigan: Wayne State University Press.

Borovsky, V. M., 1980: The drying out of the Aral Sea and its consequences, Scripta Publishing Co. (from *Izvestiya Akademii Nauk SSSR, seriya geograficheskaya*), **5**.

Burton, I. and K. Hewett, 1974: Ecological dimensions of environmental hazards. In: F. Sargent (ed.), *Human Ecology*, 253–83. Amsterdam: North-Holland.

Critchlow, J., 1991: *Nationalism in Uzbekistan: A Soviet Republic's Road to Sovereignty* Boulder, Colorado: Westview Press.

Davis, J. H., 1956: Influence of man upon coast lines. In: W. L. Thomas Jr. (ed.), *Man's Role in Changing the Face of the Earth*. Chicago, IL: University of Chicago Press.

Döös, B. R., 1994: Why is environmental protection so slow? *Global Environmental Change*, **4**, No. 3, 179–84.

Downs, A., 1972: Up and down with ecology – the 'issue-attention cycle', *The Public Interest*, **28**, 38–50.

Ellis, W. S., 1990: The Aral: a Soviet sea lies dying, *National Geographic* (February), p. 83.

Elpiner, L. I., 1990: Medical-ecological problems in the eastern Aral region. Paper presented at University of Indiana conference on 'The Aral Sea Crisis: Environmental Issues in Central Asia' Bloomington, Indiana: mimeo.

Feshbach, M. and A. Friendly Jr., 1992: *Ecocide in the USSR: Health and Nature under Siege*. New York: Basic Books.

Frederick, K. D., 1991: The disappearing Aral Sea, *Resources* (Winter issue). Washington, DC: Resources for the Future, 11–14.

Goldman, M. I., 1972: *Environmental Pollution in the Soviet Union: The Spoils of Progress*. Cambridge, Massachusetts: The MIT Press, 234–5.

Gorodetskaya, M. Ye. and A. S. Kes, 1978: Alma-Ata conference on the environmental impact of a drop in the level of the Aral Sea, *Soviet Geography*, **19** (10), 728–36.

Hoffer, E., 1952: *The Ordeal of Change*. New York: Harper and Row.

Kazakhstanskaia pravda, 1969: Kazakhstan periodical, February 6, 1969, p. 2.

Kuznetsov, N. T., 1992: Geographical and ecological aspects of Aral Sea hydrological functions. *Post-Soviet Geography*, **33**(5), 324–31

Lélé, S., 1994: Sustainability, environmentalism, and science. *Pacific Institute Report*, Spring, p. 5.

Létolle, R. and M. Mainguet, 1993: *Aral*. Paris: Springer-Verlag.

Micklin, P. and W. D. Williams (eds.), 1996: *The Aral Sea Basin*. NATO ASI Series, The Environment, Vol. 12. Berlin: Springer-Verlag.

Mnatsakanian, R. A., 1992: *Environmental Legacy of the Former Soviet Republics*. Edinburgh, Scotland: Center for Human Ecology, University of Edinburgh, p. viii.

NRC (National Research Council), 1987: *Confronting Natural Disasters: An International Decade for Natural Hazard Reduction*. Washington, DC: National Academy Press.

Palm, R. I., 1990: Chapter One: Introduction to the study of natural hazards. In: R. I. Palm, *Natural Hazards*, 1–17. Baltimore, Maryland: Johns Hopkins University Press.

Partridge, E. (ed.), 1981: *Responsibilities to Future Generations: Environmental Ethics*. Buffalo, NY: Prometheus Books.

Radzinsky, V. Ye. (ed.), 1994: *The Health of Mothers and Children in the Aral Region of Turkmenistan*. Kiev: Zdorovya Publishing House.

Redclift, M. R., 1987: *Sustainable Development: Exploring the Contradictions*. New York: Methuen Press.

Reteyum, A. U., 1991: Letter in Overview Section, *Environment*, 33(1), p. 3.

Rudenko, B., 1989: 'Solenye Peski Aralkum' (The salty sands of the Aral), *Nauka i zhizn*, 10 (October), p. 44.

Sitarz, D., 1993 (ed.): *Agenda 21: The Earth Summit Strategy to Save Our Planet*. Boulder, CO: Earth Press.

Sotsialisticheskaia industriia, 1970: Russian periodical, August 15, 1970, p. 2.

Soviet Geography: Review and Translation, 1969: Russian periodical, No. 3, p. 146.

Smith, D., 1994: Change and variability in climate and ecosystem decline in Aral Sea Basin deltas, *Post-Soviet Geography*, 35(3), pp. 142–65.

Smith, K. R., 1988: Overview: Natural hazard reduction. *Environment*, 30(6), 2–4.

Tsinzerling, V. V., 1927: *Irrigation in the Amudarya Basin*. Moscow: Izd. Upravleniya vodnogokhozyaistva Srednei Azii (Publishing House of the Water Management Board of Central Asia).

Tsutsui, H., 1991: *Some Remarks on the Aral Sea Basin Irrigation Management*. Nara, Japan: mimeo.

UNEP (United Nations Environment Programme), 1992: *Diagnostic Study for the Development of an Action Plan for the Aral Sea*. Nairobi, Kenya: UNEP.

WCED (World Commission on Environment and Development), 1987: *Our Common Future*. New York: Oxford University Press.

2   Ecological disaster linked to landscape composition changes in the Aral Sea basin

ELISABETH A. VOSTOKOVA

The first time I managed to see the Aral Sea and the Amudarya was in 1952, when large-scale studies on the northern route of the proposed but never developed Main Turkmen Canal were being developed. The Amudarya's delta with its blue- and green-colored lakes and vegetation contrasted sharply with the grayish, yellow-green desert areas of the Kynyadar'inskaya ancient alluvial deltaic plain and the saline soils (e.g., solonchaks) of the Sarykamysh depression. Nothing noticeable was taking place in the natural environment at that time that warned us of the adverse environmental changes that would be observed after 1961.

In fact, I had noticed the first alarming symptoms of irrational use of water and of the adverse changes in the vegetational mix of desert landscapes even earlier, in 1958, while carrying out field studies of the Karabil'skaya fresh water lens in the southeastern part of the Karakum Desert. At that time the depressions between the ridges in the desert sands were beginning to flood as a result of the discharge of Amudarya water into the Karakum Canal which was then under construction. My concern about these environmental changes intensified in 1963–64, when I observed the filling of the Sarykamysh depression with waste water released through the Daryalyk channel. An eyewitness (Dr G.S. Kalenov) confirmed that, during the first year of water discharge into the Sarykamysh depression, the water disappeared into and filled *with a rumble* numerous underground karst cavities. Only after those underground cavities were filled did Lake Sarykamysh begin to appear in the Sarykamysh depression.

With the extensive use of satellite data in my research carried out after 1974 in all regions of the Aral basin (Tajikistan, Uzbekistan and Karakalpakstan), I was later able to see the spatial scale of landscape changes in the Aral Sea basin. Visiting the Amudarya Delta in 1975, where vast areas were occupied by the groves of lifeless reeds and individual carabarak bushes, I was forced to think about the ecological disaster that had befallen this region. It was evident that the drying out of reed bogs had taken place quickly. The reeds, having had roots near the surface under the formerly favorable conditions, had dried out as a result of their exposure to the sun, which resulted from the rapid recession of the sea's water. The reed is an adaptable species that can grow successfully with groundwater levels at a depth of 3–5 m

Table 2.1 Development of the author's personal view about ecological changes of the natural environment in the Aral Sea basin (and their assessment)

Year of observations	Regions observed	Pattern of landscape changes noticed	My awareness of changes
1952	Aral Sea, Amudarya Delta, Kunyadarya Plain, Sarykamysh		
1955	Dzhanadar'inskaya Plain		
1958	Southeastern Karakum	Groundwater rise in the depressions of sand ridges, caused by the construction of the Karakum Canal	Concern for the irrational use of water resources
1963	Sarykamysh, Daryalyk	Lake formation, channel processes	
1974–1979	Fergana, Tajikistan, the region of Bukhara, Amudarya Delta	Groundwater rise and increase of salinity of soils; replacement of hydromorphic landscapes by halo-hydromorphic and automorphic ones	Desertification and ecological disaster for biota
1980–1989	Southern Aral Sea region, Sarykamysh, western and southwest Kyzylkum	Filling of Sarykamysh; formation of automorphic landscapes	Ecological catastrophe for biota

from the surface and with a level of water mineralization (i.e., total dissolved solids, TDS) of up to 18 g/l.

In addition, the monitoring and observations for the southern Aral Sea region, based on satellite information, provided us with an overview of desertification processes that were taking place in the region. This change was witnessed by the transformation of hydromorphic landscapes (i.e., vegetation dependent on various sources of water) into automorphic ones (i.e., vegetation dependent only on precipitation) typical of ancient deltaic and alluvial plains. My personal account of ecological changes in the Aral Sea basin are summarized in Table 2.1.

After the 1970s, a large number of papers appeared which considered different aspects of the Aral Sea problem. Although a great many were prompted by the situation that developed before 1985, these problems were later identified in the media as an 'ecological catastrophe'. In the 1970s and 1980s the majority of researchers dealing with the Aral Sea problem focused their investigations on various aspects of the destabilized environment in the southern Aral Sea region. Special meetings were held during that time. Centers for the investigation and recovery of the Aral Sea were established after 1985 and international studies of the problem began in 1991.

A retrospective analysis of Aral Sea literature shows that during the first 4 to 5 years (i.e., until 1966–67), primary attention was paid to the changes of the natural environment related to the construction of the first stage of the

Table 2.2 Awareness of ecological crisis in the Aral Sea region by the scientists and public

Time interval	Status
I The 1960s to the middle of the 1970s	Lack of attention to emerging problems related to the extension of irrigation in the region; a government priority of immediate economic benefits; little regard for adverse environmental impacts.
II The 1970s to the middle of the 1980s	Gradual increase in awareness by the scientists of the acute ecological situation; defining of the necessary tasks to improve the situation; forecasts of and projects for improvement of the ecological situation were developed, but the main causes remained uncovered; only the Aral Sea region and the Aral Sea proper were considered; measures were offered related only to specific problems; adverse water management practices continued in full scale; emphasis was given to the proposal for a diversion of the flow of Siberian rivers to the south.
III The mid-1980s to the early 1990s	Awareness of general and particular causes of environmental changes; acknowledgment of the crisis situation by some scientists; broad discussion and investigation of the problem by scientists and the public; gradual acknowledgment of the need for a partial restriction of irrigation, an increase of the efficiency of the canals, an improvement of drainage networks; new projects for the conservation and rehabilitation of the Aral Sea, including the enhancement of the ecological, social and economic situations.
IV The early 1990s to 1996	Acknowledgment by the worldwide public of the ecological situation in the Aral region as a global ecological catastrophe; interruption of inter-republic relations which had existed until 1991 with the establishment of sovereign states and the dismantling of the Soviet Union; gradual development of international cooperation; in the absence of actual concrete measures, deterioration of the ecological situation continues.

Source: Time intervals I to III, Glazovsky (1990); IV, Vostokova.

Karakum Canal from the Amudarya to the Murgab River. Changes in the natural environment in the lower reaches of the Amudarya and Syrdarya were discussed only at the end of the 1960s, with the appearance of challenging information. In 1975, Kuznetsov (1991) justified the need to investigate the whole Aral Sea basin as an integrated system.

In his monograph on the Aral crisis that reviewed its origin and its disastrous consequences for the natural environment, Glazovsky (1990) identified three periods of scientific and public awareness of ecological change in the Aral Sea region (Table 2.2). I have added a fourth period: 1991 to 1996. During periods I and II, little attention was paid to statements of changes in the natural environment which had exposed changes in certain components of the landscape and other processes. Little attention was also paid to their forecasts, including forecasts of the drying out of the Aral Sea. Periods III and IV differed from earlier ones, because there was an increase in the number of publications considering various ways to resolve the Aral crisis. Most recently, the Aral Sea problem has become more regionalized in Central Asia and more politicized (Zonn, 1993; Zonn, this volume).

An analysis of numerous scientific publications on the Aral Sea and water-resource use in the Aral basin provides the basis for the following conclusions:

- The uses of water resources in different parts of the Aral Sea basin were carried out without regard for the possible consequences either for (a) parts of the landscape or for (b) the landscape as a whole. There was little concern that such adverse changes could occur in the other parts of the basin as well.
- Most researchers focused either on specific problems (e.g., soil changes, specific vegetative changes, the water and salt regimes) or on narrowly delimited local areas predominantly along the Karakum Canal. Thus, the focus was on the eastern and southern Aral Sea region.
- The use of satellite data prompted (and enabled) an integrated approach to the investigation of anthropogenically induced changes in the natural environment. It stimulated the development of a series of interrelated maps that reflected ecological conditions (both adverse land-use violations and conservation measures). Monitoring schemes were developed.
- Disciplinary (especially economic) approaches dominated political thinking for a long time (e.g., comparison of profits from the exploitation of biological resources of the Aral Sea and the use of the sea for navigation versus the generation of profits which could result from irrigated agriculture using Amudarya and Syrdarya waters) (Geller, 1969). Political leaders pinned their hopes on solving all economic development problems in the Aral Sea region by the proposed diversion of a part of the northward-flowing Siberian rivers to the south. Such hopes significantly impeded the awareness and understanding of political leaders of the emerging ecological catastrophe in the Aral Sea basin.

Thus, the Aral Sea basin is where the effects of gradually accumulating human-induced ecological changes have become fully exposed. Those changes are most pronounced in the plains in the form of human-induced desertification, identified by drastic alterations in the desert landscape.

In a schematic diagram of the Aral Sea basin (Figure 2.1) one can identify the following: (I) the region where river flow is formed: the Pamir-Alai and Tien Shan mountains; (II) the zone of the first use of water from the rivers: the piedmont plains and intermountain depressions with ancient oases; many canals (of which the Karakum Canal is the most important); drainage networks and artificial lakes made up of collected waste water (mainly water that is saline and polluted by agricultural chemicals); (III) the zone of the secondary use of water drawn from the rivers: the valleys in the middle and lower reaches of the Amudarya and Syrdarya, their deltas and ancient alluvial-deltaic plains; also with multiple irrigation facilities and artificial waste-water collector lakes, of which Lake Sarykamysh is the largest; (IV) the Aral Sea proper and the adjacent dried bottom areas with active eolian (wind-driven) processes; (V) desert areas where the anthropogenic changes in the landscape's vegetation composition are not directly caused by water management practices (e.g., the construction of water reservoirs, irrigation and drainage networks, extension of irrigated agriculture) or by the decline of the Aral Sea level.

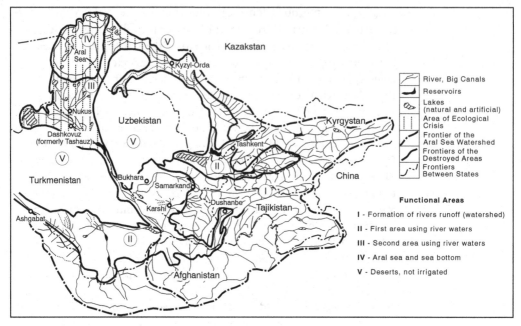

Figure 2.1 Scheme of the Aral Sea basin.

The anthropogenic changes of the landscape's composition worsen increasingly as one goes from the headwaters to the river deltas and the sea. It is the latter location where those changes acquire the characteristics of ecological disaster.

Where the rivers' waters form, water management activity is locally unique. A small number of dams and reservoirs have been constructed there. This has led to the flooding of certain landscapes, particularly in the mountains which cover an area of about 400 km². Viewed as a whole, however, the mountain landscapes have not experienced significant anthropogenic impacts because of the use of water feeding the rivers that eventually flow into the Aral Sea. Anthropogenic disturbances of the natural environment in the mountains have other origins that are primarily connected to livestock grazing activities, geological exploration and mining. Already, in the broad intermountain valleys where irrigated agriculture has been developed, the landscape's composition has changed considerably with natural landscapes being replaced by partially transformed natural and agro-landscapes. Irrigated lands in this area are generally in fairly good condition and, therefore, the impacts of changes in the landscape here can, for now at least, be viewed as positive. The rivers' regime, as they flow out of the mountains, changes only slightly. Their total discharge is about 120 km³/year with water mineralization levels ranging from 0.25 g/l to 0.6 g/l.

Piedmont (foothills) plains with its southern margins along the Kyzylkum Desert and the Ferghana Valley are more than 90% occupied by anthropogenically transformed landscapes. Only insignificant areas, totalling less than 2% of the total area, where the composition of zonal landscapes has been mostly preserved, are still represented by natural ecosystems, including low mountainous areas and certain sandy desert areas. The use of river water in this part of the basin has been essential for carrying out human activities. Long ago, rivers such as the Zarafshan, the Karshi and other smaller rivers, could not reach the Amudarya River and local water withdrawal was implemented primarily for irrigation. The ancient oases of Central Asia are located in this piedmont strip: Maryisky, Karakul'sky, Bukharsky, Samarkandsky, Tashkentsky, Ferghanasky, including the younger oases, such as Chardzhousky, Tedzhensky, and Karshinsky. Most of the important modern-day cities and industrial complexes in the region are also concentrated at the sites of these ancient oases. A dense irrigation network was developed and several large and small water reservoirs were constructed.

Before 1960, the largest canals, using water from the Syrdarya, were constructed in the Ferghana Valley and in the Mirzachul (the Golodnaya Steppe). The largest water reservoirs, using Syrdarya water, are the Kairakkumskoe and the Chardar'inskoe. Amudarya water is withdrawn by the Karshinsky Canal, the Amudarya-Bukharsky Canal (to irrigate the lands of the Karshinsky and Karakul'Bukharsky oases), and the Karakum Canal. Of these, only the first was hydraulically sealed over its entire length (of its transit part), and therefore had no adverse effects on adjacent areas. The Amudarya-Bukharsky Canal, while passing through the ancient deltaic plain of the Zarafshan, formed a narrow zone (up to 3 km) in which the groundwater level increased, where hydromorphic landscapes (sparsely populated with hygrophytes and phreatophytes) formed on the irrigated areas. While a portion of irrigation drainage water is discharged back into the rivers, the greater part flows into natural, drainless depressions such as Lake Arnasai, the Dengizkol', and the Soleno.

New hydromorphic landscapes with water-loving ecosystems formed at these locations, replacing solonchak deserts with halophyte vegetation. The decoding of satellite images provided an opportunity to reflect on changes in landscape composition, which appeared on thematic maps. These images also allowed us to monitor changes in regional hydrogeological conditions (Figure 2.2). According to Zaletaev (1989), all such regions of interactions between anthropogenic landscapes of oases and the bedrock of deserts exhibit a destabilizing pattern.

The Karakum Canal, which withdraws up to 13 km³/year from the Amudarya, is a major factor in the alteration of the land cover in the western part of this section of the Aral Sea basin. By 1989 it is estimated to have accounted for 18–24% of the total water withdrawal from the Amudarya (Kirsta, 1989). Water is generally withdrawn from the Aral basin – as even drainage waters from the Kopet dag and Tedjen-Murgab oases are discharged either directly into the Karakum sands (disturbing pasture ecosystems) – or is

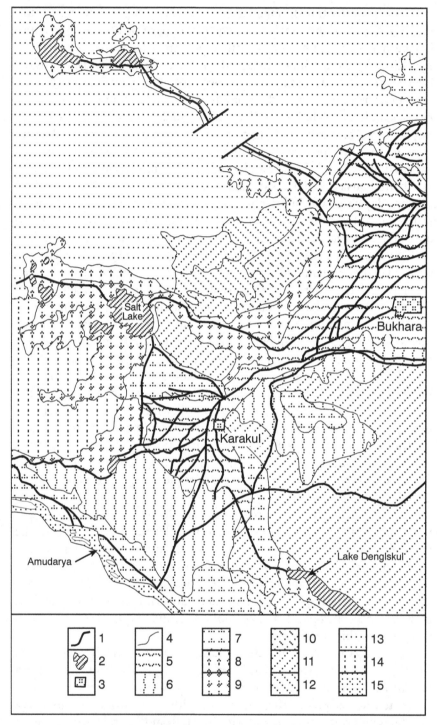

Figure 2.2 Landscape and ecology around the Buchara and Karakul oases. 1–3: Hydrography: 1, canals, collectors; 2, artificial lakes with drainage waters; 3, river bed of Amudarya. 4: Ecological frontiers. 5–14: Landscapes changed by water use around oases: 5, irrigated area; 6, ruderal (weedy) vegetation; 7, halophylic vegetation with annual soljankas and *Alhagi* spp.; 8, halophylic bushes and low bushes; 9, tamarix communities and *Alhagi* spp.; 10, artificial plantation of *Haloxylon aphyllum*; 11, sandy-clayey and loamy; 12, sandy-stony and clayey- stony; 13, clayey and a little sandy; 14, sandy. 15: Towns.

diverted by a collector main drain into the basin of the Caspian Sea (Dukhovny *et al.*, 1984). The waters of the main channel of the first stage of the Karakum Canal passes through the valley of the Kelif Uzboi (an ancient river bed) and the hilly ridge sands of southeastern Karakum. Numerous lakes have been formed from seepage into depressions and spacious subcanal lenses. Seepage losses from the Karakum Canal within the first five years alone were estimated to have been about 25 km^3.

The composition of natural landscapes in this area has been especially strongly disturbed (e.g., Grave and Grave, 1983; Gerasimov, 1978). Disturbances of natural landscapes surrounding the first stage of the Karakum Canal encompassed about 500 km^2. Where the canal crosses the sandy southeastern Karakum desert, its influence on the surrounding landscape reached more than 30 km in width. An intricate psammophytic and hydromorphic complex formed in this zone, combining the elements of zonal landscapes of shrubbery-psammophyte communities, related to ridges and newly formed lacustrine and tugai hydromorphic communities in the interridge depressions. This ecological complex was further changed near the canal by the formation of meadow–solonchak ecosystems and altered soil surfaces (e.g., takyrs and residual solonchaks) from the 'muck' from canal dredging activities to clear out overgrown vegetation.

The third and fourth stages in the development of the Karakum Canal, passing along the Kopet-Dag foothills, have already caused other kinds of landscape changes because of the rise of the groundwater level and because of the discharge of drainage water into the desert. In particular, the discharge of drainage water along the boundary of the piedmont plain and the sandy desert also resulted in the formation of a lacustrine-meadow complex, with solonchaks in the depressions and with psammophytic vegetation in higher areas. By 1989 such a complex occupied 200 to 250 thousand hectares (Glazovsky, 1990; Kirsta, 1989). As a result of the extension of irrigation at the Tedjen and Murgab oases, the area of saline lands increased. This has been confirmed by satellite images (Vostokova *et al.*, 1988; Skaterschikov, 1972).

Human-induced changes in the natural landscapes, similar to those which were observed in the Amudarya, were also noted in the Syrdarya basin. Secondary salinization of lands in the zone of influence of the Kairakkumskoe Reservoir (i.e., the Fergana Valley) and the Chardar'inskoe Reservoir provides an interesting contrast. The impact of the Kairakkumskoe Reservoir was especially negative for the landscapes upstream from the dam and adjacent fields of the Kokand oasis because of the increase in groundwater level from the side of mountains, as a result of reservoir backwater. This resulted in the formation of solonchak ecosystems; that is, widespread secondary salinization, in the fields. The Chardar'inskoe Reservoir, along with the Arnasai system of lakes (formed as a result of discharges of excess flood waters in 1969 into the Arnasai Depression), still have a significant environmental impact on the surrounding landscape.

There has been an increasing trend in the mineralization of river water, when compared with its salt content prior to 1961 – up to 0.5 g/l in the Amudarya and up to 0.8 g/l in the Syrdarya. Such an increase in mineralization rearranges flood plain ecosystems and destroys tugai (river delta ecosystem) vegetation on the islands and on the flood plain.

Thus, in the geographic zone of first water use in the Aral basin, landscape changes tend toward desertification. According to Babaev and Zaletaev (1990), the first signs include the development of hydromorphic ecosystems with simplified structures.

The second zone of water use for irrigation includes the lower reaches of the Amudarya and Syrdarya and their deltas. This zone has been subjected to considerable changes in natural conditions. It encompasses ancient oases such as Khorezm, Khiva, Kunya-Urgench, and the newer ones such as Nukus, Kungrad and Chimbai. Beginning in 1961, drainage waters of oases on the left bank were discharged through the Daryalyk and the main collector (Ozernyi) into the Sarykamysh Depression. However, a significant portion of these waters is supplied from the interridge depressions in the sandy desert by way of small collectors. In these depressions an unstable zone of lakes and bogs with hydromorphic ecosystems has formed.

Drainage waters of the right-bank area are discharged into the depressions of the ancient Akhchadarya alluvial deltaic plain and into the former bays of the Aral Sea. New hydromorphic solonchak landscapes are also formed in the desert here.

From the 1970s on, the southern and southeastern Aral Sea region has been the subject of detailed studies by various specialists and since 1985 scientists have intensified their investigations of ecological disaster (e.g., Anon., 1988; Glazovsky, 1990; Rafikov, 1982; Chernovskaya, 1988). The processes altering the landscape's composition in this region are fully described in numerous publications. It should be noted that the availability of satellite data has provided an opportunity to develop a variety of small- as well as large-scale special landscape maps. These maps reflect the extent of anthropogenic impacts on the landscape (e.g., Alekseeva et al., 1991; Bakhiev and Novikova, 1990; Vostokova, 1991; Vostokova et al., 1989; Vostokova et al., 1988; Glushko, 1991; Khodzhaev et al., 1985; Baknai et al., 1991; Popov and Vinogradov, 1982). Space-mapping and monitoring of water bodies and land-surface ecosystems have been implemented in the region (Vinogradov, 1984; Vitkovskaya et al., 1985).

When analyzing the landscape composition and its changes for this section of the Aral Sea basin, attention should be paid to the transformation of hydromorphic landscapes into the formation of new but depleted landscapes, i.e., desertification. The formation of desert landscapes apparently passes through the same stages as the formation of ancient alluvial deltaic plains of Kunyadarya, Akhchadarya and Dzanadarya. Ultimately, this process will result in the establishment of a wormwood–biyurgun–keurek desert integrated with takyrs and residual solonchaks (with the exception of low-mountain areas and the uplands [Sultanuizdag, Belitau]). These are automorphic desert landscapes which are independent of the waters of the Sea and rivers. Anthropogenic disturbances of these landscapes are connected with livestock grazing on rangelands, prospecting, and so forth.

At the same time, perennial water flow into the Sarykamysh depression completely destroyed the former black saxaul areas, sands and solonchaks, and resulted in the formation of a new ecosystem of a saline lake and a solonchak-boggy complex of partially submerged lands. The ever-increasing amounts of silt in the wastewater caused the mineralization of the lake's water to reach 13 g/l by 1990. The introduction of such chemicals gives one good reason to anticipate the near-future biological destruction of this human-created fresh-water body. Lake Sarykamysh is not the only water body fed by drainage water from irrigated fields of the Tashauz, Nukus and Chimbai regions.

The increasing level of hydromorphism in this region has led to the regeneration of processes that undermine karst both on the ancient alluvial deltaic plain of Kunyadarya and on the Akhchadar'inskaya plain and in the Amudarya Delta.

Two areas within this region exhibit a specific instability in the trends of landscape change. These trends are associated with the irregular discharges of drainage water from irrigated fields: Sudoch'e Lake which depends on water discharges from rice fields in the Raushan area and Dzhiltyrbas Bay, into which drainage water is supplied from the eastern part of the irrigated fields of the Nukus and Chimbai oases.

The Aral Sea proper, along with the newly formed deserts on its exposed dried bottom, cover an area of more than 40000 km² and represent the fourth zone of landscape transformation. Entirely new ecological conditions (and landscapes) have formed on the dried and drying bottom of the Aral Sea with pronounced tendencies toward the development of desert complexes. This zone has captured the attention of scientists. Scientists have identified the main phases of landscape changes and have made forecasts of the Aral Sea's status. Different ways for rehabilitation and stabilization have been proposed.

The main stages of change in these areas depending on the composition of soil deposits, sea bottom morphology, rates of desiccation, wind regime, and so forth, have been identified through research. The continuous, gradual retreat of the sea from year to year has created a zonal distribution of the various stages of development of desert landscapes on the dried sea bottom.

The largest zones of exposed seabed are found in the eastern and southeastern areas of the sea. In this area the sea has retreated by tens of kilometers. The initial differences of natural conditions provided an opportunity to identify different landscape zones on the dried bottom to the south Aral Sea and to make projections about the formation of new landscape compositions on the exposed seabed. A map of the locations where salt has been removed from the newly exposed surface because of wind action was also developed for the eastern and southeastern parts (Khabarov, 1991).

The areas of slightly altered landscapes in the Aral Sea basin include the Karakum and Kyzylkum deserts, the Ustyurt, the dried Aral seabed and the northern Aral Sea region. They differ significantly in physical and geographical conditions and landscape composition, and are grouped together here, only because each completely lacked appropriate water management.

The Kyzylkum area, where the water conduit from the Amudarya to the city of Zarafshan was constructed by disturbing the landscape, represents an exception. The decline in the level of the Aral Sea has had an indirect effect on the artesian waters of the Syrdarya basin, which discharged into the sea (Sadykov, 1983). Basically, the landscape of this area has remained as before, although it is subject to aridity.

The use of remote sensing, based on thematic maps of space imagery data and information systems, is important in research on the dynamics of landscape changes in the Aral Sea basin. Different flow sheets of remote monitoring are already developed, but have not been made available until recently (Babaev and Babaev, 1994; Vinogradov, 1984; Vostokova, 1991; Glushko, 1992). The creation of data bases is of great value for the purpose of monitoring from space and for modeling landscape formation processes (Vinogradov and Frolov, 1989; Glushko et al., 1994; Borodin et al., 1987; Novikova and Trofimova, 1994; Raskin et al., 1993).

Considering the Aral Sea basin as an integrated system, it is possible to trace the beginning of this ecological disaster as well as the stages of its development. Desertification in the basin has been the result of both natural factors (annual increase of the debit part of the water budget of the sea against the credit part, low water years) and water management activities aimed at increasing the area of irrigated lands, the establishment of new water reservoirs, numerous hydraulic structures, irrigation canals, etc. All this has been aggravated as well as intensified by bureaucratic interests in the distribution of water resources and by the regional interests of the various republics. The zone of river-flow formation that eventually supplies water to the Aral Sea – the mountain regions of Tien Shan and Pamir-Alai – is almost completely located within the republics (now sovereign states) of Tajikistan and Kyrgyzstan. The zone of the first use of surface waters – mainly the piedmont oases – includes parts of the republics (now sovereign states) of Uzbekistan, Turkmenistan, and Kazakstan. The lower reaches of the rivers and the sea proper belong to Uzbekistan (more correctly, Karakalpakstan) and to Kazakstan. This latter territory is of relatively little importance for Kazakstan, except for the Kyzyl-Orda and Kazalinsk oases and the city of Aral'sk. It is mostly desert with a very sparse population. For Karakalpakstan, however, it is the main area of habitation for its people.

The differences in the needs and in economic orientation of each of these republics which were under pressure from Moscow (until 1985) to expand cotton production, and the expansion of inefficient irrigation networks and the resultant natural reduction of water supply in the region, led to the irrational use of water and land resources and the creation of desertification processes. Such processes involved the replacement of natural ecosystems by human-induced landscape changes. It led to an ecological disaster, if not a catastrophe (Glazovsky, 1990).

At the same time, the destabilization of natural conditions has generated its own ecological problems in connection with the impact of Karakum Canal water on the natural environment (in particular, secondary salinization of fields), groundwater level increases in the foothills, salinization and the waterlogging of desert pastures, as a result of the uncontrolled discharge of drainage water into deserts. To many, the formation of Lake Sarykamysh has been a welcome event, because of the commercial fish catches. In Uzbekistan many people consider Lake Arnasai as a convenient setting for recreation and amateur hunting. Kazakstan has its own set of urgent problems not related to the Aral Sea (e.g., the Caspian Sea and Lake Balkhash). Thus, in reality Karakalpakstan remains alone to cope with its urgent ecological, social and economic problems. This situation has entered the realm of 'ecological politics'.

Concluding comments

The main conclusions of this analysis of the anthropogenic changes in landscape composition in the Aral Sea basin are as follows:

Anthropogenic changes in the landscape of the Aral Sea basin are found in all zones along the rivers' courses. The extent of the variability in landscape composition and the proportion of newly formed landscapes are increasing in all the zones along the river courses.

The ecological catastrophe in the Aral Sea basin is a visible example of creeping environmental changes in ecological conditions, including landscape alterations, both in space (from the mountains to the sea) and in time. More than 30 years have passed since the beginning of intense anthropogenic impacts in the region to the exposure of the Aral catastrophe (as noted in Table 2.3).

The anthropogenic modification of landscapes (i.e., when one or two aspects are altered) has been confirmed by numerous studies of the successive changes of vegetation, soil, and animal populations. New natural territorial complexes and the development of zonal landscape formation have become established on the dried sea bottom. The development of desertification processes (e.g., eolian, karst-undermining, salt removal and redistribution) was widespread in the third zone of the basin.

Unstable landscapes with complex infrastructure, representing combinations of newly formed lacustrine–boggy–solonchak complexes in the

Table 2.3 Chronological review of the state of natural environment in the Aral Sea basin in connection with intensification of anthropogenic impact

Years	Governmental resolutions, water management, scientific and organizational and research works	Certain data on the state of natural environment of the Aral Sea basin and its water area	Notes
1950	Governmental resolution on surveying along the routes of the proposed Main Turkmen Canal and the Karakum Canal	Surface flow of 120 km^3/year is formed in the mountains; the Aral Sea is at the mark of 53.5 m above sea level; water volume in the sea is	
1951	Surveying along the routes of the canals	1050–1077 km^3; water area is about 66 000 km^2 (64 490–69 531 km^2, without islands – 66 900 km^2); river flow into the Aral Sea is 52–56 km^3/year; water mineralization in the sea is 10–12%; natural decline is 0.6 cm/year; number of lakes in the deltas is 2570, total area is 1679 km^2; Sarykamysh – sandy- solonchak desert at the mark above sea level	
1952	Kattakurganskoe Reservoir is constructed[a]	An area of 84.5 km^2 is flooded	
1954	Resolution on the construction of the 1st stage of the Karakum Canal		
1954–1958	Construction of the 1st stage of the Karakum Canal by the method of 'successive sections'	Signs of groundwater levels rising due to canal construction, especially at the sections of the Kelifsky Uzboi and Southeastern Karakum	
1956	Kairakkumskoe Reservoir is constructed[b]	An area of 513 km^2 is flooded; water volume in the reservoir is 40 800 km^3; water withdrawn for the canal – 0.84 km^3	
1959	Commissioning of the 1st stage of the Karakum Canal	Water withdrawn from the Amudarya River – 3.5 km^3. Formation of lakes in the Kelifinsky Uzboi and interridge depressions of sands	First publications about the zone of canal influence
1960	Chakyr[b] and Khauzkhanskoe Reservoirs (Karakum Canal)[a] are constructed: 2nd stage of the canal is commissioned	An area of 40.7 km^2 is flooded; water volume in the reservoir is 11.35 km^3. The beginning of the sharp decline of the Aral Sea level (53.35 m above sea level)	
1961	Drainage water discharge into the Daryalyk and Ozernyi main drains	The beginning of the formation of waste-water accumulating lakes in the Sarykamysh depression. Water	

Table 2.3 *(cont.)*

Years	Governmental resolutions, water management, scientific and organizational and research works	Certain data on the state of natural environment of the Aral Sea basin and its water area	Notes
		withdrawal from the Amudarya into the canal – 4.1 km^3	
1962	Beginning of the industrial development of the Sarykamysh. Pioneering canal of the 3rd stage of the Karakum Canal	Mineralization of lake water is 3–4 g/l. Aral Sea water level is 53.12 m above sea level	
1963	Chimkurganskoe^a and Karkidonskoe^b Reservoirs	An area of 49.2 km^2 is flooded; Lake Sarykamysh – 103 km^2. The Aral Sea water level is 52.72 m above sea level	
1964	Yuzhnosurkhanskoe^a and Katpassaiskoe^b Reservoirs	An area of 67.5 km^2 is flooded. The Aral Sea water level is 52.48 m above sea level	
1965	Chardar'inskoe^b and Kassanaiskoe^b Reservoirs	An area of 9011 km^2. The Aral Sea water level is 52.46 m above sea level	
1966	Naimanskoe Reservoir^b. Beginning of the construction of the 3rd stage of the Karakum Canal	An area of 6.2 km^2 is flooded. The Aral Sea water level is 52.06 m above sea level. River flow is 39 km^3; water withdrawal into the Karakum Canal is 8.3 km^3	
1967	Dzhizakskoe Reservoir^b	An area of 12.5 km^2. The Aral Sea water level is 51.55 m above sea level. Lake Sarykamysh water volume is 1.4 km^3; salt is 7 million tonne/year	
1968	Pachkamarskoe Reservoir^a	An area of 14.2 km^2 is flooded. The Aral Sea water level is 51.44 m above sea level	Final justification of economic priority of water resources use in irrigation (Geller, 1969)
1969	Discharge of flood water of the Syrdarya into the Arnasai depression	An area of 2330 km^2 is flooded; water volume is 20 km^3. Formation of the system of Arnasaiskie lakes. Nearly 11 km^3 of water were supplied to the Aral Sea; water level is 51.96 m above sea level	
1970	Charvakskoe^b, Bugunskoe^b Reservoirs, Ashkhabadskie and Geok-Tepinkoe (Karakum Canal)^a	An area of 103.8 km^2 is flooded. River flow into the Aral Sea is 7–11 km^3. The Aral Sea water level is 51.36 m above sea level. The Syrdarya River does not reach the efficiency of the Karakum Canal of 52–74% (Kirsta, 1989)	The first alarming information appears in the press

Table 2.3 *(cont.)*

Years	Governmental resolutions, water management, scientific and organizational and research works	Certain data on the state of natural environment of the Aral Sea basin and its water area	Notes
1971– 1973		Water withdrawal by the Karakum Canal – 8.9 km^3/year; seepage losses – 1.3–2.9 km^3/year. An area of 69.5 km^2 is flooded in the zone of influence of the 1st stage; an area of 220.3 km^2 is subjected to ground-water rise; water mineralization of the reservoirs and canals is 1.3 g/l. Consumptive water losses in the lower reaches of the Amudarya are about 13.6 km^3/year. Aral Sea water level is 51.28 m above sea level. Lake Sarykamysh water area is about 1020 km^2	Low water years. Many scientists are concerned about the ecological situation in the deltas
1974	Toktagul'skoe[b], Akhangaranskoe[b] Reservoirs; the Karshinsky Canal (hydraulically sealed)	The area of 292.5 km^2 is flooded. Aral Sea water level is 50.14 m above sea level. Zone of the Karakum Canal influence is 5000 km^2, out of it 40 km^2 are solonchaks; seepage losses from the Karakum Canal within 5 years made up 25 km^3	The beginning of widespread use of satellite data
1975	Nurekskoe Reservoir[a]. Commissioning of the 3rd stage of the Karakum Canal. Conclusion of the Provisional Scientific and Technical Commission on the assessment and changes of the Aral Sea level on the environment	The area of 98 km^2 is flooded. Sarykamysh – lake area is 1450–1750 km^2; waste-water collecting lakes and bogs of the Akhchadarya – more than 100 km^2. Water withdrawal into the canal – 13.5 km^3. The Aral Sea region: dust-salt storms 6–7 times a year, with	Justification of the necessity of studying the 'basin-sea' system (Kuznetsov, 1991)
	First Working Coordination Meeting on the Aral Sea Problem, Tashkent. Meeting on the influence of interbasin redistribution of the river flow on natural conditions, Moscow. Resolution of State Committee of the USSR for the Science and Technology (GKNT) on the studies of the Aral Sea problem	dust transfer at a distance of 350–400 km. Intense incision of Daryalyk, intensification of karst-undermining processes on the Kunya- and Akhchardar'-inskaya Plains, in the lower reaches of the Amudarya; groundwater level decline from 10 to 52.5 m; width of the dried zone in the southeast of the sea is 40–50 km. Aral Sea water level is 49.34 m above sea level	Especially acute discussion on the Aral Sea problem
1976–	The works in the Amudarya Delta	Aral Sea water level is 48.52 m in	Integrated component

Table 2.3 *(cont.)*

Years	Governmental resolutions, water management, scientific and organizational and research works	Certain data on the state of natural environment of the Aral Sea basin and its water area	Notes
1985	and the Aral Sea region on the program of GKNT at sections-profiles	1976, 47.9 in 1977, 47.31 m above sea level in 1978	study of changes; forecasting
1977–1978	Establishment of Scientific Council for integrated study of Nukus deserts. Second Working Meeting on the Aral Sea problem, Alma-Ata	Development of thematic series of maps of natural resources of Tajikistan and Uzbekistan on the basis of satellite data	
1979	All-Union Meeting 'Scientific fundamental of measures for the prevention of negative effects of the decline of the Aral Sea level', Moscow: VII Congress GO (Sec. 5, p. 5 – On the Aral Sea Problem)	Aral Sea water level is 47 m above sea level; water mineralization is up to 22%; area of dried bottom is 64 500 km^2. Consumptive water losses in the Syrdarya basin are 22–24 km^3/year, in the Amudarya basin (Kerki) – 7 km^3/year	
1980	Andizhanskoe Reservoir[b]. 3rd Working Coordination Meeting on the Aral Sea problem and out-of-town Session of the Presidium of the Academy of Sciences of UzSSR on the problem of the Aral Sea and the Amudarya Delta	The area of 59 km^2 is flooded. Seepage losses of the Karakum Canal – 1514 km^3; water withdrawal – 10.7 km^3. Dust storms in the Aral Sea region – 9–10 times a year. The area of Lake Sarykamysh is 2300 km^2. Aral Sea water level is 46. 5 m above sea level; river flow into the sea is actually nonexistent	The data for feasibility report on the Aral Sea problem are prepared
1981–1982	Water discharge through the Kuvandarya and Zhanadarya Rivers[b]. Coordination meeting on the organization and planning of works on the Aral Sea problem, Moscow	Aral Sea water level is 44.9 m above sea level. Water mineralization in the lower reaches of the Amudarya is 1.5–2.2 g/l; Syrdarya is 2–2.5 g/l. Seepage losses of the Karakum Canal is 1.2 km^3	
1983	All-Union Coordination Meeting on the program of GKNT, Moscow. All-Union Meeting on the diversion of a part of the river flow to the south. Memorandum on the problem of degradation of the Aral Sea ecosystems, the Amudarya and Syrdarya Deltas and anthropogenic desertification of the Aral Sea region, caused by withdrawal of the flow of Central Asia rivers aimed at intensification of irrigated	Aral Sea water level is 44.8 m above sea level; water mineralization from 17–18 to 22%. Impact of groundwater rise is up to 200 km. The area of Lake Sarykamysh is 2800 km^2. Total discharge of drainage water into outlying lakes is about 10 km^3/year. Changes and the dynamics of soils and vegetation in the deltas and dried bottom of the Aral Sea. Existing and forecast water-salt budget of the Aral Sea is	Negative attitude to the Memorandum of the Ministry of Land Reclamation and Water Management of the USSR and a number of academicians of VASKhNIL (Glazovsky, 1990)

Table 2.3 *(cont.)*

Years	Governmental resolutions, water management, scientific and organizational and research works	Certain data on the state of natural environment of the Aral Sea basin and its water area	Notes
	agriculture (IG AN, with participation of SOPS and Soyuzgiprovodkhoz)	calculated (Chernenko, 1983; Shultz and Shalatova, 1968)	
1984	Construction of water conduit Tyuyamuyun – Nukus	Aral Sea water level is 43.5 m above sea level; the area of dried bottom is 18 825 km². Natural territorial complexes in the zone of the 1st stage of the Karakum Canal are changed in an area of 533 km². The discharge of drainage water into outlying lakes made up 3 km³, in the Bukharsky oasis is 3 km³	
1985–1986	The reservoirs: Tamarzhanskoe (Karshi)[a], Tyuyamuyunskoe[a], Andizhanskoe[b], Takhiatashskoe[a]. Final report of the results of works on the GKNT subject. VIII Congress GO with consideration of the Aral Sea problem	The area of 720 km² is flooded. Aral Sea water level is 43.2 m above sea level in 1985, 41.41 m above sea level in 1986	
1987	Governmental Commission on Ecological Situation in the Aral Sea Basin (Chairman, Dr. Yu. A. Izrael)	Aral Sea – water level of 40.3 m above sea level; water mineralization is 28–30%; the area of dried bottom is 41 000 km². Lake Sarykamysh area is about 3000 km², water mineralization is 12.1–13 g/l	Data of the Commission are not published. Participation of foreign specialists (Dr. P. D. Micklin, USA)
1988	Resolution of the government 'On measures for radical improvement of ecological and sanitary situation in the Aral Sea region, increase of the efficiency of use and conservation of water and land resources in its basin'	Lateral erosion of Daryalyk is about 1000 m³/year. Drainage water discharge into peripheral lakes (without Sarykamysh) – 7.5–8 km³/ year. Irrigation system is over 700 000 km in length; irrigation and drainage network	Widespread interest of scientists and public (Special issues of Journals 'Izv. AN SSSR, ser. Geogr.', 'Izv. VUZ', 'Vestnik MGU, ser. Geogr.', 'Probl. Osvoen. Pustyn', etc.; public journals 'Novyi Mir', 'Pamir', etc., collections.) Acknowledgment of ecological disaster
1989	The Uzbek Committee for the Aral Sea Salvation, Kazakh Public Committee on the Aral Sea and Balkhash Problem, Public Aral	Aral Sea water level is 39.5 m above sea level, the sea area is reduced by 40%; mineralization is 30%. Lake Sarykamysh water mineralization	Actual measures on improvement of ecological situation were not taken.

Table 2.3 *(cont.)*

Years	Governmental resolutions, water management, scientific and organizational and research works	Certain data on the state of natural environment of the Aral Sea basin and its water area	Notes
	Movement are established. Meeting 'The Aral Sea Problems', Scientific Symposium on the Concept of Social and Economic Development of the Aral Sea Region, Shavat. International projects for the analysis of situation in the Aral Sea region are developed	is 13 g/l, Arnasai is 1.0–13 g/l, the area is from 1750 to 2330 km^2. Irreversible transformations of delta ecosystems into desertification. The dynamics of neolandscapes of dried bottom is established depending on soil conditions and time of drying	Acknowledgment of ecological disaster
1990	Scientific and Research Coordination Center 'Aral' IG AN SSSR with affiliated branch in Nukus is established. International sub-projects within the projects: 'Critical zones of the world', 'Conservation of arid ecosystem', 'Global changes', etc. The Global Infrastructures Fund (GIF), Japan, with Committee on Improvement of Environment (Problems of the Aral Sea) is established		
1991	'Fundamentals of the concept for conservation and rehabilitation of the Aral Sea region, normalization of ecological, sanitary-hygienic and medico-biological situation in the Aral Sea region' are developed. Workshop on the problems of monitoring in the Aral Sea region. The beginning of works under the aegis of GIF, Japan	Aral Sea water level is 37.5 m above sea level. Volume of waste-water collecting lakes is 34 km^3, Lake Sarykamysh is 26 km^3. The withdrawal of water by the Karakum Canal is 10–12 km^3/year. Further deterioration of ecological situation and extensive development of desertification processes	The concept is based on basin approach. Widespread publication of the results of scientific studies and forecasted projects. Acknowledgment of the Aral Sea catastrophe by the international community
1992	Cessation of the activity of the 'Aral' Center, weakening of consolidation of scientific studies. 'The agreement between the countries of Central Asia and Kazakstan on the cooperation in the field of joint control of the use and conservation of water resources of the interstate sources' is developed. Symposium on the Aral Sea problem (GIF, Tokyo)	It is noted in the Resolution of the Symposium in Tokyo that continuing deterioration of the environment in the Aral Sea basin is one of the most serious ecological problems in the world	The Aral Sea problem becomes more regional and politicized (Zonn, 1993)

Ecological disaster and changes in landscape composition

Table 2.3 (cont.)

Years	Governmental resolutions, water management, scientific and organizational and research works	Certain data on the state of natural environment of the Aral Sea basin and its water area	Notes
1993–1995	The law 'On social protection of citizen who suffered due to ecological disaster' is The Interstate Coordination Water Management Committee is established (for the basins of Syrdarya and Amudarya) Consortium of 5 republics of Central Asia, Russia, the Global Infrastructures Fund, Japan, USA.	Aral Sea water level is 36.5 m above sea level. Lake Sarykamysh water volume is about 45 000 km³. Total area of lands flooded by water of the reservoirs and waste-water collector lakes is more than 20 000 km². Intense desertification of deltaic ecosystems; processes of salinization, eolian transfer of salts and dust, eolian formation of topography, karst-undermining, erosion are developed.	

Notes: [a]Reservoir in the Amudarya basin; [b]Reservoir in the Syrdarya basin

depressions and sandy or stony-clay deserts on elevated areas, have formed on the edges of the oases, especially on the boundaries of the newly irrigated areas and along irrigation canals and main drains. The increase in the areal extent of landscape covered with hydrophyte and phreatophyte communities results in the increase of water consumption resulting from evapo-transpiration in the middle reaches of the Amudarya and Syrdarya.

Landscape changes under the impact of human activities have been mirrored in satellite imagery data. Satellites enabled the monitoring of changes in various parts of the Aral basin, and also allowed for the detection of different types of changes in the region.

An ecological (e.g., holistic) approach to resolving regional environmental problems underscores the need for interstate agreements on the use of transboundary water resources and on the implementation of regional monitoring, surveys, and studies. Clearly, short-term economic benefits over a long term may cause irreversible adverse alterations of the natural environment, and alterations in the composition of landscapes. The benefits to certain states or people may result in catastrophic situations for others in the basin. This possibility dictates the necessity of implementing reasonably 'weighted' equitable ecological policy among the basin states.

References

Alekseeva, N. N., E. V. Glushko, T. I. Kondrat'eva and A. V. Ptichnikov, 1991: Integrated geoecological studies of the zones of ecological disaster by space images. *Izv. VUZ, Geodez, i Aerofotos'emka*, 1, p. 40.

Anon, 1988: Aral – my hope. *Essays, Poems, Articles*. Nukus, Karakalpakstan, p. 178.

Babaev, A.G. and V.S. Zaletaev, 1990: Standard objects of ecological monitoring in the arid zones. *Problems of Desert Development*, **5**, p. 3.

Babaev, A.M. and A.A. Babaev, 1994: Aero-space monitoring of the dynamics of geosystems with water supply, *Problems of Desert Development*, **1**, p. 21.

Bakhiev, A.M. and N.M. Novikova, 1990: Mapping of recent vegetative cover of the lower reaches of the Amudarya River. In: *Floristic and Ecologo-geobotanical Studies in Karakalpakia*. Tashkent: FUN, **3**, Ch. VI, p. 153.

Baknai, G.G., V.M. Dubkova and N.I. Konstantinova, 1991: Integrated nature-conservation mapping of the northern part of the Aral and Aral Sea region with the use of space imagery data. *Izv. VUZ, Geodeziya i Aerofotos'emka*, **1**, p. 115.

Borodin, L.F., V.N. Bortnik and V.F. Krapivinl, 1987: Regarding the changes, development of the models of functioning and remote monitoring of aquatic geosystems in the Aral Sea basin. *Problems of Desert Development*, **1**, p. 71.

Chernenko, I.M., 1983: Water and salt budget and use of the drying Aral Sea. *Problems of Desert Development*, **3**, p. 18.

Chernovskaya, R., 1988: *The Aral Sea Fate*. Tashkent: Mekhnat, p. 224.

Dukhovny, V.A., P.M. Razakov, I.B. Ruziev and K.A. Kosnazarov, 1984: The Aral Sea problem and nature conservation measures. *Problems of Desert Development*, **6**, p. 3.

Geller, S.Yu (ed.), 1969: *The Aral Sea Problem*. Moscow: Nauka, p. 175.

Gerasimov, I., 1978: *The Karakum Canal and Alteration of Natural Environment in the Zone of its Influence*. Moscow: Nauka, p. 232.

Glazovsky, N.F., 1990: *The Aral Sea Crisis: Causes of Origin and Ways of Elimination*. Moscow: Nauka, p. 136.

Glushko, E.V., 1991: Integrated geoecological mapping of the Aral Sea Region and Kyzylkum Desert by space images. *Vestn, MGU*, Ser. 5 – *Geografiya*, **1**, p. 21.

Glushko, E.V., 1992: Program of aero-space monitoring of nature use and geoecological situation in the Aral Sea region. *Problems of Desert Development*, **2**, p. 25.

Glushko, E.V., A.A. Pochapinskii and V.S. Tikunov, 1994: Modeling of landscape dynamics of the south Aral Sea region by space images. *Problems of Desert Development*, **2**, p. 9.

Grave, M.K. and L.M. Grave, 1983: Impact of large canals of Central Asia on desert ecology, *Problems of Desert Development*, **4**, p. 25.

Khabarov, A.V., 1991: *Establishment of Maps of the Sources of Salt Removal and Regions of Feasible Sedimentation of Salt Dust, Landscape and Ecological Fundamentals of Nature Use and Nature Arrangement*. Tselinograd, p. 38.

Khodzhaev, S.A., L.E. Markova and V.N. Poltavchenko, 1985: Regarding the procedure of integrated thematic mapping of the south-western Aral Sea region based on space imagery data. *Problems of Desert Development*, **5**, p. 18.

Kirsta, B.T., 1989: The Aral Sea problem and the Karakum Canal. *Problems of Desert Development*, **6**, p. 10.

Kuznetsov, N.T., 1991: Actual geographic aspects of present state of the Aral Sea problem and the Aral Sea region. *Problems of Desert Development*, **2**, p. 10.

Novikova, N.M. and G.Yu. Trofimova, 1994: Data base system of ecological trend for the river deltas of arid areas of Central Asia. *Problems of Desert Development*, **2**, p. 68.

Popov, V.A. and B.V. Vinogradov, 1982: Small-scale landscape mapping of the South Aral Sea region. *Problems of Desert Development*, **3**, p. 40.

Rafikov, A. A., 1982: *Natural Conditions of Drying Bottom of the Aral Sea*, Tashkent: FAN, p. 147.

Raskin P., E. Hansen, Zh. Zhu and D. Stavitsky, 1993: Modeling of water supply level and water requirements in the Aral Sea region. *Problems of Desert Development*, 3, p. 28.

Sadykov, Zh. S., 1983: Ground Water and salt flow in the Aral Sea basin. *Problems of Desert Development*, Alma-Ata.

Shultz, V. L. and L. I. Shalatova, 1968: The Aral Sea level in 1961-1966. *Problems of Desert Development*, 3, p. 19.

Skaterschikov, S. V., 1972: Investigation and mapping of anthropogenic landscapes by the materials of space photograph surveying and compilation of nature-conservation maps. *Tr. Gostsentra 'Priroda'*, Moscow, 2, p. 33.

Vinogradov, B. V., 1984: Aero-space monitoring of ecosystems. Moscow: Nauka, p. 320.

Vinogradov, B. V. and D. E. Frolov, 1989: Dynamic ecogeoinformation system with the use of aero-space data base. *Nature and Resources*, XXV, 1–4, p. 68.

Vitkovskaya, T. P., M. Mansimov and L. G. Shekhter, 1985: Dynamics of Sarykamysh Lake based on satellite survey data. *Problems of Desert Development*, 6, p. 38.

Vostokova, E. A., 1991: *Mapping Provision of Space Monitoring of Ecological Conditions, Landscape and Ecological Fundamentals of Nature Use and Nature Arrangement*. Tselinograd, p. 14.

Vostokova, E. A., V. A. Suschenya and L. A. Shevchenko, 1988: *Ecological Mapping on the Basis of Space Imagery Data*. Moscow: Nedra, p. 200.

Vostokova, E. A., I. K. Abrosimov, T. S. Kozlova and M. V. Kirpikova, 1989: Use of space imagery data for compilation of maps of phytoreclamative conditions of the south Aral Sea region. *Problems of Desert Development*, 6, p. 27.

Zaletaev, V. S., 1989: *Ecologically Destabilized Environment*. Moscow: Nauka, p. 148.

Zonn, I. S., 1993: The Aral Sea problem: The problem of great importance. *Problems of Desert Development*, 3. p. 21.

3   Alteration of water level and salinity of the Aral Sea

VITALYI N. BORTNIK

Background up to the twentieth century

The Aral Sea is the largest inland drainless water body with specific marine and lacustrine features located in the Central Asian deserts – in the Turanskaya Lowland, near the eastern edge of the Ustyurt Plateau. The water and salt budgets, the level and areal extent of the sea, the water salinity level, as well as other characteristics of the sea, are completely determined by the streamflow of Central Asia's two major rivers, the Amudarya and the Syrdarya. The amount of surface water flow into the Aral Sea depends on (a) the impacts of climate-related fluctuations of the natural water supply of these rivers and (b) on the steadily growing anthropogenic demands on the rivers' waters during the twentieth century (e.g., the consumptive withdrawals of river water mainly to meet the demands of irrigated farming).

Changes in the regime of the Aral Sea are determined primarily by alterations of two of its key characteristics: its height above sea level and the average level of mineralization of its water (i.e., salinity). The sea's level depends on the volume of water that accumulates in its basin. The average salinity of the sea's water is governed by the relationship between the mass of salts dissolved in the water and the volume (rather, the mass) of sea water.

Paleo studies indicate that there have been significant, age-old cyclical alterations of Aral Sea level and salinity throughout its history. For example, the range of sea level fluctuations during the Holocene apparently exceeded 20 meters. Fluctuations of Aral Sea level are connected with alternating phases of general humidity of Eurasia and the continental land mass of the Northern Hemisphere, the duration of which, according to Shnitnikov, covers about 1950 years. Three long-term periods of high sea level were distinguished by Shnitnikov (1959; 1961) for the Aral Sea region within the last 4500 years: the first period extended from the first half of the third millennium until the end of the second millennium BC; the second period from the first millennium BC until the beginning of the first millennium AD; and the third period from the middle of the second millennium AD to the present. Dry, low-water periods occurred between humid, high-water periods. The filling of the Sarykamysh basin, located near the Aral Sea and serving as an indicator of water supply of the basin by the Amudarya, took place during high-water periods and formed Lake Sarykamysh. The overfilling of Lake Sarykamysh enabled water to flow through the Uzboi channel (which is dry at present) into the Caspian Sea. The range of fluctuations of Aral Sea level within the last 4000 years reached 6 m.

In the course of recorded history, perhaps even earlier, anthropogenic factors also influenced changes in Aral Sea level, not only as a result of irrigated agriculture, which existed in the Aral basin since ancient times, but from periodic changes in the direction of flow of the Amudarya away from the Aral Sea to Lake Sarykamysh. These alterations took place during wars, as a result of the destruction of irrigation facilities (e.g., in the fourth to sixth centuries, the thirteenth century, and the fourteenth century) (Kes' and Klyukanova, 1990).

Fluctuations in Aral Sea level during the last 200 years are covered in more detail by Berg (1908) and L'vov (1959) using proxy data such as historical, literature, and cartographic data. During the decade of the 1720s, the sea's level was about 53 m above sea level. Then there came the period of rapid decline of its level and, by the beginning of the 1820s, it declined to 50 m above sea level. The rise in level by nearly 2 m was noted by the middle of the century, and by the 1880s it was lowered again to 50 m above sea level. From 1885 to 1905 the level of the Aral Sea increased rather quickly by almost 3 m and reached close to 53 m (Figure 3.1). The range of secular fluctuations of Aral Sea level during that period reached 3 m. The rise in sea level at the turn of the twentieth century was the result of quasi-cyclic fluctuations of climate of the whole Northern Hemisphere, which caused a gradual increase in streamflow (L'vov, 1965). In the opinion of Baidal (1972a,b), the periods of catastrophic low-water level of the Aral Sea coincide with the phase of minimum solar activity, and

Figure 3.1 Water-level fluctuations in the Aral Sea. Dashed line, from reconstruction; solid line, from instrumental data.

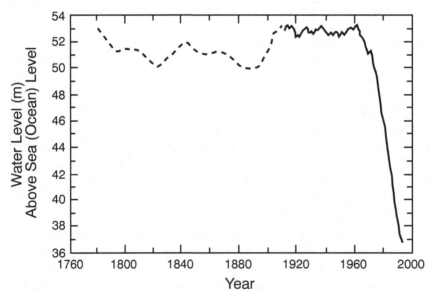

with meridional circulation prevailing in the region of Central Asia. The periods with high-water levels correspond to the periods of maximum solar activity with prevailing meridional circulation.

The Aral Sea, as a unique regional water body, attracted the attention of many scientists by the end of the nineteenth century and in the beginning of the twentieth century. Some scientists considered its existence in the desert as a 'mistake of nature'. Well-known Russian geographer Voeikov (1908) wrote about the expediency of using Amudarya and Syrdarya water for irrigation. He suggested that the drying out of the Aral Sea would not cause any significant ecological problems. He also suggested that, as a result of irrigation development, the regional hydrologic cycle in Central Asia would be intensified: there would be increased moisture supply to the mountain regions, and higher precipitation, which would increase the water supply of the rivers of Central Asia. This assumption was supported by certain scientists for more than half a century as justification for the continued extensive development of irrigated agriculture in the Aral sea basin (Molchanov, 1955; Dunin-Barkovsky and Kunin, 1965). Other scientists had expressed different opinions in a number of papers. At the same time, the beginning of the twentieth century, attention was drawn to the view that the continued development of irrigated farming in the river basins of Central Asia would undoubtedly result in the decline of Aral Sea level (Berg, 1908; Tsinzerling, 1927).

Events after 1900

The first instrumental observations of Aral Sea level were organized in 1900; casual observations were carried out from 1901 to 1910; and systematic observations were started from 1911 in the Saryshaganak Bay. The maximum development of the network for water level observations in the Aral Sea occurred from the 1940s to the 1970s, when 6 to 10 water-level stations and water-stage gauges were simultaneously in operation (Figure 3.2). With sea-level decline, their number was reduced; recently the sea-level observations were carried out at two island stations, Lazarev Island and Barsakelmes. However, these stations were closed in 1992 and by the end of 1993, respectively. Systematic observations of Aral Sea level were terminated in the mid-1990s.

The first information about salinity and salt composition of the Aral waters was obtained during analyses of the discrete samples of sea water taken in the 1870s to 1880s and in the beginning of the twentieth century (Berg, 1908).

Separate surveys of sea-water salinity were made during the 1930s and 1940s by the Aral scientific fishery station. Systematic studies of coastal water salinity were carried out by the network of hydrometeorological stations, from 1941 up to its closure. Systematic seasonal surveys of surface area and

Figure 3.2 Former and current coasts of the Aral Sea. Solid line, pre-1960 coast; dashed line, position of modern coast of Maloe (Small) Sea at 39.5 m and the Bol'shoe (Large) Sea at 37 m above sea level. Solid triangles indicate location of proposed monitoring stations (see text).

depth of the sea started in the early 1950s, after the establishment of a 'secular' network of oceanographic stations for the open sea in the Aral Sea. These surveys were carried out until the beginning of the 1980s and their data were used by the majority of investigators in their calculations of salinity and salt mass of the sea and in studies of their long-term dynamics.

Early Aral Sea investigators tried to explain the 'geographic paradox' of the Aral Sea – the apparent discrepancy between (a) its geological age and (b) the time of salt-mass accumulation, calculated by average annual ionic flow of the rivers. Thus, Berg (1908) noted that the entire salt mass of the sea could have been established during a period of only 320 years because of salt flow of the rivers. The amount of carbonaceous salts, when mixing sea and river water, does not explain the 'paradox', but only increases the amount of accumulation of current salt reserves by 2 to 3 times (Blinov, 1955).

Major assumptions about the causes of the relatively low level of salinity of water in the Aral Sea can be reduced to the following:

1. Periodic overflow of the Aral Sea in the past discharged salts into the Sarykamysh depression and sometimes through the Uzboi into the Caspian Sea.
2. Sedimentation of salts during low-water periods of the sea and their subsequent coverage by other sediments inhibited the dissolving of salts during periods when sea level had increased.
3. Sea water and salts infiltrated into the soil along the coastline and the sea bottom during periods of relatively high sea level.
4. Sea-water salinity and local sedimentation of soils increased in several shallow-water bays, coves and coastal lakes.
5. Sedimentation of salts occurred as a result of biohydrochemical processes.

Two basic periods are distinguished within the time of instrumental observations of changes in the Aral Sea level and in the Aral Sea regime as a whole: the natural period (i.e., quasi-stationary period from 1911 to 1960) and the current, sharply nonstationary period of active anthropogenic impact on the regime of the sea (from 1961 to the present).

The relative stability of the sea regime is typical for the first period. There was an approximate parity in the pluses and minuses of the Aral Sea water budget (Table 3.1) and an absence of significant long-term trends in their variability (Figure 3.3). There were insignificant fluctuations of sea level near the

Table 3.1 Average annual values of the main components of water budget and morphometric characteristics of the Aral Sea for different periods

Period (years)	River flow	Precipitation	Evaporation	Average annual increment of the sea level (cm)	Characteristics of the sea by the end of the period		
		$km^{3\,a}$ (m)			Level (m above sea level)	Area (thous. km^2)	Water volume (km^3)
1911–60	56.0 (0.847)	9.1 (0.138)	66.1 (1.000)	0.1	53.4	67.1	1083.0
1951–60	58.4 (0.873)	9.2 (0.138)	66.0 (0.991)	6.7	53.4	67.1	1083.0
1961–70	43.3 (0.685)	8.0 (0.127)	65.4 (1.035)	−21.8	51.2	60.2	950.6
1971–80	16.7 (0.293)	6.3 (0.110)	55.2 (0.968)	−57.6	45.4	50.8	628.4
1981–89	4.2 (0.132)	5.5 (0.143)	39.0 (1.050)	−76.3	38.6	36.5	328.6
1994	Small Sea				39.5	3.0	20.3
	Large Sea				36.9	30.1	256.2

Note: ^{a}Upper figure in table refers to annual volume measured in cubic kilometers. The lower figure in brackets is the thickness of the layer of water spread over the whole sea (so that the bottom figure multiplied by the area equals the upper figure).

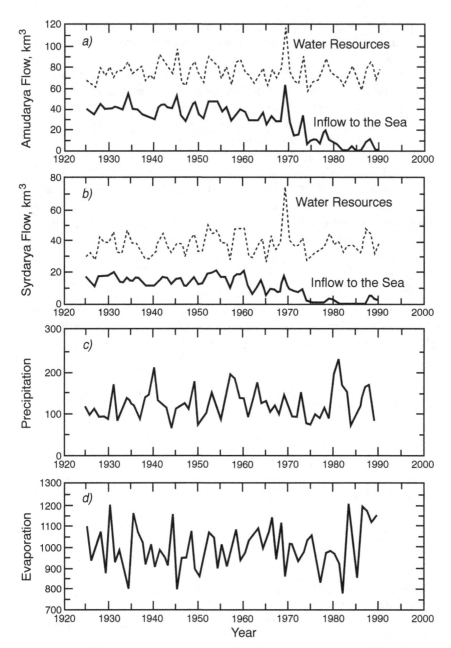

Figure 3.3 Annual changes in water resources and inflow of the (a) Amudarya water and (b) Syrdarya water to the sea, (c) precipitation, and (d) evaporation from the sea area.

mark of 53 m above sea level, which was assumed to be the average annual level. In this period the range of fluctuation of average annual levels did not exceed 1 m (Figure 3.1).

At 53 m above sea level the sea's total surface area, including islands, made up about 68 300 km² (of this the water surface area was 66 100 km²). The sea's volume of water was 1064 km³ and its average depth was 16.1 m. In this situation, the areal extent of the relatively isolated northern part of the Aral Sea, the so-called Maloe More (Small Sea), was 6000 km² and its volume was nearly 80 km³ (9.1% and 7.5% of the total sea, respectively).

During the period of the quasi-stationary regime of the sea, the relatively large annual volume of river flow was about 1/19th of the sea's volume and the peculiar salt composition of Aral waters was distinguishable from the salt composition of other closed and semi-closed seas by its high content of carbonates and sulfates. The salt budget of the sea was also relatively stable. As a result of river flow 24–26 million tonnes of salts entered the sea annually. The main volume of salts was the result of the mixing of sea and river water (due to the supersaturation of Aral water with calcium carbonate) and was deposited in shallow water areas, bays, coves and, due to seepage, the lakes of the northern, eastern, and southern coasts of the sea (Table 3.2). The mean salinity level of sea water was changed only insignificantly within that period (between 9.6 and 10.3‰).

The Amudarya and Syrdarya basins are the regions of ancient irrigation networks which had their effects on river flow and water budget and on fluctuations of the Aral Sea level over time. At the beginning of the twentieth century nearly two million hectares in the Aral basin were irrigated. By the 1960s, 5.1 million ha were irrigated, and by the beginning of the 1990s the area of irrigated lands had increased to 7.2 million ha. However, not all of this land was in production at the same time. The volumes of consumptive withdrawals of flow fluctuated insignificantly until the 1950s and was 20–25 km³/year.

During the 1950s, despite the increasing amounts of withdrawals of river flow (up to 40 km³ annually on average), there was an increase in water supply of the Amudarya and Syrdarya rivers to the sea and a reduction in natural channel losses of flow. Thus, the inflow of river water to the sea was somewhat higher than the average annual inflow for 1911–60 (Table 3.1). A small rise of the sea level was noted in 1951–60, approximately 0.5 m, and in 1960 the sea was at 53.4 m above sea level, the maximum level for the period of instrumental observations (Figure 3.1).

This period was relatively cool and humid which, apart from increased water supply from the Amudarya and Syrdarya, was also characterized by reduced evaporation and increased precipitation over the sea itself (Figure 3.3). This combination of climatic factors was most favorable and led to the excess of plus components of the water budget over minus components.

REGIONAL AND GOVERNMENT PLANNING

The first quantitative assessments of expected changes in level and in salinity of the Aral Sea go back to the late 1940s and late 1950s, with the development of the 'Plan for Nature Transformation' in Central Asia. This plan envisaged the construction of the Main Turkmen Canal with water withdrawal from the Amudarya, the construction of water reservoirs along the

Table 3.2 Average annual values (million metric ton) of the Aral Sea salt budget for typical periods

Credit	1911–1960	1961–1970	1971–1980	1981–1985	Debit	1911–1960	1961–1970	1971–1980	1981–1985
Ionic flow of the rivers	23.79	26.89	12.26	2.23	Sedimentation of salts with displacement of the river and sea waters	10.94	7.94	2.68	0.39
Salts supplied with groundwater inflow	1.40	1.40	0.70	0.70	Salts lost due to infiltration of sea waters	1.50	1.50		
Salts supplied with precipitation	0.36	0.36	0.44	0.59	Salts removed with evaporating water	0.17	0.16	0.13	0.12
Salts supplied with atmospheric dust	0.40	0.38	0.61	0.69	Wind removal of salts from water surface	0.39	0.38	0.36	0.33
					Wind removal of salts from ice	3.29	2.80	3.85	3.95
					Precipitation of salts in shallow waters and salt sedimentation on the shores	12.62	4.85	2.04	4.82
					Precipitation of carbonate and sulfate salts with increasing salinity of sea water	0	2.39	4.34	59.10
Closing error						−2.96	9.51	−0.19	−64.50

Source: Tsytsarin and Bortnik (1991).

Syrdarya and Amudarya, and a significant extension of irrigated farming areas in the Aral basin.

Preliminary assessments (Zalikov, 1946; Blinov, 1953, 1955; Samoilenko, 1955) have shown that, with an increase of withdrawals from the rivers of 10 to 20 km³/year, the level of the Aral Sea could be reduced by 5 m and 10 m, and salinity would be increased by 3.3‰ and 14‰, respectively. In the first scenario (10 km³/year), the sea's surface would be reduced to 54 000 km², and in the second it would be reduced to 39 000 km². As these calculations were further refined, they unambiguously pointed to forthcoming negative changes of the sea's regime if the planned water management policy in the basin was instituted.

Thus, the destiny of the Aral Sea was predetermined by regional economic development planning. All subsequent resolutions of the Communist Party and the Soviet government envisaged the continuation of extensive growth of irrigated farming, which depended on the increase in consumptive withdrawals of river flow and, therefore, a major reduction of flow into the sea. They understood that a drying out of the sea and an increase in salinization of its waters would occur. However, the various ecological, economic and social consequences of this occurrence were not forecast in that period because they were viewed as very remote possibilities.

Changes in the 1960s and 1970s

By the end of the 1950s, no signs of forthcoming changes in the Aral Sea had been observed. On the contrary, due to a temporary rise in sea level, construction had to be carried out to reinforce and protect against scouring during river-flow surges at certain sites along the coastal areas around the Aral ports of Aralsk and Muynak. Significant changes were taking place and accumulating in the sea basin, such as the extension of irrigated areas, and a growth of consumptive withdrawals of river flow. They were compensated for by natural factors and therefore had no effect at that time on the flow of river water into the sea. However, this was soon to change.

Continued growth of consumptive withdrawals of the Amudarya and Syrdarya flow during the 1960s and after resulted in upsetting the equilibrium of the water budget of the Aral Sea. This was primarily because of (a) the exhaustion of compensatory opportunities to reduce natural losses, and (b) the onset of the period of low water levels of the rivers of Central Asia. The inflow of river water to the sea began to drop rapidly and the sea level, beginning in 1961, acquired a stable downward trend. During 1969, an abnormally abundant water year when the inflow of river water to the sea exceeded 80 km^3, the decline of sea level ended temporarily, only to continue thereafter with greater intensity.

Already for the 1961–70 period, the average inflow of river water to the Aral Sea had been reduced to 43.3 km^3/year (or 77% of the average annual inflow during the quasi-stationary sea level regime). The average sea level drop was about 21 cm/year, and the total drop in that decade reached 2.1 m (Table 3.1). The sea surface area was reduced to 60 200 km^2 by the beginning of 1971 and the sea's volume to 951 km^3 (91% of the area and 89% of the sea volume at the 53 m level). Within the same period, the average salinity level increased by 1.7‰ reaching 11.6‰. In the salt budget of the sea, despite the reduction of the volume of river flow, the ionic flow and mineralization of the rivers increased significantly. On the whole, the influx of salts to the sea during that period exceeded the discharge and, as a result, a growth of salt in the sea occurred (Table 3.2).

An increase in precipitation over the sea was noted during the first five years of the 1960s. It then changed to a phase of reduced precipitation, which continued until the end of the 1970s. The relatively humid preceding period was followed by one that was more warm and dry, a phase of increased evaporation from the sea (Figure 3.3). Such a combination of climatic factors abetted the recession of the sea. If water use in the 1961–1970 period could have been limited to the amounts used in the 1950s, i.e., in the absence of the growth in usage, the Aral Sea level at the beginning of 1971 would have been higher by 0.9 m than it actually was.

Of course, these adverse changes in the sea's regime were known. The sea level decline between 1961 and 1965 had already reached 1.3 m, exceeding the 1-m range of fluctuations for the preceding 50-year period.

The record decline in sea level and the increase in salinity of the sea had not yet affected the sea's main physical and chemical processes and its ecological situation. However, by the late 1960s and early 1970s, negative trends in the regional economy became apparent – in navigation and in the fishery. The first of them – navigation – was connected with existing shoaling of the approaches to the ports and the need to carry out dredging works. The second negative trend – the drying of coastal shallow water areas – adversely affected spawning grounds for fish populations and led to increased salinity in these areas (reaching 14‰). These negative trends affected biota and the ability of commercial fish to reproduce. Fish catches dropped rapidly.

Forecasts of further changes of the Aral Sea regime published in the late 1960s and early 1970s (Azarin, 1964), based on the planned increases in the consumptive use of water in the Aral basin, pointed to a continuation of the desiccation of the sea. They differed only in their assessments of its intensity. 'Fruitful discussions' on the opportunity to use the areas exposed by the receding Aral Sea for farming appeared in a number of papers of that period (Geller, 1969).

Between 1971 and 1980, the annual flow of river water into the Aral Sea made up only 16.7 km³/year (or 30% of the average annual value). The average rate of sea level decline increased to 57.6 cm/year during those years, adding up to a 5.8 m drop in a decade. Its total decline since 1961 reached 7.9 m (Table 3.1). By the beginning of 1981 the surface area of the sea was reduced to 50 800 km² and the water volume had dropped to 628 km³ (77% of the area and 59% of the volume of the sea at the 53 m level). The average salinity within the same period increased from 6.5‰ to 17‰.

During the 1970s and 1980s, an extended low-water period occurred. The reduction of precipitation and of evaporation during the 1970s was also noted (Figure 3.3); and the so-called visible evaporation (evaporation minus precipitation) was close to the average for the quasi-stationary period. Thus, the total impact of the various climatic factors, mainly the low supply of river water, was unfavorable for maintaining the water budget of the sea. Under the conditions of limited water supply and use in the Aral basin (and assuming the volume of the sea that existed at the end of the 1950s), the sea's level in 1981 would have been 51.0 m above sea level. In other words, the level would have been reduced by 2.3 m within 20 years because of climatic conditions. The remaining 5.6 m of its recession are connected to the cumulative adverse effect of continued growth in the consumptive use of river flow.

Basic changes also took place in the sea's salt budget (Table 3.2). Between 1971 and 1980 the total inflow was reduced by half, including the total discharge of salts. As a result, the salt budget of the sea was close to equilibrium. The reduction of ionic flow was inadequate in comparison with the reduction of water flow due to an insignificant increase in the mineralization of river water (up to 0.7 g/l on average for the Amudarya and to 1.3 g/l for the Syrdarya). The increase in mineralization of the rivers was accompanied by an abrupt change in its ionic composition. From calcium carbonate, the waters of the Amudarya and Syrdarya were transformed into sodium sulfate. These changes resulted from the increase in the amount of highly mineralized waste and irrigation drainage waters that was returned to the rivers.

During the first half of the 1970s, the level of the Aral Sea during its recession exceeded a 2–3 m range of fluctuations that had occurred in the preceding two centuries. Already by the middle of the 1970s, the level of the Aral Sea had declined to 48–49 m above sea level (a 4–5 m drop from the quasi-stationary period). In addition, average salinity of the sea water exceeded 14‰, a threshold for the biota of fresh-water and brackish origin, including commercial species of fish. Above this concentration, reproduction ceases. Sea water quality deteriorated significantly, as a result of pollutants entering the sea from the rivers. Due to considerable shoaling, the Aral Sea completely lost its value for transport. Fish catches dropped sharply. Inflow from the Syrdarya to the sea actually terminated from 1974 until 1987.

The drying out of the Aral Sea also began to have a negative effect on the natural environment in the region. The groundwater level declined. The processes of desiccation and degradation of natural ecosystems of the Amudarya and Syrdarya deltas were initiated. The dried bottom of the sea became the source of dust and salt storms. The fall in sea level adversely changed economic activities in the sea (e.g., the loss of commercial fishing activities) and prompted migration of a part of the population away from the coastal area.

REACTION TO THE PROBLEM

The problem of the drying out of the Aral Sea and its possible consequences was realized by the majority of Soviet scientists and specialists by the mid-1970s. The scope and number of scientific studies and publications increased.

Negative environmental changes related to shortages, as well as the low quality of drinking water, and the impairment of ecological conditions in the sea basin continued. Thus, awareness of the problems of the Aral Sea expanded to include all aspects of adverse ecological, social and economical changes in the entire basin.

Concern for the future of the Aral Sea was displayed in the preparation of a scientific report on the problem of the Aral Sea, commissioned by the State Committee for Science and Technology (GKNT) and the Presidium of the Academy of Sciences of the USSR (Korenistov *et al.*, 1972). The changes which were taking place in the basin and in the regime of the sea were reported, and the inevitable drying out of the sea in future was forecast. The diversion of Siberian river flow to the arid lands of Central Asia was proposed. Moreover, additional studies were planned for the near term.

In the second half of the 1970s a number of proposals appeared in the Soviet press, aimed at decelerating the rates of decline of the Aral Sea level. They proposed ways to intervene in the process of degradation, such as the partial conservation of its relatively isolated water areas, e.g., the Small Aral Sea, the western or eastern part of the Large Aral Sea, and the major bays – Bol'shoi Sarychaganak, Adzhibai, Dzhiltyrbas (e.g., Kuznetsov and Nikolaeva, 1975; Bortnik, 1977, 1978; L'vovich and Tsigel'naya, 1978). The essence of these scientific and technical proposals focused on reducing the loss of moisture evaporating from the sea by reducing the sea's surface area. With sea level decline, there appeared an opportunity to control its water budget and the sea regime by the separation of its various parts and the maintenance not of a single unit, but of several water bodies connected to each other. Such an opportunity is provided by the morphological features of the Aral basin.

Changes after the 1970s

From the end of the 1970s water flowing into the sea was intermittent. In the 1980s, the rates of drying of the Aral Sea intensified even further. During low or average water years, the rivers' waters were used almost completely for economic needs in the basin and did not reach the sea. Large volumes of compulsory sanitary water releases are almost fully used for water supply of the delta areas of the Amudarya and Syrdarya. River flow into the sea in 1982–1983 and in 1985–1986 was nonexistent; the average inflow to the sea within the 1980s was only 4.2 km³/year. The average rate of sea level decline within that period increased to 76 cm/year, and the total decline of the sea level of the Large Aral Sea (from 1961) reached 15.5 m. During certain years with a complete absence of the river flow, the annual decline in sea level reached 89–109 cm. Climatic conditions in that period were characterized by a continuation of low water levels in the rivers of Central Asia and by an increase in precipitation and evaporation in the region of the sea (Figure 3.3; Table 3.1).

Changes also took place in the salt budget of the sea (Table 3.2). With insignificant water flow (despite the continued increase in the mineralization of the Amudarya water to 1.4–1.6‰ and the Syrdarya to 1.6–2.0‰), the ionic flow was sharply reduced. With the increased salinity of sea water, the processes of sedimentation of carbonate and then sulfate salts directly onto the sea bottom were intensified. In those years the earlier trend of increase in the salt mass was changed in the opposite direction – its reduction was observed.

In the beginning of the 1980s the cumulative changes in the Aral Sea regime reached a new 'threshold level' – the decline of sea level by 15 m. This decline exceeded the range of fluctuations (6–7 m) during the last several millennia. This process of decline was aggravated by a complete absence of water and salt flow in some years. Thus, the Aral Sea was losing some of its most important hydrological and ecological functions as the main receptacle of water and salt for the whole basin.

By the mid-1980s the waters of the Aral Sea were converted from gradations of brackish water to gradations of saline waters. This resulted in a change in a number of important physical and chemical processes in the sea proper, such as the autumn/winter convective mixing of sea water, which determined the pattern of salt sedimentation (Anon., 1990). Negative trends in this period were intensified and continued in the Aral region. All these changes are indicative of the transition of the Aral Sea's environmental problem into a crisis.

Scientific assessments and the design of economic development activities in that period both forecast the continued drying out of the sea. They differed only in their forecast of the intensity of that process.

By the end of the 1980s the drop in sea level led to the separation of the sea into two water bodies, the Small Sea (Maloe More) (the northeastern part) and the Large Sea (Bol'shoe). Each has a separate river feeding it, the Syrdarya and Amudarya, respectively. As early as 1987, the level of the Aral Sea reached a 'critical' level of 40 m above sea level, when it separated into the Large and Small Seas. At the time, the Large Sea's surface area was 36 700 km² and its volume 360 km³; and the Small Sea's surface area was 3100 km² and its volume 22 km³. Their total area and water volume by the time of separation made up only 60% and 32%, respectively, of the area and volume of the sea at the 53 m level. Average water salinity had increased by then to 27‰.

The separation into two of the Aral Sea in 1988–1990 was unstable due to wind-caused and eustatic (volumetric) fluctuations of the level of the Small Sea. During those years, the Small Sea overflowed due to the temporary increase in flow of the Syrdarya, resulting in a partial discharge of water from the Small Sea into the large one.

ACTION FOLLOWING REALIZATION OF THE CRISIS

The separation of the Aral Sea into two water bodies indicated that a new 'threshold level had been crossed,' the most demonstrative landmark change in the desiccation of the sea and in the realization of the extent of the Aral crisis. There was a realization of the need to take action to prevent either new or continued adverse changes in the natural environment, in the economy and in society. This resulted in the formation in 1987 of a Soviet governmental commission on the Aral Sea in order to prepare a detailed report for the leaders of the country. The main provisions of the report were published (Izrael' et al., 1988). The data in this report were used as a basis for the Resolution of the Central Committee of the Communist Party of the Soviet Union and the Council of Ministers of the USSR, dated 19 September 1988 'On the measures for the radical improvement of ecological and efficient use and intensification of the conservation of water and land resources in the basin'. In this resolution the Aral Sea and the Aral Sea region (i.e., the delta areas) were recognized for the first time as direct consumers of water, and it proposed a progressive increase in water supply from 9 to 15 km³/year by the year 2000 and to 22 km³/year by 2010. A number of measures to secure safe inflow in the basin were aimed at the conservation of the sea but with a reduced surface area.

Of course, such a proposed volume of inflow did not address the problem of restoration of the sea to its 'original' level or even its maintenance at the 1980s level. However, the proposed inflow would give a chance to conserve the residual water bodies in the Small and Large Seas and would establish 'green zones' in the ante-deltas of the rivers.

Prior to the disintegration of the Soviet Union in late 1991, under conditions of *glasnost* and the formation of the 'green' movement, stories of the Aral problem had not left the front pages of many publications. Because many of the articles of the Resolution of 1988 were not realized, the governmental commission was re-established in 1990 for the development of measures on the restoration of ecological equilibrium in the Aral Sea region and control of its realization. This commission conducted the All-Union open competition for ideas about the conservation and rehabilitation of the Aral Sea including the normalization of ecological, sanitary, medical, social, and economic conditions in the Aral Sea region. Such ideas were submitted to the Supreme Soviet and Council of Ministers of the USSR by the working group. Basic provisions of these ideas were published in 1991 (Anon., 1991). The 'Resolution on the Aral Sea' was adopted by the Supreme Soviet of the USSR on 4 March 1991. Concerning the future of the Aral Sea, the idea was put forward for the 'conservation of a unified (integrated) brackish water body with the level not below 38 m'. An alternative to establishing a number of regulated water bodies was not completely excluded as a possible measure for improving the Aral's ecological situation.

A number of projects were developed in 1989–1991 and a feasibility report prepared for establishing small artificially regulated water bodies in the former bays of the Aral Sea in order to create 'green zones' near the 'coastal' cities of Aralsk and Muynak. In fact, the project for enhancing the water supply of small water areas near Muynak was the only one to have been implemented.

After the disintegration of the Soviet Union, the burden of solving the Aral crisis fell to the newly independent states of Central Asia and Kazakstan. Economic difficulties and the absence of financing precluded the implementation of many of the proposed measures. The Interstate Coordination Water Commission (ICWC) was established in 1992 by the five Central Asian states. Its purpose was to implement the plan and to operationalize activities to control and distribute water resources with regard for the interests of the Aral Sea and the Aral Sea region. The Interstate Council for the Aral Sea basin (ICAS) and its Executive Committee were established at a meeting of the heads of state of the Central Asian Republics and Kazakstan. A number of new documents to resolve Aral Sea basin problems were prepared.

Present situation

The level of the Large Sea continued to decline between 1988 and 1995, reaching 36.9 m above sea level in 1994. The sea's surface area was reduced to 30 900 km² and its volume to 268 km³. In the middle of 1995 the level of the Large Sea was at about 36.5 m. Thus, the total decline of the Aral Sea (Large Sea) between 1961 and 1995 was about 17 m. The average salinity of the Large Sea is close to 40‰, making the continued existence of biota in the sea nearly impossible.

At the same time, an adequate supply of water had stabilized the regime of the Small Sea. Between 1990 and 1994, its level fluctuated between 38.5–39.5 m above sea level and during periods of excess, water was discharged through a channel into the Large Sea. The surface area of the Small Sea is about 3000 km² and its water volume approximately 20 km³. Its average salinity reached 24–27‰ with significant spatial variations.

The changes in Aral Sea level after 1961 were caused in large measure (about 80%) by anthropogenic factors, e.g., the growth of consumptive withdrawals of river flow, and by climatic factors, e.g., a relatively low water supply in this period. If consumptive withdrawals of river flow from the Amudarya and the Syrdarya had been maintained at the level of the late 1950s, the level of the Aral Sea would have declined by only 3.4 m because of natural factors and would have made its present level about 50 m above sea level. The reduced atmospheric supply of precipitation in the region during the last 30 years resulted primarily from changes in the general circulation of the atmosphere in Central Asia. Prior to the beginning of the 1960s, the latitudinal (i.e., zonal) and combined forms of atmospheric circulation probably prevailed here, but within recent decades the meridional forms of circulation have persisted (Anon., 1990).

Future changes in the state of the Large and Small Seas will depend primarily on the volume of inflow from the rivers and from other sources of water. However, the economic use of water resources of the Amudarya and Syrdarya will mainly be governed by human factors: water releases and discharges of river and drainage waters. Only during high-water years can the insignificant supplementary volumes of unused river flow reach these water bodies.

To maintain the Small Sea at 39–40 m, at least 3 km³ of the Syrdarya water should be supplied to it annually. With larger volumes of water to feed the regulation of the Small Sea regime it would be possible to discharge the excess water and salts through the former Gulf of Berg into the Large Sea. A volume of inflow of 5–6 km³/year would in principle allow a rise in the level of the Small Sea up to its previous level of 53 m above sea level, if the construction is undertaken of a dam in the Gulf of Berg. With an input of not less than 3 km³/year, the level of the Small Sea would be reduced to a 'level of equilibrium', i.e., when credit and debit parts of the water budget would be in balance. In this case, the Small Sea would separate into eastern and western parts below a sea level of 38 m. The surface area of water in the eastern part of the Small Sea (into which the residual flow of the Syrdarya will go) at this level will be 1700 km², its water volume approximately 9 km³, and the required annual volume of water feed to maintain it will be 1.5–2.0 km³.

To maintain the Large Sea at its present level, close to 37 m, the required flow of the Amudarya would have to be 28 km³/year. Apparently, this value is unrealistic for either current or future high levels of water use of the Amudarya. Therefore, the continued sharp decline of 0.4–0.6 m/year in the level of the Large Sea can be expected in the near term (Anon., 1990; Bortnik *et al.*, 1991).

At current amounts of annual water flow into the Large Sea, its level would be reduced to 31 m above sea level by 2005–2010. At that point the Large Sea would separate into a smaller (in area) but deeper western part and a more spacious but shallow eastern part. The surface area of the eastern part of the Large Sea, at the 31 m level, would be about 12 000 km² and its volume at 45 km³, the necessary amount of water inflow to maintain the main body of water at 31 m would be 11–12 km³/year. The water in the western part of the Large Sea would have the following characteristics: surface area, 5000 km²; water volume, 76 km³, and the volume of inflow required to maintain that level, 4–5 km³/year. With lower volumes of water flowing into the sea, the drying out of these water bodies would continue until corresponding new, lower 'levels of equilibrium' were reached.

Separation of the Large Sea into two water bodies creates an opportunity in the future to regulate the water regime of one of them by regulating the flow and discharge of some water and salts into another terminal water body, which would serve as a salt receiver. From a geo-ecological point of view, in order to reduce the total exposed area of dried sea bottom and to arrest the removal of salts by wind action, it would be more expedient to establish such a system of flow in the eastern part of the Large Sea with a sea level of 31–32 m above sea level. However, a larger volume of river flow would be required, approximately 18–20 km³/year, in comparison with only 8–10 km³/year for the western water body at the same 31 m level.

The implementation of the rational use and, therefore, saving of water resources in the Aral basin and the purposeful limitation of consumptive use of water could secure a 'safe' inflow to the sea of a volume of about 20–25 km³/year. This could create opportunities in the future to conserve or rehabilitate a significant water area and regulate the regime of its isolated areas (e.g., Small Sea, eastern and western parts of the Large Sea, and separate water bodies in the former isolated bays of the sea).

Conclusion

Changes in the level of the Aral Sea during the twentieth century were the result of the combination of climatic and anthropogenic factors. Before 1960 the influence of climatic factors prevailed. At that time the constant growth in consumptive withdrawals of river flow to a considerable extent was compensated for by increased river flow and reduced natural losses. In the last 30 years or so, anthropogenic factors had the dominant effect on the level of the Aral Sea. As a result of a catastrophic decline in sea level, the Aral Sea separated into two water bodies, Bol'shoe and Maloe Seas (Large and Small Seas).

Several stages can be distinguished with regard to the problem of changes in the level and salinity of the Aral Sea (Table 3.3).

Prevention of the problem could have taken place at the beginning of the 1950s, when the designing of water use for irrigation began, and following the expansion of the areas of irrigation farming within the Amudarya and Syrdarya basins.

Essential changes in the sea's regime were already noticed in the middle of the 1960s, beginning with change in level and salinity, when they began to affect economic activities associated with the sea, i.e., navigation and fishery.

Awareness that a problem existed dated from the 1970s, when the drying out of the sea began to influence the natural environment of the Aral Sea region. These influences had negative ecological, social and economic consequences.

In the beginning of the 1980s further changes in the sea's regime and the aggravation of consequences caused by changes in the sea proper and in the region surrounding it resulted in the escalation of the problem into a crisis.

It was only by the end of the 1980s that Soviet leaders realized that there

Table 3.3 Stages reached during gradual changes in water level and salinity in the Aral Sea basin

	Thresholds				
Prevention of the problem possible	1 Realization of the changes	2 Realization of the problem	3 Realization of the crisis	4 Realization of the necessity of action	5 Action
Beginning of the 1950s	Middle of the 1960s	Beginning of the 1970s	Beginning of the 1980s	End of the 1980s	Non-existent at present

was a need to act. Primarily, these actions were aimed at mitigating the negative consequences of the crisis, certain aspects of which warranted the label of catastrophe in the Aral Sea region. The region was designated as a zone of ecological disaster, and plans were made to address the sanitary-hygienic, medical, and biological problems in order to create acceptable conditions of life for local populations.

Until the present time actions with respect to the sea proper have been absent. On the one hand, the lack of action is connected to the fact that any growth or even stabilization of the inflow to the sea are governed by the societal measures related to water use. On the other hand, the 'rehabilitation of the sea' would depend on any one of the following frequently suggested scenarios: the separation of the Aral Sea into large and small Seas; the feasible stabilization of the Small Sea at a certain level; the further drying out of the Large Sea with its subsequent separation into its western and eastern parts. It is probable that the conservation of the Aral (Large) Sea in its present state is not possible in the near term because it is not seen as a high-priority task on the extensive list of ecological, social and economic aspects of the Aral Sea crisis.

References

Anon., 1990: Hydrometeorology and hydrochemistry of the seas of the USSR. In: *The Aral Sea*. Leiningrad: Gidrometeoizdat, 195 pp.

Anon., 1991: Fundamentals of the concept for conservation and rehabilitation of the Aral region, normalization of ecological, sanitary-hygienic, medico-biological, social and economic situation in the Aral Sea region. *Izv. AN SSR*, ser. Geograf., 4, 8–21.

Azarin, A. E., 1964: Level regime of the Aral Sea when developing water use in the River Basins of Syrdarya and Amudarya. *Tr. Gidroproekta*, 12, 211–21.

Baidal, M. Kh., 1972a: Types of atmospheric circulation stipulating fluctuations of climate of the Aral Sea, *Tr. KazNIGMI*, 44, 21–9.

Baidal, M. Kh., 1972b: Relation of secular fluctuations of the Aral Sea level with ratios of solar activity and macrotypes of atmospheric circulation. *Tr. NIGMI*, 44, 62–6.

Berg, L. S., 1908: The Aral Sea: Scientific results of the Aral expedition. *Izv. Turkest. otd. Russk. Geogr. Obschestva*, St. Petersburg, 5(9).

Blinov, L. K., 1953: Alteration of the Aral Sea salinity in connection with water withdrawal from the Amudarya River flow. *Tr. GOIN*, 12, 77–97.

Blinov, L. K., 1955: Salt balance of the Aral Sea in connection with partial water withdrawal from the Amudarya River flow. *Tr. GOIN*, 20, 167–83.

Bortnik, V. N., 1977: Current and future changes of water level and salinity of the Aral Sea. *VNIIGMI – MTsD, Ekspress-informarsiya, ser. Okeanologiya*, 3(43), 1–8.

Bortnik, V. N., 1978: *Present and Future of the Aral Sea*. Obninsk: Girdrometeoizdat, 11 pp.

Bortnik, V. N., V. I. Kuksa, and A. G. Tsytsarin, 1991: Present state and feasible future of the Aral Sea. *Izv. AN SSSR, ser. Geograf.*, 4, 62–8.

Dunin-Barkovsky, L. V. and V. N. Kunin, 1965: Nature transformation of the deserts of Central Asia. *Izv. AN SSR, ser. Geograf*, 5, 70–5.

Geller, S. Yu., 1969: Certain aspects of the Aral Sea problem. In: S. Yu. Geller (Ed.), *The Aral Sea Problem*. Moscow: Nauka, 5–24.

Izrael', Yu. A., A. L. Yanshin, P. A. Polad-zade *et al.*, 1988: Present state and proposals for cardinal improvement of the ecological, sanitary and epidemiological situation in the Aral Sea region and lower reaches of the Amudarya and Syrdarya rivers. *Meteorologiya i Gidrologiya*, **9**, 5–22.

Kes', A. S. and I. A. Klyukanova, 1990: On the causes of level fluctuations of the Aral Sea in the past. *Izv. AN SSR, ser. Geograf.*, **4**, 78–86.

Korenistov, D. V., S. N. Kritskii, M. F. Menkel' *et al.*, 1972: The Aral Sea problem. *Water Resources*, **1**, 138–62.

Kuznetsov, N. T. and R. V. Nikolaeva, 1975: Morphological and morphometric fundamentals of the Aral Sea. *Proc. VI Congress VGO SSSR: Environment and Man.* Leningrad, 43–5.

L'vov, V. P., 1959: Fluctuations of the Aral Sea level within the last 100 years. *Tr. GOIN*, **46**, 80–114.

L'vov, V. P., 1965: The Aral Sea level and solar activity. *Tr. GOIN*, **85**, 91–172.

L'vovich, M. I. and I. D. Tsigel'naya, 1978: Water budget control of the Aral Sea. *Izv. AN SSSR, ser. Geograf.*, **1**, 42–54.

Molchanov, L. A., 1955: Hypothetical changes of climate and hydrological conditions of Central Asia when developing irrigation in its southern half. *Izv. Uzb. Geograf. Obschestva*, **1**, 34–40.

Samoilenko, V. S., 1955: Status of the problem of the water budget and fluctuations of Aral Sea level. *Tr. GOIN*, **20**, 127–66.

Shnitnikov, A. V., 1959: The past and future of the Aral Sea from the positions of significant rhythms of climate. *Proceedings of III All-Union Hydrological Congress, Vol. 7.* Leningrad: Gidrometeoizdat, 47–57.

Shnitnikov, A. V., 1961: Dynamics of water resources of the Aral basin in the light of its climatic transgression. *Proceedings of the Laborotory of Limnology, Vol. 14.*, 10–88.

Tsytsarin, A. G. and V. N. Bortnik, 1991: Currents problems of the Aral Sea and the perspectives to solve them. In: *Monitoring of the Environment in the Aral Sea Basin*, ed. U. A. Izrael, pp. 7–28. St. Petersburg: Hydrometeoizdat.

Voeikov, A. I., 1908: Irrigation of the Transcaspian lowland from the point of view of geography and climatology. *Izv. Russk. Geograf. Obschestva*, **44**(3), 131–60.

Zalikov, B. D., 1946: Current and future water budget of the Aral Sea. *Tr. NIU Gidrometsluzhby*, ser. IV, issue 39, 25–59.

4 Desertification in the Aral Sea region

ASOMITDIN A. RAFIKOV

The Aral Sea region is a classic one for investigation of the processes of anthropogenic desertification in the arid parts of Central Asia. The rapid dynamics of those processes, sparked by the decline of Aral Sea level, and the pollution of river and sea waters among other problems, have taken place in plain view during the span of one generation. Desertification is manifested in the intense degradation of natural resources up to the point of their complete depletion in a given area. Desertification processes observed on the dried part of the newly exposed seabed are especially highly dynamic, having changed within a relatively short period of time. Comprehensive studies of regional desertification processes are of special importance to the development of the concepts and methods needed to plan for the control of those processes on the basis of scientific research.

Scale of desertification

The most negative effect on the environment of the drying out of the Aral Sea is in the delta plains of the Amudarya and Syrdarya, and in a radius of influence of the atmosphere on relative humidity and the temperature regime at a distance of 150–200 km in the southwestern direction. At the same time, the impact of the drying sea has been insignificant on the composition of coastal sediments on the Ustyurt Plateau and in the eastern part of the sea's periphery (the northwest Kyzylkum part of the Aral Sea region).

The intense impact of the declining sea on the degradation of the natural environment in the delta plains is expressed by the reduced influence of the high groundwater table due to flat topographical features, by the gradual reduction of the volume of flow into the lower reaches of the Amudarya and Syrdarya, and by the increased pollution (primarily mineralization) of river water. Therefore, changes in the natural environment in the delta systems have been affected by changes in both the sea and in the rivers. Intensification of desertification processes in the deltas since the 1960s can be explained by the following facts.

Nearly 1.3 million ha of agricultural lands in the Amudarya delta were subjected to desertification, 58% of which was occupied by reed pastures and hay fields. According to Starodubtsev et al. (1978), the reed and brushwood tugais in the lower reaches of the Syrdarya were subjected to desertification in the 1970s. This area increased to 2 million ha in the 1980s, because of the additional degradation of 846 000 ha of pastures and tugais. Along with degradation of vegetation due to progressive salt accumulation in the ground aeration zone, the extensive salinization of alluvial-meadow, meadow-boggy, and bog soils was observed. According to the data of Uzbek soil scientists, solonchaks covered an area of 84 000 ha prior to 1961 and, according to our studies, this area has expanded by four times. The area of the dried sea bottom, depositing hundreds of thousands of metric tons of salt dust on the adjacent plains of the Aral Sea region as a result of wind erosion, encompasses 3.5 million ha. Together with the area of degraded Amudarya and Syrdarya deltas, the area in this region subjected to desertification processes exceeds 6 million ha.

Factors influencing desertification development

Contemporary desertification is the result of the impact of natural and anthropogenic factors and, as is known, the latter is of decisive importance. In fact, under conditions of intense human impact on the environment, desertification develops at an accelerated pace. This is attested to by examples drawn from many regions of the world of the formation and intense degradation of the natural environment (Zonn and Orlovsky, 1984).

The development of desertification processes in the Aral Sea region is directly connected with economic activity, mainly with the withdrawal of streamflow from the Amudarya and Syrdarya for irrigation purposes. As is known, stable development of the intrazonal hydromorphic ecosystems in the Amudarya and Syrdarya deltas prior to the 1960s was based on two factors: (a) the annual flood regime of the region's rivers, the Amudarya and Syrdarya, and (b) the pressure of high sea level on the groundwater flow from the side of the deltas. With the weakening or termination of their impact, a sharp alteration of the hydromorphism of the delta ecosystems became inevitable. This fine-tuned mechanism of nature was disturbed in the beginning of the 1960s in the Aral Sea region. This in turn disturbed the ecological equilibrium in the region, which has been progressing since that time, worsening and extending in space. Its ecological and socio-economic after-effects have also grown with time.

It is known that disturbance of the water budget of the deltas was not only the cause of the initiation and formation of desertification, but that its acceleration was prompted by the following factors: (a) regulation of the hydrological regime of the deltas; (b) decline of the level of groundwater and increase of its mineralization; (c) salt accumulation in the soils; (d) evolution of hydromorphic soils; (e) degradation of vegetative communities; and (f) technogenic impact on ecosystems. In their potential for adverse impacts, these factors do not act independently; they are interdependent and mutually reinforcing. In other words, the development of certain processes of desertification is initiated not by one factor alone, but represents the collective impact of all factors. At any given time, however, one or two factors could be dominant.

The decline of the groundwater level in the Amudarya and Syrdarya deltas promoted a significant transformation of almost all components of the living and the physical environment. In particular, it contributed to the trend of transition from hydromorphic ecosystems into semi-hydromorphic and automorphic ecosystems typical of a desert zone. It also contributed to the accumulation of considerable amounts of salts in the zone of aeration (from groundwater table to surface). The increase of salts in the root layer of soil, especially in the 0–2 cm horizon (from 1%–10% or more), was accompanied by the change of vegetative cover, because those plants (reed, etc.), which had developed in the past under conditions of slightly mineralized groundwater, began to perish or to wither as a result of the accumulation of salt and an invasion of plants that were more salt- and drought-resistant. Such plants are typical in solonchak deserts. Yulgun, annual saltwort, and carabarak communities became dominant in place of hydrophytes and hygrophytes.

Certainly, the succession of vegetation was accompanied by the reduction of productivity of pastures and hay fields. The total reserve of reeds in boggy soils (an area of 400 000 ha) with an average yield of 20 000 kg/ha of all surface mass, produced 800 000 metric tons. Mowing the reed prior to the generative phase of a plant's development provides the best quality and quantity of hay (the yield of hay fields averaged 40–50 centners/ha). The yield of newly emerged vegetative communities, on the other hand, is considerably lower in comparison with reed brushwood and is not used for pasture or hay purposes.

The irrigated zones of the Amudarya and Syrdarya have been subjected to irrigation-related desertification since ancient times, because of the specific character of secondary salinization of soil in the deltas. Owing to the lack of natural drainage and the inadequate development of artificial drainage, gradual salt accumulation occurred in the ground aeration zone. However, wintertime leaching and an insignificant level of land use, when set against the background of random drainage, made it impossible for large volume of salts to accumulate in the root zone.

Since the 1960s, the salinization of cultivated lands has intensified. This resulted from an increase in land use, an expansion of the area of irrigated lands, the gradual increase in river water salinity and drainage discharge from the oases. The inadequate water supply to the zone of drainage runoff also promoted salinization. A positive salt budget prevailed for a long time under such conditions. This phenomenon developed especially quickly between the 1970 and 1990s. During low-water periods, the rate of accumulation of salts in the irrigated zone has been the highest. The average salinity of Amudarya water during those years reached 0.6 g/l–1.2 g/l in summer in the Chatly river section, a value exceeded during winter.

Varying degrees of salinization of soils are typical for irrigated land in the delta plains of the Amudarya and Syrdarya. There are spots of slight, medium and high salinization surrounded by nonsaline areas and by solonchak spots as well. The area occupied by such spots is about 0.10 to 0.5 ha. Often, neighboring spots fuse into one, forming areas with continuous salinization to varying degrees. The causes of 'spot' salinization are usually governed by differences in the topographic features of these irrigated areas.

The main factors of desertification in the development of the irrigation zone of the deltas of the Aral Sea region are as follows: (a) the stable occurrence (1–3 m) of groundwater close to the surface and an inability to drain the soils artificially; (b) a complete absence of drainage; (c) an inadequate leveling of lands prepared for irrigation; (d) the irrigation of agricultural crops with high water demand; (e) the availability of excessive salts in the zone of aeration.

Recently, the problems of the Aral Sea and desertification have been discussed increasingly by foreign scientists (e.g., Micklin, 1990a,b; UNEP, 1992; Létolle and Mainguet, 1993; Glantz *et al.*, 1993). Such discussions underscore the high level of international concern about the Aral Sea problem. Glantz, for example, noted that desertification in the Aral Sea region resulted from slowly increasing, long-term, cumulative processes. In fact, the 35-year history of desertification development in the region indicates that the gradual accumulation of ecological changes has been directly connected to widespread irrigation in the Aral Sea basin.

Desertification in the delta plains of the Aral Sea region

Desertification in the Aral region intensified in stages. An analysis of the dynamics of the various processes within the 1961–1995 period reveals the main stages and trends of their development.

Prior to 1961, 8 km³ of river water was supplied to the Amudarya delta (Shults and Shalatova, 1964) and 3 km³ was supplied to the lower reaches of the Syrdarya (Borovsky and Pogrebinski, 1958). Gradually the water supply to certain arms and branches in the deltas began to shrink and dry out. This resulted in the drying and intense shoaling of certain lakes and the drying out of bogs and previously waterlogged areas. Considerable changes took place in the hydrography of the central and eastern parts of the delta of the Amudarya. The desiccation in this area caused a significant reduction in the productivity of pastures and hay fields.

The general drying out and drainage of ecosystems negatively affected existing ecological conditions. Within these geographic areas the formation of desertification was first observed in the form of soil salinization, pasture degradation, and the exogenous succession of brushwood tugais. The actual centers of formation and establishment of desertification were mainly on the slopes of interchannel depressions, in the watersheds between water bodies, along the periphery of natural levees and lakes, and along the arms and branches of deltas where river flow was discontinued (Figure 4.1).

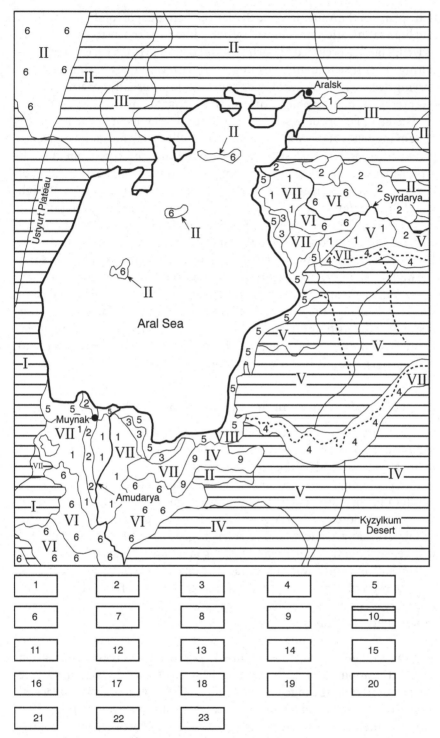

Figure 4.1 Basic stages of desertification development in the Aral Sea region (1961–73). For the legend, see text, pp. 71–2.

The legends for Figures 4.1–4.4 are as follows:

1. shoaling of lakes, drying of bogs;
2. decline of ground water level;
3. slow drying out of soils due to recession of groundwater and lowering of the Aral sea level;
4. sharp changes in groundwater table and drying up of soils;
5. appearance of sandy plains on the dried sea bottom;
6. without change;
7. overgrown with reeds, salt accumulation in some places;
8. salt accumulation, extension of the areas of meadow-takyr soils and xerophytes;
9. intense salt accumulation and extension of halophyte areas;
10. intensification of techno-erosion, deflation of substratum and cutting of shrubs;
11. salt accumulation in irrigated lands;
12. salt accumulation of the dried bottom of the sea and lakes;
13. degradation of reed pastures, salt accumulation on the bed of dried lakes, deflation of substratum on channel levees;
14. intensification of eolian processes, desiccation of land surface;
15. deflation of sandy soils, appearance of eolian land forms;
16. deflation of typical solonchaks, desiccation of topography, transition of active solonchaks into residual;
17. cutting of brushwood, deflation and deep erosion of slopes;
18. drying up of brushwood tugais, extension of areas of xerophytes and halophytes;
19. deflation (wind erosion) of seabed sands;
20. combined effects of watering in relatively high-water years and relatively low-water years;
21. stabilization in the regularly and sporadically water-supplied arms, degradation of tugais in the arms without water supply and in their peripheral areas, intensification of eolian processes, takyr formation;
22. appearance of typical eolian landscape forms; and
23. deflation of salt deposits.

Natural complexes of the Aral Sea region are as follows:

I. Flat Ustyurt Plateau, formed by limestone, sandstone and clay with a combination of plant communities, dominated by *Anabasis salsa, Artemisia terrae-albae, Salsola gemmascens*, with addition of *Galoxylon phyllum* on gray-brown soils.
II. Elevated plains, formed by clay and sandstone, with combination of plant communities of *Artemisia semiarida, Anabasis aphylla, Artemisia terrae-albae, Anabasis salsa, Salsola arbuscula, Salsola orientalis, Atriplex cana* on brown, gray-brown soils, and on typical takyrs.
III. Eolian sandy plains with solonchak deflation depressions growing up near plant communities of cereal, *Artemisia arenaria*, Buxus, mixte herbosa on hilly sands.
IV. Elevated flat plains formed by clay and sandy deposits with *Haloxylon persicum, Ephemerosa, Calligonum* spp., and *Haloxylon aphyllum* in depressions.
V. Complex of hilly ridge sands (formed from river deposits) with plant communities of *Haloxylon persicum, Artemisia terrae-albae, Ephemerosa*, accompanied by

Salsola richterii, Calligonum aphyllum, Ammodendron spp. on desert sandy soils and takyrs with algae.

VI. Deltaic alluvial plains formed by a combination of sand, clay, and loamy deposits transformed by irrigation for cotton and rice fields with old and recent irrigated meadow soils.

VII. Deltaic alluvial plains with degraded tugai woody bush and grassy plant communities on differently salinized meadow-takyr soils in combination with typical solonchaks and takyr soils.

VIII. Slightly inclined sandy and sandy loam-clayey sea plains with hilly barkhan (dune) sands with differently fixed complexes of *Haloxylon aphyllum, Ephemerosa, Salsola richterii*, and *Tamarix* spp. In some places there are residual solonchaks.

IX. Slightly sloping loamy-clayey-silty dried seabed with annual saltworts, in some places with *Tamarix* spp. and *Halostachys belangeriana* on residual and typical solonchaks.

Stages of desertification development

In research on the history of desertification in the Aral Sea region its initial stage – 1974–1977 – is of importance. Widespread serious alterations of the natural environment of the region began from that time.

Beginning in 1972, flooding ended in the contemporary delta of the Syrdarya. Water withdrawals in the upper and middle reaches of the river had established extremely difficult conditions for irrigation in the lower reaches of the river, especially during low water years. The termination of the supply of flood water to the delta (about 1–1.3 km³) caused the drying out of almost all of its territory (Kievskaya *et al.*, 1979).

The Takhiatash hydraulic power system was commissioned in 1974 and, since that time, Amudarya water has been distributed in the delta primarily via irrigation canals. The remaining water resources have been supplied to the pastures and hay fields in interchannel depressions and, occasionally, to the other ecosystems. Providing water to delta ecosystems under these conditions became more complicated and more difficult. During certain years the expected water supply to some areas could not be guaranteed. The frequent low-water periods in the Amudarya basin along with the increase in the area of irrigated land contributed to this shortfall of needed water resources.

Cessation of flow through almost all arms resulted in the drying up of the deltas' lakes and bogs (during this period a sporadic flow existed only in the Akdarya, Kipchakdarya, Akbashli, Kazakhdarya rivers and seldom through the Raushan, Taldykdarya, Kunyadarya rivers). By 1975, 100 000 ha of lakes had ceased to exist. The bogs and waterlogged ecosystems also dried up for good. During the 1970s in the lower reaches of the Syrdarya, the Kara-Uzyak and the Aksai-Kuvandarya systems of lakes with an area of 50 000 ha each disappeared; water mineralization increased in many of the remaining lakes (Starodubtsev *et al.*, 1978).

The sharp alteration of the surface water regime of the delta directly affected the changes in the groundwater regime. Depending on the specific local granulometric and geomorphological conditions, the groundwater regime changed in different ways. In particular, on the elevated areas (channel levees, watersheds between depressions) the groundwater level was reduced more rapidly than on the slopes of drainless depressions or in the beds of depressions. This groundwater level decline determined the salt accumulation regime in subsoils. Generally, moisture was used in part for evaporation and in part for groundwater outflow (Figure 4.2). This initiated the deposition of salts in soil profiles in large quantities (5%–28% of dry residue with a chlorine ion content of 2.50%–12.75%).

The evolution of hydromorphic soils in the trend toward the formation and development of semi-hydromorphic and eluvial soils was accompanied by a declining groundwater level and widespread salt accumulation. As a result of soil changes, those areas composed chiefly of meadow-takyr, meadow-desert, typical and meadow solonchaks were significantly extended. A 1977–78 salt survey of soils of the western and central parts of the Amudarya delta identified the accumulation of a large amount of salt due to intensive salinization. Salt reserves in the 0–2 m layer in an area of 1.1 million ha was estimated at 268.3 million metric tons.

The humus content in these soils is rather high (more than 2% and in some places 8–10%), which was the result of conservation of humus under strongly saline conditions of the former meadow-alluvial soils of the delta. However, at present deflation of the peaty layers of soils has intensified. Therefore, the humus horizon of soils has become degraded and has been transformed into areas no different from those with typical soils in the region.

The initial stage of desertification processes in the delta, which is connected with the general salinization of soils, has sharply changed the natural conditions in the region; ecosystems began to degrade in a trend toward halo- and xerophytization. The destruction of intrazonal ecosystems, as a result of salt accumulation in the soils, is the outcome of desertification. In order to label the specific cause(s) of the initial stage of desertification this process could be referred to as 'halo-desertification'. On the higher elevations the process is accelerated because of the salinization of soils as a result of the sharp decline in groundwater level (by 2.0–2.5 m) and also because of the drying of subsoils. This contributed to the emergence of more xerophytic communities. This type of degradation of ecosystems can be called 'xero-desertification'.

Development of desertification under eluvial conditions (1978–82)

By the end of the 1970s, the problem of water supply to the live (present-day) deltas of the Amudarya and Syrdarya became more serious. As

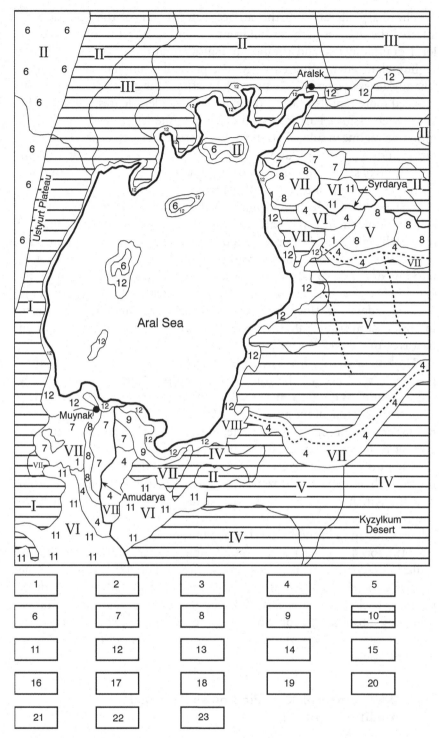

Figure 4.2 Basic stages of desertification development in the Aral Sea region (1974–77). For the legend, see text, pp. 71–2.

significant amounts of water were diverted for irrigation, the flow into the sea was greatly reduced. Under such critical water supply conditions, desertification rates became significant. Consider the periphery of Sudoch'e Lake, Makpalkul', the natural landmark Maipost, Shege, and Kazakhdarya (in the Amudarya delta); each was supplied not only with river water, but also with drainage water, and, as a result, their ecosystems developed hydromorphically (e.g., reed brushwood was used principally as pastures and hay fields by the farmers of Karakalpakstan).

At the same time, the deflation of both sandy loam and sandy soils accelerated in the remaining part of the Amudarya delta where the groundwater level had declined below 5 m, and in some places even below 7 m. Due to the general drying out of subsoils, vegetation in these areas was reduced by 20%–40% or more. The availability of mostly sandy loam soils in the upper layers in combination with sandy soils has led to the deflation of subsoils. This process was particularly intense in the old, strongly degraded tugais along the delta's arms, where brushwood communities were subjected to major changes (Figure 4.3).

The increase in the extent of eolian processes in the delta is closely connected to the pattern of succession of vegetation. Reeds first appeared after the drying out of lacustrine-boggy complexes in the northeastern part of the live Amudarya delta. Then, after a sharp reduction in soil moisture, they were replaced by mixed grass saltworts in combination with yulgunniks. By the end of the 1970s and the beginning of the 1980s herbage became extinct and saltworts remained. Wind activity was intensified because of the decrease of 10%–40% in the protective vegetation cover. This circumstance generated the formation of small hollows of deflation, followed by steepwalled hollows with depths exceeding 1 m, especially along tracks made by vehicles.

Thus, the former intrazonal ecosystems of the delta were subjected to transformation into zonal natural desert complexes under the influence of aridification processes. The dynamic desertification processes, being the catalytic factors in the alteration of hydromorphic ecosystems into eluvial ones, not only promoted the emergence of new indicators of aridity, but also reduced the productivity of the ecosystem to the level observed in deserts.

Development of desertification in combination with eluvial and waterlogging conditions (1983–95)

Under conditions of creeping cumulative changes, e.g., desertification, of the Amudarya and Syrdarya deltas, local populations back in the 1970s supplied them with water in order to maintain certain pasture areas, hay fields and brushwood tugais and some lakes. In this case the ecosystems nearest to water sources were supplied naturally with water in large quantities, whereas the more distant ecosystems were subjected to desertification.

From the end of the 1970s, the pastures and hay fields of the northern

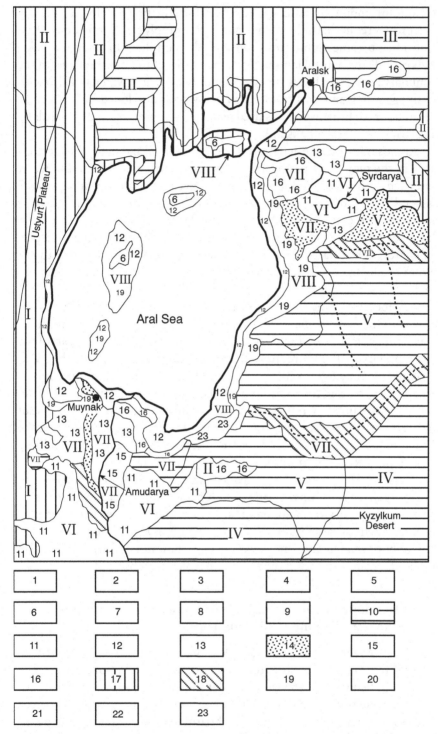

Figure 4.3 Basic stages of desertification development in the Aral Sea region (1978–82). For the legend, see text, pp. 71–2.

and northeastern periphery of Sudoch'e Lake, along with the natural landmarks of Shege, Maipost and other areas in the Amudarya delta, were regularly supplied with water. The Kamyslybas, Akshatau, Tamankol' Lakes, etc., and sometimes the Zhanadarya and Eskidaryalyk channels in the Syrdarya delta also received water. Reduction in the water supplies to large areas of reed pastures and hay fields, as well as to most of the lakes and arms of the western and central areas of the Amudarya delta during the vegetative period, also reduced the flow to the sea at that time (approximately 500–600 m³/s and more) of the Akdarya River. This reduction sharply changed the existing vegetative cover in the region, making it open to intense degradation.

Temporarily established subaqueous and superaqueous conditions differ radically from the hydromorphic conditions within the former live delta. These are brought about in the vegetative cover and, to a limited extent, in the soil and groundwater regimes. Despite this, they temporarily halt the desertification process and its intensification and tend to create favorable conditions for the development of ecosystems. At that time, desertification in the Amudarya delta covered only those areas where water supply had disappeared. These areas were mainly the peripheral strips of the channel levees of the arms (e.g., the left bank of the Taldykdarya River from the Akbashli channel to the Shagyrlyk settlement; the right and left banks of the Erkindarya River mostly in the northern part; the Gedeuzek basin; a number of areas in the region of the old arm of the Inzheneruzek, etc.), and the right bank to the north of the Kazakhdarya channel (Table 4.1).

During years of water shortages, the ecological situation in the deltas becomes especially tense: eluvial and semi-hydromorphic conditions dominate in some areas. The area covered by saline soils and by xerophytic and halophytic plants becomes extended and the importance of eolian processes increases. This phenomenon intensifies during a second and third consecutive year of low water. During the first year, degradation is not yet obvious because of the availability of moisture in the subsoils. In this year, reeds can be found in the same areas where they existed during abundant water years, but they are stressed. Reeds wither as the groundwater level drops and mineralization increases. Abruptly, the importance of annual saltworts and yulgunniks increases, whereas under ordinary conditions they are sparse in the vegetative cover. Large quantities of salt (more than 3%) concentrate in the root zone, especially in its upper horizon (Figure 4.4).

Main stages of desertification development on the dried-out Aral Sea bottom

The exposed dried seabed has been subjected to desertification processes over a 30-year period. Therefore, its different parts are in various stages of progress, i.e., from initial stages to shifting sands. The dried-out part of

Table 4.1 Present state of desertification of the Amudarya and Syrdarya Deltas

| Ecosystem | Desertification | | | Productivity (centner/ha) | | |
| | | | | Irrigated lands | | Pastures |
	Class	Type	Cause	Raw cotton	Rice	
Interchannel depressions	Desertification is unavailable	Salinization	Normal water supply	–	–	10–50
	Moderate		Water supply is unavailable			1–4
Elevated parts of plains	Severe	Salinization	Groundwater use for evaporation	–	–	1–3
Channel levees and their periphery	Severe	Degradation of tugais	Termination of regular water supply of the arms	–	–	3–5
Flat lakes plains	Very	Salinization	Groundwater use for evaporation	–	–	0–1 and less
Sea plains	Very	Salinization	Groundwater use for evaporation	–	–	0
Subareal part of the deltas	Weak	Deflation	Technoerosion	–	–	1/3
Artificially drained flat plains	Moderate, severe in some places	Salinization	Close occurrence of groundwater level and low efficiency of drainage system	2–20	20–40	–
Denuded uplands	Moderate	Deflation, technogenic desertification	Technoerosion deflation, gully erosion of slopes	–	–	1–3

the sea bottom has been expanded annually by hundreds of thousands of hectares. There, desertification is most dynamic and intense and can be found in all stages. The intensification of desertification from one stage (in particular, from the stage of marsh solonchaks) to another (typical solon-chaks) can be identified by qualitative changes in the natural properties of desertified ecosystems. Investigation of the dynamics of desertification on the exposed seabed over time is important for an improved understanding of the processes of desertification.

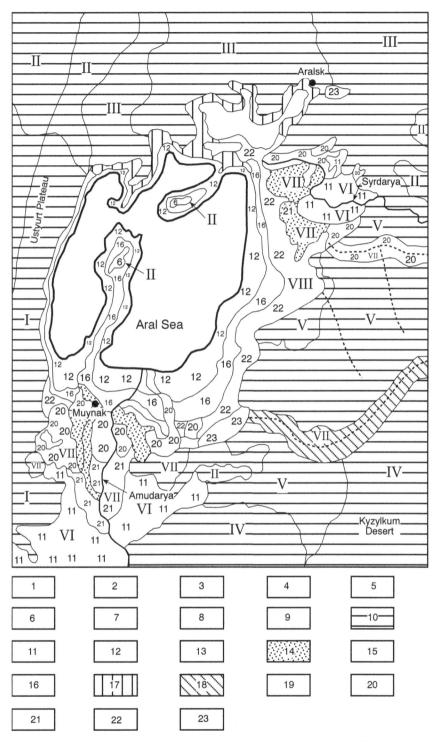

Figure 4.4 Basic stages of desertification development in the Aral Sea region (1983–95). For the legend, see text, pp. 71–2.

The area identified by the 0–5 m isobath of the sea dried completely between 1961 and 1976 (when the sea level dropped from 53.3 m to 48.4 m). A considerable part of the Aral seabed became exposed in the bays: Adzhibai, Muynak, Rybat, Bozkol', and the zone of Akpetkinski Archipelago areas. This part of the dried sea formed as a result of sandy bottom sediments deposited by the rivers. Beneath these sediments are mainly silty clays, silts, loamy-sandy loam soils interspersed with aleurites and sands. The exposed sands are subjected to wind erosion. In the beginning of the 1980s typical eolian land forms were created: barkhans, hills, bush-trapped sands, deflation hollows, etc.

As a result of intense deflation processes, this 0–5 m strip is the zone of the eolian processes, with no apparent differences between Kyzylkum and Karakum sand. The main feature of this strip is that eolian land forms are constantly shifting, influenced by north and northwest winds, whose velocities sometimes exceed 20 m/s.

The extremely dynamic nature of sand forms resulting from wind action is also affected by the fact that the sands are inadequately fixed by vegetation (psammophytes). In some places the vegetation is so sparse that there are large areas of shifting sands. Evidently, this condition is perpetuated by the constant movement of sand. During the earliest stages of drying of the sea, the annual vegetation (e.g., saltworts) prevailed. They completely disappeared within 2 to 3 years, thereby opening the area to invasion by other types of vegetation. Vegetation was absent here for a long time. Only from the late 1970s did sparse, withered yulgunniks begin to emerge. By the end of the 1980s, the distribution of cherkez and other types of psammophytes were noted. Yulgun, cherkez, and the rare black saxaul can now be found, although still in insignificant numbers.

There is a narrow contact zone where a combination of solonchaks and eolian land forms can be found. The zone is between the strips of sand and solonchak plains on the dried sea bottom. The development of sarsasan solonchaks with eolian land forms has been noted here with phytogenous sands. The topography becomes flat with an insignificant gradient seaward. It is disturbed only by sparse, low, flat-topped and rounded hills of sarsasanniks, a semi-bush halophytic vegetation, which is inedible to cattle. Therefore, this zone is similar to other parts of the dried sea bottom – ecosystem productivity is very low.

The isobaths (lines of equal depth) of 6–10 m correspond to the open part of the dried sea bottom beyond the strip of dried bays and islands, where the seabed becomes more uniformly flat and spacious. This 6–10 m strip dried completely between 1977 and 1984 (in this period the sea had dropped from 47.7 m to 43.1 m above sea level). During those years the soil in the strip was primarily composed of typical solonchaks. The generation of salt formations after the recession of the sea is directly connected to (a) the flatness of the dried seabed, (b) the predominance of heavy-textured soils which tends to reduce the outflow of groundwater, and (c) drainless areas. Because of the significant amount of groundwater evaporation, a concentration of large volumes of salt built up in the active zone of aeration. Under conditions of an extremely slow horizontal outflow of groundwater, a gradual increase in mineralization occurred, reaching values as high as 50–80 g/l and more.

With a deepening of the groundwater table, in particular below 5 m from the surface, the accumulation of salts in the upper layers of subsoils gradually ended. When the level declined below 7 m, the accumulation of salts in the root layer ended. During that period, a stabilization of salt accumulation in the soil took place. The desalinization of soils was observed in elevated areas in contact with atmospheric precipitation. In the absence of drainage, salts that had been washed out from the upper layers and substratum accumulated in the lower horizons of the soil. In certain locations (i.e., in elevated areas) this resulted in the development of solonchak subsoils.

An analysis of the dynamics of desertification on the dried sea bottom at 6–10 m shows a trend toward halophytic and xerophytic vegetation. In the beginning of the seashore's regression, hydrohalophyte desertification (meadow and marsh solonchaks) took place, followed by halomorphic desertification (typical solonchaks), and then (nowadays) by xeromorphic and halomorphic desertification (residual typical solonchaks).

The seabed encompassed by the 11–17 m isobaths emerged as an exposed seabed in 1985 (42.3 m above sea level). As the seashore receded progressively each year, so too did the area of this strip. At the same time, due to the decline of groundwater level below 5–7 m from the surface, it gradually acquired natural properties of the previous isobath. Therefore, the ecosystem at this isobath is changing considerably (Table 4.2).

Active typical ring-shaped solonchaks over the whole area of the dried part of the Aral Sea bottom are characteristic of that strip. As a rule, marsh solonchaks originate from the water's edge, followed by the strips of meadow (from 1–4 km and more), crust, and puffed solonchaks. Progressive salt accumulation is observed in this strip. Desertification acquires apparent hydrohalomorphic properties, which actively develop until the groundwater level declines below 7 m. Therefore, huge amounts of chloride-sulfate salt accumulate in subsoils (from 1200 to 4000 metric tons/ha in the 0–1.5 m layer), and this will determine the vegetative cover and type of solonchaks that develop. In any event, one should expect to see very sparse halophytes in withered condition, as well as most denuded areas covered completely in some places with chloride-sulfate salts.

The strip of sands of river origin, highly enriched by salts, is outcropped in places in the southwestern part of the dried sea bottom. The groundwater level close to the surface continuously provides the sandy layer with salt crystals. As a result, the crystals are compacted and are not easily subjected to deflation. But, with the future transition from hydromorphic solonchaks to eluvial activity, one should expect the emergence of eolian land forms, strongly enriched with salts. The vegetative cover will, therefore,

Table 4.2 Natural factors, determining desertification of the dried part of the Aral Sea bottom (within the boundaries of Uzbekistan to isobath of 12 m, 1986)

Isobath (m)	Topography	Lithological composition of sediments	Groundwater			Soil	Vegetation (%)	Area (thous. ha)	(%)
			Depth (m)	Mineralization					
				Extent (g/l)	Type				
12–12.1	Gently sloping plains with uplands and depressions	Fine-grain sand, interbedded with loams and sandy loams	0–0.5	30–50	C–S	Marsh solonchaks	Seablight Salicornia, projective cover 0–20	20–24	2.60
12–10.5	As above	As above	0.5–2.0	30–50	C–S	Meadow solonchaks and wet solonchaks	Ditto, projective cover 20–40	171.67	22.14
10.5–7.5	As above	As above	2.0–3.0			Sand solonchaks (typical)	Unavailable	381.02	48.86
7.5–2.5	Sloping plains with uplands and depressions	Sandy loam interbedded with sands and loams	3.0–5.0 5.0–7.0	30–50 10–30	C–S	Residual solonchaks in complex with typical	Annual saltwort yulgunniks, projective cover 30–40, 50–80	27.71	3.58
1.0–3.0	Drainless depressions	Clays interbedded with silts and loams	0–0.5	100–500	S–C	Sors	Unavailable	3.46	0.45
3.0–3.5	Gently sloping plains	Loams interbedded with sandy loams and sands	0–0.5	50–100	C–S	Boggy solonchaks	Sparse annual saltworts, projective cover 10–25	48.94	6.30
2.5–0	Hilly barchan sands	Sand underlayed by loams and sandy loams	Below	10–30	C–S	Sands	Black saxaul yul gunniks in complex with cherkez, projective cover 40–90	122.18	15.80
							Total:	775.26	100.00

Note: C–S, chloride-sulfate; S–C, sulfate-chloride.

develop a complicated composition that includes psammophytes and halophytes.

Trends of desertification development

An analysis of desertification dynamics reveals basic trends. In the eluvial complexes of the deltas the processes are linked to the deflation of substratum and the evolution of the soils (the formation of takyr soils in the areas of meadow-takyr soils where the level of groundwater is below 7–10 m). The development of at least two processes can be noted: eolian processes and the evolution of soils in the dry-valley areas. This is supported by the pattern of environmental change in the eastern part of the Amudarya delta and the northern and southern parts of the Kazalinskaya delta of the Syrdarya. This happened as a result of the cessation of a regular water supply to these parts of the deltas. The interbasin plains of the Amudarya delta (e.g., Muynaksko-Kinkairskaya, Akdar'inskaya, Kundarar'inskaya) have been transformed (at present) into an eluvial stage.

Desertification processes are noted in superaqueous ecosystems; in particular, salt accumulation, the degradation and succession of vegetation, and the transformation of soils. The development of desertification processes under subaqueous and superaqueous conditions is a function of the regularity of a water supply.

On the dried sea bottom there are salt-accumulating processes. Solonchaks appeared first, but were later changed as a result of eolian processes, forming barkhans. This sequence will be stable as long as the Large Aral Sea continues to dry up.

Aerospace monitoring of desertification

The Aral Sea region is a natural laboratory for studies on desertification processes. Since the development of desertification is occurring over such a large area, traditional methods of investigation are rather complicated, requiring large financial investment. Thus, the use of remote sensing in desertification studies is considered efficient. Satellite images, covering large areas, at a given point in time can provide valuable information on the status, trend and dynamics of changes in surface processes. Analysis of the images of certain areas over a period of several years enables researchers to determine desertification trends. Theoretical approaches and individual mapping developments have been used by a number of specialists (e.g., Vostokova et al., 1988; Glushko, 1992).

For aerospace monitoring of desertification in the Aral region, the following blocks covering certain subregions should be distinguished: agroecosystems of the developed zone (e.g., rice and cotton plantations), ecosystems of the drying sea (its sand solonchak parts), and the water surface of the sea (e.g., the southern, central, northern and western parts of the Aral Sea, including the Small Sea). It is necessary to implement regular monitoring activities in the future using space imagery data.

The regular collection and processing of data will provide valuable information on the status of desertification processes in different subregions in the Aral Sea area. Such information is needed to justify measures proposed for desertification control. Certain activities are irregularly carried out by specialists of SANIGMI and KazNIGMI for aerospace monitoring. The program is restricted and does not cover all of the aforementioned ecosystems. These intermittent, limited activities provide the required scientific underpinning for revealing desertification in the region and for determining short-term, seasonal, annual, and long-term changes in the dynamics of desertification processes. Aerospace data and images must be supplemented by surface survey data (i.e., groundtruthing), such as the salt content in soils, the mineralization of surface and groundwater, the status of the vegetative cover, agricultural crops, and pasture productivity.

The dried part of the sea bottom is the source of salt storms. There is no agreement, however, on the extent of these processes. According to Kosnazarov (1990), the total amount of dust-salt aerosols in the southern Aral Sea region reached 1.5–6.0 metric tons/ha per year on average, of which dissolved salts made up 170–800 kg/ha. According to the *Ecological Map of the Aral Sea Region* (1992), dust-salt aerosols in the Aral Sea region made up 20–600 metric tons/km² per year. The author of *Hydrometeorological Problems of the Aral Sea Region* (Chichasov, 1990) wrote that the total volume of salt removed from the dried bottom of the Aral Sea was about 50–70 000 metric tons/year. Rubanov and Bogdanova (1987) determined that the annual removal of salts from all dried parts of the Aral sea bottom amounted to 39–42 million metric tons/year. According to Semenova *et al.* (1991), in the Kazakstan portion of the dried sea bottom alone, the annual removal of salts with 50% probability reached 10–20 000 metric tons/year on average. Thus, it is necessary to develop a research program to determine the amount of the removal and accumulation of salt dusts and salts within the Aral Sea region. The development of the *Maps of Desertification of the Aral Sea Region and the Aral Sea* on a scale about 1:500 000 should be undertaken as a result of aerospace monitoring of desertification processes, including the mapping of certain subregions on a scale of 1:100 000–1:300 000 (e.g., irrigated ecosystems). Such maps should be updated at three-year intervals.

A retrospective analysis of the development of desertification from the time of its origin until the present has revealed certain typical stages of evolution of desertification in the region and has identified the main trends of changes in the future. The problem of desertification in the Aral Sea region is a complicated one, with a variety of processes occurring in various locales and subregions throughout the area. The adverse impacts of these processes require the immediate attention of political leaders if such processes are to be arrested, if not reversed.

References

Borovsky, V. M. and M. A. Pogrebinski, 1958: *The Ancient Delta of the Syrdarya River and Northern Kyzyl Kum: Volume 1.* Alma-Ata: Izd-vo AN KazSSR, 514 pp.

Chichasov, S. N., 1990: *Hydrometeorological Problems of the Aral Sea Region.* Leningrad: Gidrometeoizdat, 278 pp.

Ecological Map of the Aral Sea Region, 1992: Scale 1:100 000, Kazakhstan: Leninsk.

Glantz, M. H., A. Z. Rubinstein and I. S. Zonn, 1993: Tragedy in the Aral Sea basin: Looking back to plan ahead? *Global Environmental Change,* 3(2), 174–98.

Glushko, E. V., 1992: Program for the aerospace monitoring of the use of nature and the geoecological situation in the Aral Sea region. *Problems of Desert Development,* 2, 25–37.

Kievskaya, R. Kh., N. F. Mozhaitseva, and V. P. Bogachev, 1979: Alteration of natural conditions of the modern delta of the Syrdarya River in connection with regulations of its flow. *Problems of Desert Development,* 4, 11–17.

Kosnazarov, K. A., 1990: *Importance of Eolian Transfer in the Salt Regime of the Southern Aral Sea Region.* Author's Abstract, Canad. of Agricultural Sci. Tashkent: SANIIRI, 22 pp.

Létolle, R. and M. Mainguet, 1993: *Aral.* Paris: Springer-Verlag, 358 pp.

Micklin, P. P., 1990*a*: Drying of the Aral Sea: Water management catastrophe. *Land Reclamation and Water Management,* 5, 16.

Micklin, P. P., 1990*b*: Drying of the Aral Sea: Water management catastrophe. *Land Reclamation and Water Management,* 6, 12.

Rubanov, I. V. and N. M. Bogdanova, 1987: Quantitative assessment of salt deflation on the drying bottom of the Aral Sea. *Problems of Desert Development,* 3, 9–16.

Semenova, O. E., L. P. Tulina, and G. N. Chichasov, 1991: About changes of climate and ecological conditions of the Aral Sea region. In: Yu A. Izrael and Yu A. Anochin (eds.), *Monitoring of Natural Environment in the Aral Sea Basin.* St. Petersburg: Gidrometeoizdat, 150–1760.

Shults, V. L. and L. I. Shalatova, 1964: Water budget of the Aral Sea. *Proceedings Tash. GU. Hydrology and Physical Geography,* 269, 3–14.

Starodubtsev, V. M., T. F. Nekrasova, and Yu. M. Popov, 1978: Aridization of soils of delta plains of the southern Kazakhstan in connection with river flow regulation. *Problems of Desert Development,* 5, 14–24.

UNEP (United Nations Environment Programme), 1992: *Diagnostic Study for the Development of an Action Plan for the Aral Sea,* 6. Nairobi, Kenya: UNEP.

Vostokova, E. A., V. A. Suschenya, and L. A. Sherchenko, 1988: *Ecological Mapping Based on Satellite Imagery Data.* Moscow: Nedra, 220 pp.

Zonn, I. S. and N. S. Orlovsky, 1984: *Desertification: Strategy of Control.* Ashkhabad: 'Ylym', 320 pp.

5 Climate fluctuations and change in the Aral Sea basin within the last 50 years

ALEXANDER N. ZOLOTOKRYLIN

Introduction

More than a hundred papers have been published within the past 50 years about climate and atmospheric processes in the Aral Sea basin. A thorough review of them would require the publication of a separate book. Thus, in this chapter only one key aspect of climate in the Aral Sea basin is investigated: a diagnosis of regional climatic change and creeping environmental change.

The following sections are presented in general terms: natural changes in regional climate characteristics, temporal and spatial scales of change and their interactions, the rate of climate change and local anthropogenic impacts on climate. The data presented in this chapter are important for a retrospective identification of threshold levels in the development of gradually accumulating ecological changes in the Aral basin.

Basic works dedicated to the diagnosis of climate changes in the Aral Sea basin can be divided between those of the 1960s and 1970s and those of the 1980s. The steady decline of the Aral Sea level became obvious in the 1960s, thereby stimulating scientific studies. Another reason for the growth in the number of Aral-related publications was the fact that, by 1970, spatially detailed meteorological observation data in the Aral basin had already covered a period of over 30 years and were in need of scientific analysis. The development of the classification of types of the synoptic processes of Central Asia (Bugaev et al., 1957) and the classification of macrocirculation processes as applied to the Central Asian Kazakstan region (Baidal, 1964) were completed in the 1950s. The application of these classifications in the diagnostics of regional climate underscores the importance of relating large-scale atmospheric circulation changes to the causes of climate changes in the region.

In the beginning of the 1970s, on the basis of the diagnosis of regional climate changes, forecasts were made of the implications of climate changes for the future condition of the Aral Sea (Baidal and Kiyatkin, 1972). It is important to note that these authors, among others who provided the most pessimistic scenarios of the possible consequences of anthropogenic effects in the region, could not foresee the acceleration of development of the Aral Sea crisis. Within the 1980s, publications appeared assessing the importance of natural and anthropogenically induced changes of climate in the creation of the Aral crisis. Among these, the following publications should be noted: Molosnova and Ilinyak (1991), Kuvshinova (1980, 1982), Baidal and Khanzhina (1986), Molosnova *et al.* (1987), and Zolotokrylin and Tokarenko (1991).

Natural changes of climate

CIRCULATION FACTORS OF CLIMATE CHANGE

Regional circulation processes, in particular in the Aral Sea basin, do not occur in isolation, but are closely connected with the general circulation of the atmosphere and with circulation processes developed in the other parts of the globe. To explain past as well as future changes of climate in the Aral basin, most authors were drawn to the classifications of large-scale atmospheric circulation (Baidal and Khanzhina, 1986; Molosnova and Ilinyak, 1991; Subbotina and Chevychalova, 1991*a*).

One of the best-known classifications that has been widely used in hydrometeorological studies and in forecasting is the classification developed by Vangengeim (1941). This classification is based on the concept of the elementary synoptic process which is established according to the distribution of the sign of altitudinal pressure and the direction of basic air transfers within the Atlantic-Eurasian sector of the Northern Hemisphere. All circulation processes in this sector are divided into three types: western (W), eastern (E) and meridional (C). The catalog of circulation types according to Vangengeim covers the period from 1891 to the present.

With type E meridional processes the pattern of distribution of altitudes of isobaric surfaces is opposite to those processes of type C. The location of the axis of the altitudinal ridge, contoured by the planetary frontal zone, between meridians of 30° E and 60° E, is the criterion for attributing the meridional process to form E. The earliest of the catalogs of daily forms of circulation by Baidal is dated 1930. Complete matching of similar forms of circulation according to Vangengeim's classification and by the classification of Baidal for the 1930–1985 period makes up 70% of the total (Subbotina and Chevychalova, 1991*b,c*).

The regional classification of atmospheric circulation of Central Asia is represented by the types of synoptic processes (Bugaev *et al.*, 1957). A daily record of these types has been kept since 1934. A summary of the types of synoptic processes is listed in Table 5.1. As seen in Table 5.1, the group of cyclonic processes (types 1, 2 and 3) makes up 23.3% in the cold half-year (November–April) and only 6.6% in the warm half-year (May–October). The days with active cold intrusions (types 5 and 6) occupy 24.3% in the cold half-year and 33.7% during the warm half. The anticyclonic position makes up another basic group of processes, with a high percentage of recurrence (29.6%

Table 5.1
Table 5.1 Percentage of days with various types of synoptic processes over Central Asia during the cold (November–April) and the warm (May–October) half year

	Processes	Type number	(May–October)	(November–April)
I	Cyclonic breaches:			
	South-Caspian	1	4.3	10.8
	Murgabsky	2	0.7	8.4
	Verkhne-amudarinsky	3	1.6	4.1
II	Extended warming	4	0.6	1.6
III	Cold intrusions:			
	northwest	5	22.6	15.9[a]
	northern	6	11.1	8.4[a]
IV	Wave activity	7	1.7	6.1
V	Slightly mobile cyclone in the north of Central Asia	8	0.6	2.4
VI	Anticyclone periphery	9–9a	24.5	29.6
VII	Western intrusion	10	24.8	12.7
VIII	Thermal depression	11	–	7.5

Note: [a]15.9 and 8.4 equal 24.3 as indicated in the text.

Source: Bugaev et al. (1957).

during the cold and 24.5% during the warm half-year). The recurrence of western intrusions is higher during the warm half year (24.8%) than during the cold half-year (12.7%).

Precipitation in Central Asia during the cold half-year is determined by warm southwest flows of the free atmosphere (types 1–4, 7) and cold northwest intrusions. The frequency and intensity of these cold intrusions and the cyclonic activity preceding the intrusions determine the level of accumulation of winter snowpack in the mountains and its distribution of these reserves in various altitudinal zones. With cold intrusions taking place against a low temperature background, the main mass of snow tends to fall at low altitudes (up to the level of 1.5–2.0 km). With the cold intrusions taking place against a higher temperature background, the bulk of snow falls at higher altitudes (from 2 to 4 km) (Bugaev et al., 1957).

Cold intrusions during the warm half-year (types 5–6) are often accompanied by precipitation in the mountains at altitudes above 3 km. In this case, cold intrusions regulate the melting of snow and ice in the mountains. During the periods with frequent cold intrusions in spring and summer, when a prolonged spring and a cool summer occur, river flow is more uniform. Only certain types of synoptic processes affect the increase in river water discharges. During winter and spring, flooding is induced by the outbreak of cyclones from the southwest (types 1–2) and wave activity in the southeastern part of Central Asia (type 7). In the summer, the large floods are caused either by western intrusions (type 10) or by cold intrusions (types 5, 6) or with summer thermal depressions (type 11) (Bugaev *et al.*, 1957).

EPOCHS OF ATMOSPHERIC CIRCULATION

The extended periods of increases or decreases of any circulation form are referred to as the epochs of atmospheric circulation. They are indicators of global as well as of regional climate fluctuations and change. There are several methods for distinguishing circulation epochs, but most often epochs are identified by the deviation of a number of days with certain circulation compared with an average. Since 1930, two major circulation epochs have been identified in large-scale circulation: (a) 1930–1960 and (b) after the 1960s (Molosnova *et al.*, 1987, 1991; Subbotina and Chevychalova, 1991*a,c*). An excess of the western W(L) form (where L refers to latitudinal) compared with the average form of circulation and the meridional form E are characteristic features of the first circulation epoch. Meridional forms C and E circulations prevailed during the second epoch of circulation. The circulation epochs defined by Baidal and Khanzhina (1986) are the shortest. From 1930 to 1941 and from 1973 to 1985, Baidal identified epochs with prevailing form E circulation. Combined forms of circulation C+L, E+C and E+L prevailed, respectively, within the periods of 1941–53, 1954–63, and 1964–72. The shift in circulation patterns between the end of the 1950s and the beginning of the 1960s, noted by many researchers, is of great importance for the identification of the origin of the natural contribution to the crisis of the Aral Sea. From that time, the rate of recurrence of the main types of synoptic processes – responsible for positive anomalies of precipitation in the mountains and water supply level of the rivers – became lower.

TRENDS OF ATMOSPHERIC CIRCULATION IN THE PERIOD
OF THE ARAL SEA CRISIS

To assess the importance of natural factors in the decline of the Aral Sea level, one needs to know the trend of change in atmospheric circulation. Assessments of linear trend rate and the contribution of the linear trend for different intervals of time are presented in Table 5.2.

In the period 1956–85, the trend signs of circulation form recurrence (as defined by Vangengeim) were not changed compared with the period 1981–85. The recurrence of meridional C and eastern E forms of circulation maintained the sign of their trend also in 1966–85. However, the trend sign of the recurrence of the western form W changed in 1966–85: for the first time during the twentieth century the trend of growth of this form of circulation took shape. Essential changes as suggested by Baidal in the values of the trend of circulation form recurrence in 1976–85 point to the formation of a new circulation epoch. A considerable reduction of the meridional form C and an increase of the latitudinal form L was observed in that 10-year period.

Noticeable trends in the development of synoptic processes of Central

Table 5.2 Assessments of the parameters of linear trend of recurrence of different forms of atmospheric circulation according to the classifications of Vangengeim

Forms of atmospheric circulation	Period	B (days/decade)	a (%)
I Western, W	1956–1985	−6	6
	1966–1985	11	20
Meridional, C	1956–1985	−16	34
	1966–1985	−13	14
Eastern, E	1956–1985	21	27
	1966–1985	2	0
II Latitudinal, L	1956–1985	−9	16
	1966–1985	−9	7
	1976–1985	15	5
Meridional, C	1956–1985	5	2
	1966–1985	3	5
	1976–1985	−25	17
Eastern, E	1956–1985	5	5
	1966–1985	3	1
	1976–1985	9	1

Note: (I) Molosnova and Ilinyak (1991) and (II) Baidal, Subbotina and Chevychalova (1991). B, velocity of linear trend (days/10 years); a, relative contribution of linear trend into general dispersion of the series (percent ratio).

Asia in the 1935–89 period are typical for cold intrusions and for anticyclonic situations. The recurrence of northern and northwest intrusions (types 5, 6) was reduced sharply within that period, and the recurrence of anticyclonic fields increased sharply (types 9, 12). Although the statistical importance of trends of these synoptic types is not high, they can be shown to have had an influence on the reduction of precipitation in the mountainous part of the Aral Sea basin (Subbotina *et al.*, 1991; Molosnova *et al.*, 1991).

CHANGES OF AIR TEMPERATURE AND PRECIPITATION

Air temperature During 1936 to 1985, observations of air temperature were carried out at 60 stations in Central Asia. A continuous series of observations started in 1936 and at some of the stations between 1943 and 1946.

During the decades of 1931–40 and 1971–80, the most cold January (i.e., winter) temperature decades recurred. Cold decades mainly covered the plains, the piedmont areas, and the open valleys. Negative deviations from average for January temperatures reached 2–3 °C. The last decade of cold winter temperatures (1971–80) was followed by the warmest period since the beginning of observations. Positive deviations from normal for January temperatures in 1981–88 reached 3.0–3.5 °C in the Aral Sea region, 2.0–2.9 °C in large parts of the piedmont areas, and nearly 1 °C in the high mountains (Chanysheva *et al.*, 1991).

The month of July, and the entire summer as a whole, were the warmest within the decade of the 1970s. Positive deviations of July temperature from normal was on the order of 1 °C. The coldest summer seasons over the largest part of the territory occurred in 1931–40 and 1951–60 (Chanysheva *et al.*, 1991).

Spring and summer temperatures in the plains increased by 1.2–1.6 °C within 40 years (1951–90). During the autumn and winter seasons, monthly temperatures increased by 0.1–0.9 °C in this period. The trend in the increase of annual air temperatures is generally typical for Central Asia (Zolotokrylin and Tokarenko, 1991; Chanysheva *et al.*, 1991).

Thermal resources of a warm period The value of a warm period was determined to be the total effective temperatures (above 10 °C) in the plains and piedmont territories for the vegetative period of cotton (April 11–October 31). Thermal resources of the high mountain regions were considered to be the total positive temperatures (over 0 °C) during the whole warm period.

Total temperatures in the 1936–60 period over the whole region were lower on average than after 1960. The differences between the average values of the periods of 1936–60 and 1961–85 made up 100–150 °C, which is equivalent to the shifting in space of isotherms at a distance of 50 to 100 km. Fluctuations of total temperatures in the majority of cases covered a significant part of the Aral Sea basin, which is confirmed by the simultaneous fluctuations at high mountain and plains stations. The coldest decade (according to total temperatures) was the period of 1951–60 and warmest decades were in 1971–80 and 1976–85 (Chanysheva *et al.*, 1991).

Warm summer seasons contribute to the volume of streamflow in the rivers and to glacial and snow melt, because summer streamflow of these rivers and, thus, annual river flow to a considerable extent are governed by the increase of summer temperatures in the mountains. For the rivers with significant melting of glaciers and glacial feed, warm summer seasons usually provide up to 25% of the annual flow, but sometimes only 7–10%. The reserves of snow in the mountains are of main importance for streamflow of the rivers in the Aral Sea basin. In this regard, the importance of the role of winter precipitation for the annual flow significantly exceeds the role of summer precipitation (Bugaev *et al.*, 1957). Therefore, the increase of total temperatures within the 10-year periods of 1971–1980 and 1976–85 was reflected only to a small extent in the increase of streamflow of the mountain rivers.

Precipitation The precipitation data from 50 stations within the period of 1927–87 are considered here and grouped by 4 latitudinal zones of Central Asia: 44–42° N, 42–40, 40–38, 38–35° N latitude (Molosnova *et al.*, 1991; Subbotina *et al.*, 1991; Zolotokrylin and Tokarenko, 1991). Analyses have been made of year-to-year fluctuations in the annual, seasonal and monthly total precipitation of precipitation anomalies, and in the areas of propagation of these anomalies.

The trends of change of annual total precipitation on the plains and in the mountains are specified by the rate of the linear trend (mm/10 years). A weak trend in the increase of annual precipitation over the plains and the mountains is apparent in 1930–85. Prior to 1960, the positive trend of annual precipitation prevailed almost over the whole of Central Asia, and the increase of precipitation in this period was significant over the mountains (85 mm within 10 years).

The 1960–85 period is characterized by the reduction of the annual total precipitation only over the mountains (16 mm within 10 years). Especially noticeable periods in the reduction of precipitation in the mountains were 1960–85 and 1976–85. Systematic changes in annual precipitation over the plains were not observed in 1960–85.

When assessing the trends of changes in the precipitation of cold (Nov–Apr) and warm (May–Oct) half-years, the reduction of precipitation was observed at a major number of stations during the warm half-year in 1958–87, and during the cold half-year at certain mountain stations. The decade of 1978–87 is characterized by the trend of reduced precipitation in much of Central Asia during the cold half-year and an increase during the warm half-year (Molosnova et al., 1991).

The anomalies of annual total precipitation have not fully covered either the Aral Sea basin, or any latitudinal zone. A larger number of stations with a precipitation anomaly of one sign was noted to the south of 42° N latitude compared to the latitudinal zone of 42–44° N latitude. The number of negative anomalies of annual precipitation is larger on the plains than in the mountains. Positive anomalies of precipitation covered as a rule the greater part of the mountains (Subbotina et al., 1991).

The number of large-scale negative anomalies of monthly total precipitation, covering more than 70% of stations, exceeds the number of large-scale positive anomalies. An increase of both positive and negative large-scale anomalies of monthly precipitation after 1960 can be seen. Thus, the number of large-scale positive anomalies of monthly precipitation increased twice, generally during the warm half-year. The number of large-scale negative anomalies of monthly precipitation increased by 40% during the cold half-year.

THE RELATIONSHIP OF TEMPERATURE AND PRECIPITATION CHANGES TO CHANGES IN ATMOSPHERIC CIRCULATION

It was noted above that the period of 1935–60 is specified by latitudinal and meridional forms of circulation in compliance with Baidal's classification, and by a predominance of meridional forms after 1960. The prevalence of any one form in both plains and mountains is shown in Table 5.3.

The largest amount of total annual precipitation in the Aral Sea basin falls within the periods of predominance of the meridional form C of circulation. Positive anomalies of the total annual precipitation with C form of circulation on the plains and in the mountains make up, respectively, 112% and 110%.

Table 5.3 Statistical characteristics of total annual precipitation, averaged for flatland and mountain territory, with prevalence of forms by Baidal

Forms of circulation	Territory	Average (mm)	Root-mean-square deviation (mm)	Coefficient of variation	Anomaly (%)
			Statistical characteristics		
Meridional C	Plains	257	50	0.20	112
	Mountains	521	85	0.16	110
Meridional E	Plains	218	35	0.16	95
	Mountains	468	83	0.18	98
Latitudinal L	Plains	215	38	0.18	94
	Mountains	442	56	0.13	93

Source: Subbotina et al. (1991).

The prevalence of the meridional form E of circulation is connected with annual precipitation close to a standard rate in the mountains and below the rate on the plains. The lower amounts of annual precipitation in the Aral Sea basin have been observed during the years when latitudinal forms of circulation prevailed.

Alterations of the epochs of large-scale atmospheric circulation were displayed by the end of the 1950s and beginning of the 1960s in the development of synoptic processes of Central Asia. During the cold half-year after 1960, southern cyclones were reduced in number (especially between 1976 and 1985), and the recurrence of extensive warm releases increased sharply. The number of cold intrusions was significantly reduced from 1976 to 1985. These changes of atmospheric circulation were accompanied by an increase in the recurrence of warm winters and by a noticeable reduction in precipitation in the mountains.

HUMAN-INDUCED CHANGES OF LOCAL CLIMATES

Anthropogenic changes in climate in this area comprise the alteration of climatic characteristics as a result of irrigation, salinization of soils, urbanization, and the drying out of large water bodies. The numerous causes of land-surface changes are widespread in the Aral Sea basin. Visible changes of local climate have been noted in the coastal strip of the retreating Aral Sea.

The irrigation of large areas to a considerable extent changed the heat budget of climate at the surface. Aizenshtat (1991) traced the alterations in the heat budget caused by the extension of irrigated lands in the Golodnaya Steppe for the period of 1913–78. The area of the Golodnaya Steppe makes up about 850 000 ha. Irrigated lands of the Golodnaya Steppe occupied nearly 80% of the territory in 1978. Irrigation development in the Golodnaya Steppe changed the pattern of surface heat budget from the arid conditions of a desert in 1913 to the conditions of high humidity in 1978. On conversion to weighted average climate characteristics of the territory, it was found that, due to irrigation, the air temperature was reduced by 2.7 °C and relative humidity increased by 20%. The values of changes in climate characteristics of the same order are obtained by other authors for smaller-scale irrigated areas (about 10 000 ha) (Orlovsky and Utina, 1977; Orlovsky and Durdyev, 1978).

The signal of anthropogenic changes of climatic characteristics for irrigated lands is statistically important only during the warm half-year. Its capacity in July with respect to April is increased 4 to 5 times for large irrigated areas and 1.5–2 times for small ones. Interannual natural variability of mean July temperature for large irrigated areas is about 77%, governed by the interannual variability of weather conditions. The remaining share of this variability in July can be considered the result of irrigation impact (Zolotokrylin and Tokarenko, 1991).

Significant local climate changes have been observed with the reduction of Aral Sea surface area. As established by long-term records of microclimate, the modifying impact of the Aral Sea is propagated from the sea's shore for several tens of kilometers (Kuvshinova, 1980, 1982). Therefore, the influence on climatic data of the drying out of the Aral Sea may continue until the time when the dried strip exceeds 40–50 km. Further retreat of the coastal line beyond that will make it difficult to see its influence in climatic data over the long term (Zolotokrylin and Tokarenko, 1991). According to Kuvshinova (1980, 1982), because of the Aral Sea drying out, summer air temperatures increased in the period of 1970–80 by 0.1–0.4, and in the springtime by 0.5–0.7 °C. Winter and autumn temperatures were reduced by 0.2–0.6 and 0.5–1.3 °C, respectively. The daily amplitude of temperatures at coastal stations was increased and the reduction of relative humidity of the air, especially during the warm period of the year, was observed.

A greater number of dust storms and drifting snow was registered, with a maximum outbreak from April to July. The initiation, as well as the reactivation, of centers of dust storms on the dried bottom of the Aral Sea was also noted. The main regions of the fallout of dust and salts are over the Aral Sea, the eastern Ustyurt, and the land in the lower reaches of the Amudarya (Grigor'ev and Zhogova, 1992). The annual variation in the mean monthly air temperatures at the coastal meteorological station of Muynak is shown in Figure 5.1. A positive trend of spring–summer temperatures in 1970–80 was identified when Muynak station was at a distance of a few tens of kilometers from the water's edge. Because of the rapid recession of the Aral Sea by the middle of the 1980s, a new semi-desert climate with clearly defined annual variability of climate characteristics formed by the mid-1980s in the vicinity of Muynak (Zolotokrylin and Tokarenko, 1991).

The transformation of local climates on the dried territory has been accompanied by changes in the recurrence of certain types of weather. Thus, in the region of Aral'sk the frequency of sunny and very hot and dry weather increased by 15%. At the same time, the frequency of sunny, moderately humid weather conditions was reduced by 4 times. In sum, the frequency of

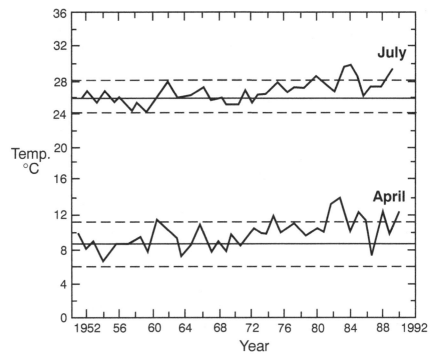

Figure 5.1 Annual variations of monthly values of air temperature in April and July at the coastal meteorological station at Muynak (Zolotokrylin and Tokarenko, 1991). Thin solid lines are average values, dashed lines are root-mean-square deviations.

more severe and unfavorable weather for the carrying out of human activities increased on the newly exposed seabed (Rusmanova and Afanas'eva, 1988).

Conclusion

This review of climate in the Aral Sea region allows us to answer certain questions concerning the climatic aspects of the identification and analysis of various thresholds associated with creeping environmental changes.

The temporal scale of changes in regional atmospheric circulation responsible for climate changes in the Aral Sea basin has been on the order of several decades. The human-induced changes of local climate and the formation of new local climates on the dried river areas and in the coastal strip surrounding the Aral Sea have taken place over several years.

The changes of atmospheric circulation are regional in pattern and have spatial peculiarities, e.g., on the plains and in the mountains of the Aral Sea basin. The spatial scale of anthropogenic changes of climate has been on the order of 100 km², and climate changes in the river deltas and coastal strip of the Aral Sea has ranged from 1000 km² to 10 000 km².

Natural changes in regional climatic characteristics are the result of a composition of short- and long-term positive and negative trends, suggesting a quasi-cyclical character of regional and local climate change. In many cases, the rates of natural change in climatic characteristics are unstable and statistically insignificant. However, the rates of human-induced changes of local climates are stable and can be detected.

The following paragraphs present my view of the threshold levels of comprehension of the climatic causes of creeping environmental changes in the Aral Sea basin.

THRESHOLD 1: AWARENESS OF CHANGES

By the 1950s, the data collected as a result of instrumental meteorological observations had already encompassed several decades. The regime changes of atmospheric circulation of climate characteristics in Central Asia were identified. Concepts were developed about atmospheric circulation 'epochs', causing relatively long-term regional climate change. The possibility for the potential degradation of the Aral Sea is discussed.

THRESHOLD 2: AWARENESS OF PROBLEM

The steady reduction in the volume of water in the mountain rivers was noted in the 1960s. At the same time there was an increase in the withdrawal of river water for irrigation. Scientific interest in the Aral Sea problem was on the increase. Climate forecasts about the state of the Aral Sea began to be published.

THRESHOLD 3: AWARENESS OF CRISIS

The negative effects of the phase of lowered atmospheric humidity in the mountainous parts of the Aral Sea basin were exposed in the 1970s and were linked to a new trend in regional atmospheric circulation. The withdrawal of river water for irrigation continued to increase. Some scientists became concerned by the decline of the Aral Sea level.

THRESHOLD 4: REALIZATION OF THE NEED FOR ACTION

During the 1980s, the trends which began in the 1960s of a circulation epoch with the phase of reduced humidity in the mountains continued. The negative effects of changes in local climates in the river deltas and in the Aral Sea's coastal area were exposed as a result of the sharp reduction in river flow. New locations of salt and dust storms emerged on the dried seabed of the Aral Sea, depositing salt and dust on the plains. The concern of Soviet scientists with the Aral Sea problem reached its peak.

THRESHOLD 5: ACTION

The Aral Sea crisis intensified in the late 1980s and early 1990s. Scientific recommendations on mitigation of the Aral Sea crisis are developed.

Compiled on the basis of this review, the retrospective perspective of the climatic problems and responses to them in the Aral Sea basin is presented in Table 5.4.

Table 5.4 Climate-related problems and responses to them in the Aral Sea basin

Period	Problems	Factors	Social response	State response
End of the 1950s	Change of the epochs of regional atmospheric circulation	Reduction of precipitation in the mountainous part of the basin	Water deficiency for irrigation on the plains; population increase in the region	Construction of new irrigation facilities
The 1970s	Human-induced changes of local climates on the plains and in the river deltas	Irrigation, desertification of lands; urbanization	Sharp growth of population; increase in agricultural production; deterioration of ecological conditions	Construction of new settlements on newly irrigated lands
The 1980s	Change of local climates in the river deltas and in the coastal area of the Aral Sea	Continued reduction of the river flow and catastrophic decline of the Aral Sea; origination of new centers of initiation of dust storms	Degradation of infrastructure in the river deltas and in the coastal areas of the Aral Sea. Deterioration of social, economic, medical and sanitation situations	Development and partial implementation of state measures to mitigate the Aral Sea crisis

References

Aizenshtat, B.A., 1991: Comparative analysis of heat budget characteristics of irrigated and nonirrigated lands of Central Asia. In: V.A. Bugaev (ed.), *Climate of Central Asia and its Variability*. Proceedings of V.A. Bugaev Central Asian Regional Scientific and Research Hydrometeorological Institute, 141(222). Moscow: Gidrometeoizdat, p. 127.

Baidal, M. Kh., 1964: *Long-term Forecasts of Weather and Climate Fluctuations of Kazakhstan*. Parts I and II. Leningrad: Gidrometeoizdat, p. 446.

Baidal, M. Kh., and D.G. Khanzhina, 1986: *Long-term Variability of Macro-circulation Factors of Climate*. Moscow: Gidrometeoizdat, p. 130.

Baidal, M. Kh., and A.K. Kiyatkin, 1972: Current and future problems of the Aral Sea. Trudy KazNIIGMI, 44, *Problems of Climate and Water Resources Fluctuations*. Moscow: Gidrometeoizdat, p. 5.

Bugaev, V.A., and V.A. Dzhordzhio *et al.*, 1957: *Synoptic Processes of Central Asia*. Izd. AN UzbSSR, p. 360.

Chanysheva, S.G., T.L. Veremeeva, E.P. Ilinyak, and N.V. Porfir'eva, 1991: Long-term fluctuations of temperature regime of Central Asia. Climate of Central Asia and its Variability. *Trudy SANIGMI im. V.A. Bugaeva*, 141(222). Moscow: p. 27.

Grigor'ev, A.A., and M.L. Zhogova, 1992: *Powerful Salt Removal in the Aral Sea Region in 1985-1990*. Doklady RAN, 324(3), p. 612.

Kuvshinova, K.V., 1980: Climate of the Aral Sea Region and its feasible changes in connection with drying of the sea. *Weather-forming Factors and Their Importance in Bioclimatology*. Moscow: Mos, Fil. GO, p. 17.

Kuvshinova, K.V., 1982: The thermal regime of the Aral Sea region in connection with anthropogenic desertification. *Data of Meteorological Studies*, 5, p. 86.

Molosnova, T.I. and E.P. Ilinyak, 1991: Climatic fluctuations in the general circulation of atmosphere and types of synoptic processes over Central Asia. *Proceedings of Central Asian Scientific and Research Regional Hydrometeorological Institute*. Climate of Central Asia and its Variability, 141(222). Moscow: Gidrometeoizdat, p. 3.

Molosnova, T.I., E.P. Ilinyak, and T.M. Chevychalova, 1991: Large-scale anomalies of atmospheric precipitation on the territory of Central Asia and their climatic variability. *Trudy SANIGMI im. V.A. Bugaeva*, 141(222). Moscow, p. 98.

Molosnova, T.I., O.I. Subbotina, and S.G. Chanysheva, 1987: *Climatic Consequences of Economic Activity in the Zone of the Aral Sea*. Moscow: Gidrometeoizdat, p. 119.

Orlovsky, N.S, and A.M. Durdyev, 1978: Influence of irrigation on thermal resources of the deserts of Central Asia. *Problems of Desert Development*, 4, p. 70.

Orlovsky, N.S. and Z.N. Utina, 1977: Influence of irrigation and water supply level on microclimate of deserts. *Problems of Desert Development*, 5, p. 3.

Rusmanova, T.S, and N.A. Afanas'eva, 1988: *Alteration of Climate and Bioclimate of the Aral Sea Coast in Connection with Decline of its Level. Complex Bioclimatic Studies*. Moscow: Moskovskii Fil. GO SSSR, p. 51.

Subbotina, O.I. and T.M. Chevychalova, 1991a: Specific features of long-term changes of atmospheric circulation on the territory of Central Asia. Proceeding of V.A. Bugaev Central Asian Regional Scientific and Research Hydrometeorological Institute. Moscow: Gidrometeoizdat, 141(222), p. 12.

Subbotina, O.I. and T.M. Chevychalova, 1991b: Concerning the cyclic recurrence of annual total precipitation on the territory of Central Asia and their long-term trend. Climate of Central Asia and its Variability. *Trudy SANIGMI im. Bugaeva V.A.*, **141**(222), Moscow, p. 113.

Subbotina, O. I. and T. M. Chevychalova, 1991c: Specific features of long-term changes of atmospheric circulation on the territory of Central Asia. Climate of Central Asia and its Variability. *Trudy SANIGMI im. V.A. Bugaeva*, **141**(222), Moscow, p. 12.

Subbotina, O. I., A. A. Zveryanskaya, and T. M. Chevychalova, 1991: Specific features of the distribution of anomalies of annual total precipitation over the territory of Central Asia. Climate of Central Asia and its Variability. *Trudy SANIGMI im. Bugaeva V.A.*, Moscow, Gidrometeoizdat, **141**(222), p. 87.

Vangengeim, G. J., 1941: The prediction of seasonal distribution of meteorological characteristics. *Isvestiya Academy of Sciences USSR. Seria (SIA) Geography-Geophysics*, **3**, 289–315.

Zolotokrylin, A. N., and A. A. Tokarenko, 1991: *About Variations of Climate in the Aral Sea Region within the Last 40 Years*. Izd. AN SSSR. ser. geograficheskaya, p. 69.

Priaralye ecosystems and creeping environmental changes in the Aral Sea

NINA M. NOVIKOVA

Introduction

The Aral Sea lies in the center of the Turan Desert in a transitional belt between the 'cold' northern and 'warm' southern parts of the desert. The area immediately around the Aral Sea is called the 'Priaralye'. Its external boundary is hypothetical and, according to the definition of Barykina *et al.* (1979), 'coincides with a border of the area where the sea affects ecosystems and land use'. The newly exposed drying-out part of the seabed is, thus, a part of the Priaralye.

The ecosystem is an historically determined formation of living (e.g., biota) and non-living (e.g., soils, groundwater, microclimate) components transformed by biota that are in constant movement and interaction and are capable of self-regulation and reproduction. Landscape conditions in the Priaralye include a gypsum desert (the Ustyurt Plateau), high desert plains in the northern Priaralye, ancient delta plains and a modern plain in the eastern Priaralye, and ancient deltas and a modern (live) delta plain in the southern Priaralye (Figure 6.1).

About 200 basic ecosystems classified as a vegetation association or as a landscape are found on Priaralye territory. The composition of the ecosystem in each landscape is specific to a region. The floristic composition of vegetation communities in the Priaralye includes about 1400 species, which is a large percentage of the total flora (about 1700 species) of the Turan Desert. There are more than 150 species of birds and 80 species of mammals. In the Priaralye, northern desert plants and animals are replaced by subtropical and tropical species.

The natural environment of the Priaralye is a harsh area for life. The climate is continental and arid with cold winters. The mean temperature in July on the northern coast of the sea is 24 °C; on the southern coast it is 28 °C. The mean temperature of the coldest month (January) is −12 °C in the north and −8 °C in the south. The zero degree C isotherm passes far to the south of the Aral. In the winter, air temperature can drop below −35 °C, while in the summer it can rise up to 46 °C, producing an amplitude of over 80 °C. Annual precipitation north of the Aral Sea varies from 60 mm to 200 mm and on the southern coast from 80 mm to 130 mm.

The combination of high summer temperatures and very low precipita-

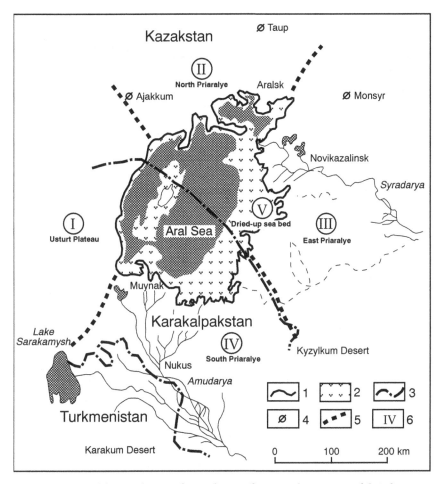

Figure 6.1 Region of the Priaralye, according to the prevailing type of ecosystems and their changes due to aridization. Region: I, Ustyurt Plateau; II, North Priaralye; III, East Priaralye; IV, South Priaralye; V, Dried-up seabed. The six designations are: (1) previous boundary of the Aral Sea; (2) dried-up seabed; (3) boundaries between Central Asian Republics; (4) meteorological stations; (5) boundaries between landscape regions, and (6) numbers of the landscale regions.

tion creates an arid climate for the Priaralye. The dry season lasts from the beginning of April until the end of September. Low winter temperatures and high aridity are limiting climatic factors which, on the one hand, protect modern ecosystems from the invasion of 'newcomers' (i.e., they prevent an invasion of mesophytes from the north and tropical varieties from the south) and, on the other hand, create a harsh 'habitat' for a limited number of species and communities which have nevertheless been able to adapt to such conditions.

Xeromorphy of vegetation is the main adaptive response to water deficit on a variety of levels. This is manifest, first of all, in the morphology of varieties. Succulence and plant surface areas are unique to plants in the area. Dwarf subshrubs and dwarf perennial plants lose many of their branches in winter (a widespread phenomenon in cold deserts), which should be considered as an adaptive feature. This enables plants to be protected under snow cover from severe frosts during harsh winters and to use moisture sparingly in dry summers. One of the most widespread varieties in the Priaralye is the xerophilic semi-shrub (*Artemisia terrae-albae*) which is capable of dropping its leaves in summer because of the reduced water supply, thereby reducing the plant's moisture needs.

Anabasis salsa is the second most widespread variety. It, too, drops its leaves; photosynthesis takes place in its green branches. An important adaptive feature exhibited by perennial and annual grasses (making up the ecological groups of ephemeroids and ephemers) is that they have a short growing period in autumn or early spring, the wettest seasons of the year.

The composition of plant communities also contributes to saving water, because the most active growth stages (i.e., blossoming and fruit-bearing) of the different plants occur at different times. In addition, they have an unequal distribution of aboveground and underground parts with, for example, an essential predominance of roots in relation to aboveground parts of plants (up to 70–80 times).

Animals of desert ecosystems either migrate northward or become dormant in summertime, hiding in holes in the ground or becoming active only at night. Desert animals get water only from food. Therefore, vegetation in desert ecosystems is vitally important as food and as a source of water.

Ecosystems are classified into two categories according to a water regime:

1. zonal automorphic (rely on water from precipitation only) and xerophilic desert ecosystems, which are confined to the Ustyurt Plateau and to the north and west Priaralye – they have an acute moisture deficit, and

2. hydromorphic mesomorphic azonal ecosystems, which develop within the Amudarya and Syrdarya deltas and within oases, and receive additional moisture from natural or artificial flooding and from groundwater whose level is close to the surface.

Apart from climate, edaphic factors (e.g., soil structure and salinity) also play an important ecological role. Under identical climatic conditions, the landscapes with various types of deserts exhibit very different ecological habitats. Thus, landscapes of sandy deserts provide a more hospitable habitat for plant species than do clay and solonchak soils. With an absence of soil salinity and with the ability of sands to accumulate moisture, a more favorable water regime exists. Thus, many more layers and a greater variety of life forms with higher annual productivity develop. The number of plant varieties on standard sandy locations is 2 to 3 times higher than in clayey locations.

Soil salinity is a serious limiting factor for vegetation. Even with sufficient moisture, as is the case with solonchaks, high soil salinity makes moisture inaccessible to plants. The Turan Desert was formed in the dried-up seabed of the ancient Tethys Sea and, therefore, soil-forming rocks contain significant quantities of water-soluble salts. Under hyper-arid desert conditions, salts are kept within landscapes by physically shifting from higher locations to lower ones, where they then accumulate, contributing to the formation of saline soils and solonchaks. The latter are usually devoid of vegetation. Phytomass and an annual increase of plant cover in saline habitats are usually much lower than in nonsaline ones.

Groups of halophytes are confined to higher salinity conditions. In view of the widespread development of saline soils, halophyte ecosystems are also widespread in deserts. Thus, practically the whole coastal part of the Ustyurt Plateau is occupied by communities formed by the haloxerophilic subshrubs (*Anabasis salsa*).

Three principal edaphic types of automorphic ecosystems can be identified in the Priaralye; xerophylic subshrubs on brown soils, xerohalophilic subshrubs on saline brown soils, and xero-psammophylic on desert sandy soils.

Under relatively stable conditions of an arid climate on vast expanses of the desert plains in the Priaralye, one can see a natural, region-wide tendency toward soil desalinization and a concentration of salts on the limited lowland solonchak sites. This tendency facilitates the increase in biological activity and the increase in the productivity of automorphic ecosystems. The natural environment of the modern delta plains of the Amudarya and the Syrdarya, where hydromorphic ecosystems exist, differs from the natural environment of the desert. Seasonal flooding in the deltas and groundwater levels close to the surface alleviate aridity and allow for the leaching of salts from the soils. This creates a more favorable habitat for plants and animals and provides a habitat that can support a variety of mesomorphic ecosystems. Also during seasonal flooding, sediments containing large quantities of nutrients for plants (such as nitrogen and phosphorus) are deposited in the rivers. These sediments provide an additional contribution to the increase in fertility of hydromorphic deltaic soils.

Ecosystems formed in river deltas are usually called 'tugai'. The species composition of tugai vegetation is not very diverse (about 40 to 80 varieties). These communities differ from desert communities because of their more efficient use of their natural environment. In forest communities, plants of the principal life forms are usually present: a wood stratum which reaches a height of 7 to 8 m; a shrub stratum at 2.5 to 3 m; perennial and annual grasses form 1 to 2 strata. Two to 3 kinds of vines reach to the tops of trees. Under conditions of sufficient heat and moisture, the phytomass reserves and the productivity of the tugai community are higher than in a desert.

Based on composition and structure, one can distinguish between two types of tugai ecosystems: meadow ecosystems and ecosystems of riverine forests and shrubs. Dominating in the wood communities in the Amudarya delta are the following: *Salix wilchelmsiana, Populus ariana, Populus pruinosa,*

Elaeagnus turcomanica, E. angustifolia; and in the Syrdarya delta – *Elaeagnus angustifolia*, and *Salix wilchelmsiana*. The shrub composition in both deltas is similar; dominating are *Halimodendron halodendron, Tamarix ramosissima, T. hispida*, and *Lycium ruthnicum*. High grasses are represented by *Phragmites australis, Glycyrrhiza glabra*, and *Trachomitum scabrum*, among others. Dominating in meadow communities are the following: *Phragmites australis, Calamagrostis epigeios, Glycyrrhiza glabra*, and *Alhagi pseudalhagi*.

Tugai are inhabited by a variety of animals. About 115 species of birds (Abdreimov, 1981) are found, of which 16 species have a permanent tugai habitat, 24 are found during migration, 18 hibernate, 57 are nesting, and one species is found only in summertime. There are not many mammals in tugai. In woody ecosystems 27 species have been found; whereas in meadow ecosystems only 16 have been recorded (Reimov, 1985). The typical species for tugai ecosystems include *Sus scrofa, Canis aureus, Felis chaus, Meles meles* among others.

With the drying out of the river network, sources of additional moisture are also disappearing. Increasing aridity enhances the impact of zonal climatic and edaphic factors. As a result, typical tugai ecosystems are replaced by solonchak, takyr and sandy ecosystems (Novikova, 1996). Such ecosystems are widespread in the ancient delta plains of the southern Priaralye located between the modern deltas of the Amudarya and Syrdarya (Figure 6.1).

The Aral Sea as an ecological factor

The Aral Sea ameliorates the region's continental, arid mesoclimate. It supports certain ecosystems because it is a source of moisture to adjacent areas within the Priaralye. Moisture transported by the air from the sea is a minor but constant additional moisture source which noticeably increases perennial plant production (by 3 to 4 times) in the desert at a distance of up to 3 km from the seashore. The Aral sea level controls the moisture supply of ecosystems by way of subsurface and surface water regimes.

The sea also functions as a barrier to erosion for delta landscapes. The sea's high level supports hydromorphic ecosystems because of a high groundwater table, river flooding, and the maintenance of numerous lakes. The lowering of the sea level would lead to the disappearance of lakes, the formation of drainage channels, and the desiccation of hydromorphic landscapes.

Natural ecosystem stability

The lifetime of ecosystems is measured from the time of their formation in a given region: (a) an absolute age and (b) a relative age, the length of time of their existence at a particular location. Both indices characterize, to a certain extent, an ecosystem's natural stability. An absolute age of modern ecosystems in the Priaralye should begin approximately 4000–6000 years ago, at the end of the last pluvial period and the beginning of an arid climate over the whole territory of Central Asia. However, some groups of plants and animals in this ecosystem have proven to be much older (e.g., the Tertiary Period). The relative age of xeromorphic zonal ecosystems is nearly equal to their absolute age, while the relative age of tugai ecosystems in the modern delta plains dates back to the first decade of their formation. Ecosystems of riverine areas and on young alluvial islands are still in the process of forming, while ecosystems of the ancient delta-plain landscapes are several centuries old. In other words, the stability of zonal automorphic ecosystems is maintained by a relatively stable desert climate, while the development of delta-plain ecosystems is constantly being affected by the dynamics of the river system. In this context, under existing climatic conditions, xeromorphic desert ecosystems should be recognized as more stable in comparison with hydromorphic intrazonal (tugai) ecosystems.

Human activity in the Priaralye has a thousand-year-old history. The vast desert plains were used as grazing lands by pastoralists. During centuries of grazing, the zonal ecosystems became semi-natural. In recent years, the historically established mix of ecosystems in some areas has changed as a result of poor grazing practices. Major changes in floristic composition have been observed in the vegetative cover; less prevalent species have disappeared and the structure of the vegetative cover has been altered if not degraded. According to Basova (1993), about 64% of the territory of the northern Priaralye is at some level of degradation. The sandy ecosystems have been most affected with moderate to severe levels of degradation having reached almost 70%.

Ecosystems in the eastern Priaralye are also disturbed because of too many livestock on rangelands. Sandy desert ecosystems are affected more than others. They lose their productivity as their protective vegetative cover diminishes and the mobility of the sand increases. The initiation of this process is very dangerous because of its longer-term implications; it has a tendency to feed back upon itself (i.e., a positive feedback). It lends itself to restabilization only with great difficulty. Hence, the constantly increasing anthropogenic pressures such as more intensive land use, and the processes of desert development, have endangered many rare species of plants and animals as well as natural complexes. Thus, beginning in the late 1960s, at least five species of big mammals have disappeared from the fauna of the Priaralye: the Turan tiger (*Panthera tigris virgata*), the cheetah (*Acinonyx jubatus*), the Kyzylkum mountain sheep (*Ovis vigney*), the Asian donkey (*Equus hemoinus*), the Bukhara red deer (*Cervus elaphus bactrianus*) and the Bukhara noble deer.

More than 20 species of mammals from the Priaralye fauna are referred to as rare and on the verge of extinction and, as such, they have been listed in the second edition of the USSR Red Data Book (Sokolov, 1984). One of these species (*Salinoglotus heptnery*) is endemic to the takyr landscapes within the region of the southern Priaralye, some are at the northern and northwestern parts of this area (*Jaculus turkmenicus, Salpinoglotus crassicauda, Felis manul*), and still others have gone beyond the borders of this region (*Felis caracal*). Within the southern Priaralye and Kyzylkum, one can find a population of gazelle – djeiran (*Gazelle subgutturosa*). The population of some beasts of prey (*Vormela peregusna, Felis caracal, Felis manul, Felis margarita*) has decreased with each passing year, because of the shrinking of their biotopes and the reduced availability of their food resources.

In the late 1960s and early 1970s, three wildlife preserves were created in the Priaralye; the Badai-Tugai in the Amudarya floodplain, the Kaplankyr on the border between the Ustyurt Plateau and the Karakum Desert, and on the Barsakelmes Island in the Aral Sea. In addition to their usual work, people in the wildlife preserves are busy reintroducing species that have disappeared from the Priaralye. For example, the Asian donkey ('culan', *Equus hemionus*), was introduced to Barsakelmes Island where its population has since grown. As another example, the Bukhara red deer, which was introduced to the Badai-Tugai preserve, is successfully breeding there and can once again be found in forests along the Amudarya banks.

The environment in the Priaralye has been greatly stressed since the beginning of the Aral ecological crisis a few decades ago. Much of this region has been destroyed as a result of the irrational use of nature. The continued lowering of the Aral Sea level and the expanding newly exposed seabed with its drifting surface sands, is responsible for incremental but cumulative changes in the ecosystems of the Priaralye. These ecosystems have in general been affected by the following: aridization of the mesoclimate, lowering of the sea level, and removal, by wind erosion, of dust and salt from the dried seabed.

In each part of the Priaralye (Figure 6.1), changes occurred following a certain progression, with varied rates of change for different parts of the region reaching crisis levels at different points in time. For ecosystems of the delta plains, the leading cause of such variations in rates of change has been the region's water supply.

Methods and techniques

The Aral Sea program was initiated and carried out by a team of researchers from the Institute of Geography of the Russian Academy of Sciences. When formulating the research concept (Anon., 1970), it was referred to as 'The Aral Sea Problem' and has remained so named until now. When formulating the goals and objectives of the study and when developing the program and the general methods of research, studies of the Priaralye environment (including regional changes in flora and fauna resulting from sea level decline) acquired a top priority alongside a study of the regularities of natural ecosystem formation and the evolution of natural processes on the recently dried-out part of the Aral seabed.

The general theoretical underpinning for environmental studies in this region was the concept of present-day desertification in the Priaralye region, i.e., a climate-related process initiated by human activity. An important approach was to identify functional units within the Aral basin (the drying seabed, the Amudarya and Syrdarya deltas, and adjacent areas) in order to carry out *in situ* long-term ecological observations using a common methodology. The principal method used has been referred to as 'integrated field profiling'. Regular observations at sites under different landscape and environmental conditions of the Aral coast and exposed sea bottom enabled the accumulation of empirical data from 1975 to 1985. These data were later used in various research activities and in monographs published in the 1980s.

When formulating the problem in 1975, researchers gave special emphases to the following: (a) forecasting future changes, (b) studying natural processes, and (c) identifying regularities in landscape and environmental conditions. Forecasts were based on landscape and genetic succession that exhibited a relative similarity of space and time sequences.

From 1975 to the late 1980s, studies of natural complexes in the Priaralye region were aimed at defining the trends in the interchange between biocomplexes and at evaluating the impacts of the drop in sea level on adjacent areas. Prior to the beginning of the human-induced decline of the Aral Sea in the 1960s, the boundary of the zone of impact was restricted to the relatively narrow geographic scope of impacts on the coastal meso- and micro-climates, estimated at 10 km from the shore of the sea. The impact zone along the low-lying coastline of the Akpektinski archipelago and the northern coast peninsulars was between 40 and 60 km.

An evaluation of the magnitude of the sea's impact on adjacent areas and on the nature of environmental changes in the region generated considerable debate, which has been summarized by Kabulov (1990). Based on his review of the literature, Kabulov concluded that the sea's effects were only of limited (i.e., regional) scope. The Aral Sea acts as an atmospheric temperature regulator in the Priaralye. This has been seen in such environmentally important processes as reduced evaporation and as changes in the frequency and duration of hot dry winds.

The drying out of the Aral Sea has been accompanied by a growing aridization in the Priaralye: an increase in extreme air temperature values, an increase in the amplitude of mean daily, monthly and annual values, a decrease in winter temperatures and an increase in summer temperatures (a pattern typical of a continental climate). The sea's influence on temperature extends out to a distance of 200 to 250 km. Previously, the sea's contribution to higher atmospheric humidity had extended out to 300–400 km. With the sharp drop in sea level, its role in the regional water budget has been reduced.

Research that summarized and evaluated the changing nature of the Priaralye appeared in the 1980s and early 1990s (e.g., Rafikov and Tetyukhin, 1981; Ptichnikov, 1994; Kust, 1993). These authors identified and assessed environmental 'thresholds' of change in the qualitative state of the ecosystems of the Aral Sea alone.

Popov (1990) and Zhollybekov (1995) forecast a transition of hydromorphic complexes into solonchak and semi-hydromorphic complexes, based on trends and rates of change of the natural hydromorphic complexes and of the soils in the Amudarya delta. Ptichnikov (1990) distinguished among several 'belts' exhibiting different rates of desertification in the Priaralye region. He mapped three belts with varying degrees and scopes of desertification and, based on the current rates of desertification, predicted an expansion of the belt with a high rate of desertification and then a relative stabilization of desertification processes in the Priaralye in the next 10 to 20 years.

The dominant Soviet economic development ethic up to the late 1980s was that of 'conquering nature'. In the 1960s, the USSR along with many other countries considered the environment as a competitor to be tamed, controlled, and 'developed'. From this perspective the disappearance of natural communities in the estuaries and deltas was considered a positive step because it was viewed as the elimination of 'non-productive' water uses in this arid region. By the late 1960s, most delta lakes had dried up. Later, when identifying environmental changes in the deltas that resulted from changes in the Aral Sea level, the loss of delta lakes was considered to have been negative, because of the lost economic value of biologically productive natural lands.

A misdirected idea was pursued in 1975 to divert a part of Siberian river runoff in order to sustain the sea's level and hydromorphic delta ecosystems. This was essentially a hypothetical solution to the Aral problem, which included sea level control in order to mitigate the negative impacts on ecosystems and the environment of a dropping sea level. This plan, however, underestimated the time and space scales of the Aral problem. The processes' rapid rates and the environmental transformations to qualitatively different states, after having crossed certain thresholds of ecological change, had been underestimated.

Identifying principles for evaluating the critical state of the environmental situation in the Aral Sea basin and identifying thresholds of regional environmental change were not on the government's agenda in the 1975–85 period. The realization of a need for such studies emerged only in the late 1980s, when the Aral crisis had reached a peak and had clearly threatened the livelihood and health of the inhabitants in the Priaralye region.

One of the first steps in exposing the crisis state of the Aral was taken by Glazovsky (1991). He proposed five scales and criteria for crisis evaluation (similar to the Richter Scale for earthquake assessment by the damage caused). A 1:2 500 000-scale map of environmental crisis situations was prepared for a detailed study of environmental processes in the Aral region. Working with this map, Zaletaev and Novikova (1995) compiled a map-scheme of the present state of biota and ecosystems at the same scale. Table 6.1 shows the evaluation criteria used to identify the severity of the regional crisis

Table 6.1 Assessment of the status of biota in the Aral Sea region

Index assessment	Changes in biota and ecosystems	Population health	Economy
Satisfactory	Reproduction and production changes fluctuate within normal boundaries.		
Tense (reversible processes)	Anthropogenic processes affect less than 10% of the area causing a regrouping of species in communities; 10–15% reduction of the original cover while still preserving the initial structure of ecosystems.		
Conflict (reversible with special reclamation)	Mosaic-like qualitative transformations of the ecosystem cover; onset of local restructuring of ecosystems; elimination of subdominant species, 20–25% reduction of the existing cover and 20–25% reduction of bioproductivity. 30% of land takes on a mosaic pattern. Extinction of climax communities.		
Grave (reversible with sophisticated reclamation)	Qualitative disturbances of ecosystem content and structure; pattern of changes affecting 60–75%. Initial cover reduced by 20–35% and useful production by 50%. Population of weed plants and rodents increases. Changes in structures in terms of ecotones.	Pandemic danger	Shrinking fodder and food resources
Disastrous (irreversible)	Irreversible transformation of ecosystem structure; extinction of dominant, rare and endemic species. Appearance of biotic badlands. Rates of degradation processes exceed adaptive capabilities of organisms.	Environmental quality is unsuitable for vital activity of population	Loss of resource potential; extinction of pharmaceutical plant species

based on the condition of the biota in the late 1980s. Based on these criteria, a situation typical of clay deserts at a radius of 70 km from the Aral Sea was considered 'critical'. Ground observations in the region showed a lower level of productivity of semi-shrub communities but no changes in their structure.

The situation on sand massifs adjoining the Aral Sea from the north and northeast was evaluated as one of 'conflict'. Mosaic-like qualitative transformations of the vegetative cover and the local restructuring of ecosystems with a disappearance of dominant plant species were reported. This resulted from the lowering of the groundwater table, increased livestock grazing pressures, and possible mesoclimate changes.

A 'grave' situation emerged in the sandy desert in a 40–50 km belt along the northeastern Aral Sea coast, where adverse changes have affected 60–70% of the area. At this stage, rodent populations increase and expand their range of activities, thereby increasing the risk of epizootic diseases. This situation could be reversed by intricate land reclamation and attempts to restore the previous environmental regimes. According to these evaluation criteria, a *crisis situation* had emerged on (a) the dried Aral sea bottom and (b) in the Amudarya and Syrdarya deltas.

The present crisis differs from previous ones in that it is now self-developing (i.e., a positive feedback situation has developed) and is not necessarily related to human activities. The falling sea level, the shrinking surface-water area, and the appearance of new dryland sites are agents of modern ecological destabilization. The following sections present the chronology of destabilization of the Priaralye's natural environment.

Changes in the Amudarya and Syrdarya deltas

The Amudarya and Syrdarya deltas have acted as barriers that slowed down the disastrous fall in Aral sea level that resulted from increasing river diversions in the 1960s and early 1970s. Nevertheless, changing environmental trends and resultant impacts on ecosystems were rather swift. With a sharp reduction in surface water from the rivers, submerged and boggy ecosystems disappeared. On cleared areas that were wet enough in the 1960–65 period, shrub, meadow and wood ecosystems were still developing. Solonchak ecosystems developed widely in 1974–80. By 1980, after the lowering of the groundwater level below 5 meters, meadow ecosystems perished, and simultaneously shrub and wood ecosystems began to degrade. By 1985 the groundwater level had dropped below 10 meters, which became critical for shrub and wood ecosystems; they began to perish *en masse*. New types of desert ecology started to become established (Novikova, 1996) (Table 6.2).

Thresholds of changes for delta plain ecosystems coincided with major stages of the changing ecological regime, as well as stages of delta landscape evolution. As is clear from Figure 6.2, changes in river runoff, water supply to the delta, and fish and muskrat catches occurred almost simultaneously.

In the mid-1980s, the modern Amudarya and Syrdarya deltas entered a

Table 6.2 Threshold values for groundwater levels in delta ecosystems

Ecosystem type	Critical groundwater level (m)	Notes
Meadow-boggy ecosystems	0.5 ± 0.5	with a regular supply of surface water
Meadow ecosystems	0.5–1.5	with periodic supply of surface water
Shrub	0.5–5	
Solonchak ecosystems	0–5	
Wood ecosystems	0.5–10(15)	

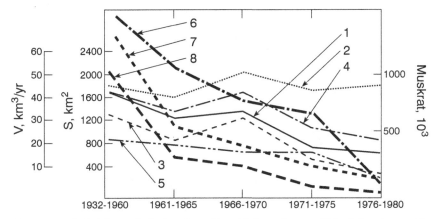

Figure 6.2 Ecological changes in the Amudarya delta with decreasing runoff. The designations are (1) actual inflow into the delta, km³/yr; (2) restored inflow into the delta region, km³/yr; (3) runoff to the Aral Sea, km³/yr; (4) runoff, using within delta, km³/yr; (5) lake area, km²; (6) flooded area, km²; (7) tugai forest area, km², and (8) muskrat skins in units.

semi-hydromorphic stage of development with takyr formation completed and the desertification of ecosystems just beginning (Threshold 3; see Table 6.3). But the artificial flooding that began in 1984–85 reversed those parts of the landscape that had already passed the hydromorphic stage back to the hydromorphic and solonchak stage. This process increased environmental destabilization in the region.

Changes in environment and in biological complexes in the river deltas have been described in detail, and the data, summarized by Kurochkina *et al.* (1991) and Zaletaev and Novikova (1995), are shown in Table 6.3. An assessment of thresholds of change over time in separate components of ecosystems is considered in the following section.

Changes in the tugai forest and bush ecosystems

River forest ecosystems (tugai forests and bushes) are very diverse in the Amudarya delta. They occur as large forest islands only along the main waterway, the Akdarya and its branches (Figure 6.3). Over 22 large forests can be found in the delta, each with its own name. Most tugai bushes and forests in the Amudarya delta were formed in the 1960s with the drying up of the central section of the delta. Desertification processes in the forests started in 1974, when annual flooding stopped. River runoff had terminated at most secondary river branches and, with an end to flooding, there was a sharp drop of the groundwater table. As a result, tugai bush and forests were rapidly transformed (in 10 to 15 years) to desert and low-productivity vegetation or were replaced by tamarisk thickets and annual *Salsola* barren lands. Where there was a slow fall of the groundwater table, tugai forests were replaced either by halophyte bushes or by barren solonchaks (Figure 6.4). Tugai forest

Table 6.3 Creeping environmental changes in pre-Aral ecosystems

Creeping environmental changes	Awareness of the problem	Threshold 1 Realization of changes	Threshold 2 Realization of the problem	Threshold 3 Realization of the crisis	Threshold 4 Realization of need for action	Threshold 5 Action
Degradation of hydrophylic tugai forest ecosystems in the Amudarya and Syrdarya deltas	1971	1940–1965 Onset of drying, extinction of lake and wetland ecosystems ___ 1969	1965–1974 Termination of floods, extinction of lake and wetland ecosystems and changes in tugai forest ecosystems ___ 1975	1974–1985 Replacement of tugai forest ecosystems by solonchaks ___ 1977–1979 Realization of the impossibility to compensate for water supply losses to the deltas; from 1974 — an artificial regulation of water supply to the Amudarya delta by the Takhiatash water reservoir and after 1981, by the Tuyamuyun reservoir; and to the Syrdarya delta by way of the Kazalin reservoir	1985–1990 Development of takyr process, introduction of desert species ___ 1977 Construction of collectors and filling dry water bodies with drainage water; construction of two dams in the Syrdarya delta, and one dam in the Amudarya delta	1990 Formation of desert biocomplexes, increasing role of hydromorphism and partial restoration of marshland and wetland ecosystems ___ 1984, 1990 Filling of dried-out water bodies with river water
Desertification of halomesophylic ecosystems of marine terraces	1979	1977–1979 Sea level fall, extinction of hydromorphic ecosystems ___ 1977	1979 Initiation of solonchak process ___ 1979	Early 1980s Development of halophylic bush communities ___ None	Late 1980s Onset of psammophyte succession ___ None	None

Process (year noticed)	Stage 1	Stage 2	Stage 3	Stage 4	Stage 5
Formation of biotic complexes on marine bay area and avant-delta — 1979	Second half of the 1970s: Drying seabed, ephemer ecosystems declining — Not reported	1979–1980: Formation of meso-psammo-halophylic bush complexes — 1979	1979: Wasteland ecosystems — 1990	1980: Wasteland ecosystems — 1979	Early 1980s: Selective phytoreclamation and the creation of artificial water bodies
Formation of biocomplexes on the newly exposed seabed — 1969	1977–1979: Retreat of the coastline by 500 m, ephemer and bush ecosystem affected — 1979; Realization of the impossibility of regulating the sea level by Siberian river diversions	Early 1980s: Drying of the first 2–5 km of the seabed, development of mesophylic complexes near the coast and of a solonchak-sandy wasteland — 1979	Mid-1980s: Rapid coastline retreat, changes in ephemer and desert ecosystems, salt-dust transport to surrounding territory	Late 1980s: Formation of a large area unfixed by vegetation, salt accumulation on the surface, eolian processes moving barkhan (dune) systems — Late 1980s – early 1990s	1981–1983 and beyond: Attempt to artificially stabilize mobile seabed substrate
Degradation of zonal ecosystems on coastline — 1979	Ecosystem of clayey deserts; Mid-1980s: Changes under impact of salt and dust transport, groundwater salinization and level decline due to fall in sea level	Sandy desert ecosystems in the northeast; disappearance of subdominants, 10% reduction in productivity — 1990	Ecosystems of a sandy desert on the Kyzylkum coast within a 40–50 km strip, 20% reduction of productivity; clayey deserts on SE site, change of perennial communities by halophylic annual communities — 1990	1990	None

Note: The first date in each category means duration of the stage of the process in every threshold. The second date (after a line) means the date when this threshold was noticed by scientists and announced (or, in some cases, this is the date when action was taken).

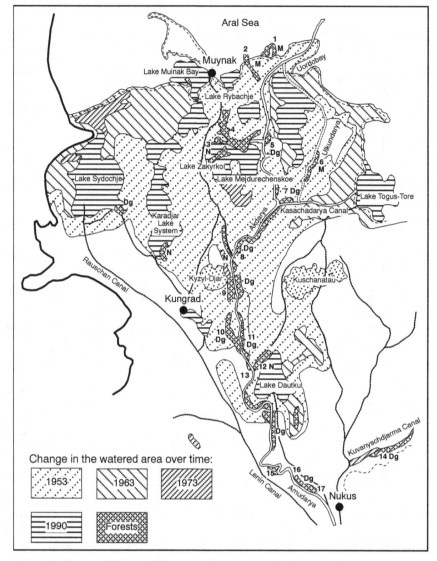

Figure 6.3 Changes in lake area from 1953 to 1990 and large tracts of forests. Designations are (1) Uordobay; (2) Archantay, Inzjenerusjak; (3) Zakirkol; (4) Shege; (5) Woroschilov; (6) Ulkun; (7) Kazakhdarya; (8) Aspantay; (9) Kyzyldjar; (10) Sajat; (11) Porlytau; (12) Erkin; (13) Naiman tube; (14) Nurumtubek; (15) Khatep; (16) Chortambay; and (17) Samambay. Status of the forests: N, normal; Dg, degraded; M, dead.

degradation was accelerated by river branch incisions that resulted from a declining sea level.

Today, the tugai forests in the Amudarya delta represent a middle-aged plantation with small areas in the drying river bed. River branches are occupied by newly forming tugai forests.

Tugai ecosystem shrinkage in the lower Amudarya started in the 1930s, increasing in scale during and after World War II. From 1930 to 1948 an area

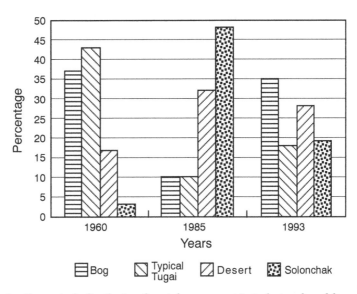

Figure 6.4 Changes in the distribution of tugai plant communities in the Amudarya delta.

of 120 000 ha of tugai forests was destroyed (Treshkin, 1990, figure 2a). According to Granitov (1970), the tugai area in the lower Amudarya in the 1950s totalled 1.3 million ha, second only to Uzbekistan's tugai forests (1.6 million ha). It included the area under tugai forests proper (100 thousand ha) and bush vegetation. According to Medetullayev (1974), the area under tugai forests proper was 52 300 ha. By 1989 that area had been reduced to 3300 ha.

In 1971 the Badai-Tugai reserve was established to protect vegetation and animal populations in the lower Amudarya. However, this designation failed to prevent a disaster. Changes in the Amudarya regime and the fall in the water level in the estuary resulted in the degradation of tugai forests in this protected area. According to Ptichnikov and Saburov (1990), the area of moisture-loving cane tugais along river terraces was reduced 1.5 times and mesophilic communities were reduced twofold from 1972 to 1982. This demonstrated that merely 'fencing' for protection is not a sufficient approach to protecting hydrophilous tugai complexes. Specialized land reclamation activities are necessary to protect floodplain forests. Such activities were designed in the 1980s; for example, the artificial irrigation of separate sites can help to restore tugai forests.

Recent changes in tugai forests resulted from aridization as well as from direct human-induced impacts which have increased because of an improved accessibility of tugai forests as parts of the delta dried up (Table 6.4).

Several threshold situations can be identified in the desertification of forest communities in the Amudarya and Syrdarya. Some of them were realized and evaluated recently, while others were identified some time ago.

Table 6.4 Damage to tugai forests resulting from different human-induced impacts (%)

Name of tugai forest	Impacts			
	Burning	Cutting	Cattle grazing	Transportation
Nurumtubek	14	36	42	8
Erkin-urai	22	44	25	9
Kyzildjar	7	40	32	21
Nazarkhan	28	38	22	12
Bekbai	24	36	30	10
Badai-Tugai	68	14	18	none
Hodjakul	4	18	74	4
Kazakhdarya	11	29	38	22
Shabbas	18	34	42	6
Samanbai	21	44	28	6

Source: After Treshkin (1990).

THRESHOLD 1

Bakhiev *et al.* (1977) were the first to realize the danger of tugai forest extinction and its replacement by bush, when the forest area in the Amudarya delta had already been reduced by half. A proposal was developed to protect all tugai forests. From 1975 to 1985 the Nurumtubek tugai was protected as a game reserve.

THRESHOLD 2

The crisis emerged in the 1980s as a result of: (a) the disappearance of willow forests, a moisture-loving plant not resistant to soil and groundwater salinization and (b) the lack of proper conditions for optimum growth and seed reproduction of the existing tugai forests. Bakhiev (1981) estimated that the ongoing xerophytization was a threat to tugai vegetation. Despite that information, the Nurumtubek game reserve lost its nature reserve status in 1981.

THRESHOLD 3

A crisis situation developed as a result of a further drying out of the area, resulting in a lowering of the groundwater table below 6–10 m; the delta landscapes entered a semi-hydromorphic stage. Bakhiev (1985) identified the second threshold of tugai forest change – the danger of delta transformation into a delta desert, because of the replacement of the tugai community by halophilic bush.

THRESHOLD 4

The expansion of species-induced solonchak desalinization and the onset of the process of takyr formation indicated that landscapes and ecosystems had entered the automorphic stage. The vegetative cover is characterized by the disappearance of bush and the expansion of annual *Salsola* and ephemers.

THRESHOLD 5

Desert-type biocomplexes (e.g., *Haloxylum aphillum* and *Ceratoides papposa* communities) formed at estuary mounds. The regular artificial irrigation of lake depressions facilitated a partial restoration of wetlands.

THRESHOLD 6

An expanding solonchak ring around the reclaimed reservoirs could serve as this threshold of change. The secondary salinization of soils and the delta's transformation to a solonchak desert signify the crossing of this threshold.

The crisis in tugai forest ecosystems resulted from the following factors:

- disappearance of glycophyllic willow forests because of the lack of favorable conditions for their growth and reproduction;
- prevailing communities with a single dominant poplar species and of a similar age of its plantations;
- termination of forest seed renewal;
- lack of conditions for the natural renewal of forests;
- unfavorable conditions for growth in all forests and the appearance of conditions leading to desertification in the forested areas (Figure 6.3);
- adverse human-induced impacts such as cattle grazing, burning, and uprooting of vegetation.

Changes in meadow ecosystems

Many researchers believe that meadow ecosystems owe their existence to human activities. For example, grazing and haymowing prevent forest and bush expansion onto meadows. Under desert river conditions, key factors that facilitate meadow emergence are tree-cutting, fires, and flooding for hay production. By cutting tugai forests, grass communities are able to develop as meadows. Their widespread appearance and their economic value necessitate their examination as a separate ecosystem of the delta floodplains. Meadow ecosystems in the Amudarya delta are represented by wetland, true meadow, and desert and solonchak variants.

Studies in the Amudarya delta by Mamutov (1991) showed that meadows are natural intrazonal ecosystems formed in floodplains under arid conditions. Their total area is 235 000 ha, of which 114 000 ha are used for pastures, and 111 000 ha for hay production.

Basic meadow communities of value as fodder for livestock are those of reed (*Phragmites australis*), thistle (*Alhagi pseudalhagi*), licorice (*Glycyrrhiza glabra*), and *Aeluropus littoralis*. These make up the bulk of the fodder crops grazed by cattle.

Successional changes of meadow vegetation under regulated runoff and extensive irrigated farming in delta regions result in halo- and zerophytization accompanied by a marked decrease of grass productivity and an increase in the variety of weed species.

From 1980 to 1985 the maximum harvest of reed communities reached 31.6 metric tons per ha; licorice, 3.6 metric tons per ha; and thistle, 2.57 metric tons per ha. The ratio of lands with different meadow types changed over time (Figure 6.3). In the 1960s typical and wetland meadows prevailed; by 1985 solonchak and desert meadows had expanded; and by 1993 marshy and desert meadows prevailed, although the amount of solonchak and typical meadows was still considerable.

Today, favorable conditions exist for meadow vegetation along reservoir margins and on lake bottoms that are regularly flooded. An expanded irrigated area in 1990–93 favored cane thickets and herbage (including *Aeluropus littoralis*). Yet, natural meadow areas are insufficient in size to meet economic demands and, therefore, these lands are irrigated to produce coarse fodder. Repeated haymowing of these lands, however, markedly reduces their productivity.

In the past, *Phragmites australis* communities that are typical of marshy meadows in the modern Amudarya delta occupied 600 000 ha. Today, despite a tenfold reduction of their area, they are most common along lake shores. Fodder stock from the total reed area in the Amudarya delta varied from 526 000 metric tons in poor harvest years to 593 000 metric tons in good harvest years. The biggest area (10 580 ha) lies in the Karadjar Lake system and Mezhdurechensk and yields a mean annual harvest in the range of 120 to 280 metric tons per ha. There are significant reed fodder areas around the Toguz-tore basin (an area of 5022 ha), producing a total phytomass of about 58 000 metric tons.

Glycyrrhiza glabra communities, typical of meadows proper, have been reduced in size by 90% in the last several decades, as a result of both aridization and commercial production for the medical industry. Today, the roots of this plant cover an area in Karakalpakstan estimated at 1716 ha (Bakhiev, 1985) and its harvest varies from 3149 to 3355 metric tons.

Alhagi pseudalhagi communities that are typical of desert and secondary meadows in the Amudarya delta occupied about 140 000 ha, prior to the reduction in runoff. As a result of aridization in the Priaralye region, thistle communities have been replaced by more salt-resistant and less moisture-loving communities with a type of fodder of lower economic value. The main thistle sites in the 1970s and 1980s were found on the beds of regularly flooded lakes. For example, a significant thistle site was located at the dry bottom of the Greater and Lesser Zakirkul lakes and along the shores of the Shege and Makpalkul lakes, a total area of 3870 ha and a fodder stock of about 3000 metric tons. Thistle vegetation perished with lake flooding. The total area of thistle communities in the delta has been estimated at about 44 355 ha. The minimum fodder stock averages 44 276 metric tons, but can reach as high as 54 073 metric tons in favorable years.

By the late 1970s, the fodder areas in the Syrdarya delta had been reduced by 60–70% and the harvest had decreased by 0.7–1 metric tons per ha (Kuznetsov et al., 1980).

An understanding of the negative changes in meadow ecosystems, particularly cane communities, came in the 1970s and 1980s with a loss of their economic production (Kuznetsov *et al.*, 1980; Bakhiev *et al.*, 1987). It was then made worse by reducing the area planted in commercially valuable licorice which had been exported from the Amudarya delta. Increased water supply to the delta in the mid-1980s ameliorated the situation to some degree. It brought about a partial rehabilitation of the cane lands, but failed to compensate for the loss of grass meadows valuable for fodder production.

Changes in bird populations

An important consequence of habitat change was the loss of trophic potential for birds and mammals. The extinction of lakes and cane areas complicated the life of fish-eating and insect-eating bird species in the wetland complex.

The bird population is an important component of tugai bush and forest ecosystems. More than 115 bird species inhabit the Amudarya delta (Abdreimov, 1981). One of the most valuable Central Asian bird species is the Khiva subspecies of pheasant (*Phasianus colchicus*). Most bird species in tugai forest and bush feed on insects and, therefore, act as a biological pest control.

Diverse vegetative cover, adequate shelters, habitat diversity and fodder abundance made lakes ideal sites as nesting areas for water fowl. A migration route crosses the lower Amudarya. Over 60 game-bird species inhabit the Amudarya delta, 30 of which are used for sport hunting. Up to 9000 birds rest at delta lakes during a migration period. Birds from northern Kazakstan and Siberia winter there. Among the delta birds, there are endangered and stressed species, many of which are in need of priority protection.

A sharp reduction of water supply to the delta and changes in the hydrologic regime have modified habitats and have severely affected bird populations in delta wetland ecosystems. We have shown that succession changes in lake biocomplexes are accompanied by modifications in bird populations (Table 6.5).

In addition, as a result of the uprooting of the tugai forest and the cutting of old, tall, and hollow trees, bird species nesting in the tugai forest are being replaced by xerophylic species coming from the desert (e.g., Abdreimov, 1981; Ametov, 1981).

The period from 1965 to 1984 was a crisis for ornithocomplexes in the Amudarya delta. With a minimum water supply, the bird population congested in the remaining basins of the river-fed Karadjar lake system and at the Toguz-tore and Koksu Lakes, which were fed by collector and drainage runoff. Lakes in the Kungrad oasis have preserved a unique ornithocomplex of about 15000 nesting bird couples including 12 rare species of Pelicaniformes and Ciconeformes.

Table 6.5 Evolutional succession of lake systems and changes in bird populations in the delta

Lake characteristics	Aquatic vegetation	Series of coastal ecotone ecosystems	Birds	Economic use
Permanently flowing transparent water, Depth up 6–8 m Transparency up to 4–5 m	High degree of overgrowth with *Potamogeton* spp. *Chara* spp.	Hydrosere *Typha* spp.→ *Phragmites australis*	*Netta rufina* *Aythya nyroca* *Phalacrocorax carbo* *Pelecanus crispus*	Fishery Hunting Grazing Haymowing
Permanently flowing turbid water Depth up 2–4 m Transparency up to 0.2–0.4 m	Almost a total absence of submerged aquatic plants, i.e., *Phragmites australis*	Hydrosere *Typha* spp.→ *Phragmites australis* Vast massifs of half-submerged reed plantations	*Larus argentatus* *Egretta garzetta* *Ardea cinerea* Anatinae	Fishery Grazing Haymowing
Floodplain lakes: (a) permanently connected with main delta branches by channels; (b) lakes fed by groundwaters during the low-flow period	Rich and diverse submerged plants: *Potamogeton* spp. *Chara* spp.	Xerosere *Phragmites australis*→ grass meadows→ bushes→ takyr	*Anas clypeata* *Anas platyrhynchos* *Netta rufina* *Sterna hirundo* Charadiiformes	Haymowing Grazing
(a) Lake systems are flowing during floods. During dry periods water is unavailable. (b) Certain floodplain lakes completely or appreciably drying in the low-flow period. Depth 1–2 m, in dry period 1 m	Diversified aquatic plants dying during the dry season *Potamogeton cristatum,* *Chara* spp.	Halosere weakly developed reed→ *Halostachys caspica*→ solonchaks	*Hymanthopus hymanthopus* *Charadrius alexandrinus* *Charadrius dubius*	Grazing
Salinized lakes fed by groundwater; Depth 1–3 m	Vegetation depends on salinity, *Zannichelia* spp.	Halosere *Salicornia herbacea* → *Aster tripolium* *Salsola* spp., *Suaeda* spp.→ solonchaks	*Tardona tardona* *Hymanthopus hymanthopus* *Charadrius alexandrinus*	
Floodland lakes with stable water supply all year	Diversified	Xerosere reed→ grass meadows→ tugai forests and bushes	*Phasanius colchicus* Anatinae *Ardea*	Haymowing Grazing Hunting

Source: Novikova (1995).

Studies by Lukashevich (1990) have identified the reasons for ornithocomplex changes in the Amudarya delta between 1964 and 1984. Although hydrophylic bird species have different requirements with regard to nesting habitats, their dynamics reflect changes in biotope structure. Hydrophylic ornithocomplexes reach maximum species diversity and population density in those delta lakes with abundant vegetation.

It is noteworthy that the human-made reservoirs fulfill functions similar to the natural ones, helping bird populations to survive during this environmental crisis. For instance, rice paddies provided shelter for ornithocomplexes typical of the initial stages of reservoir succession. Ornithocomplexes in collector and drainage reservoirs in the location of former lakes have a lower density, occupying an intermediate place between the reservoirs during earlier succession stages.

A similarity between ornithocomplexes of the delta and the discharge reservoirs indicates that the latter presently are and will be refuges for bird species belonging to the delta ornithocomplexes.

Zaletaev (1989) suggested that such changes had taken place during the periods of minimum water supply to the Amudarya and Syrdarya deltas. Nesting and congregation areas of the Pelicaniformes species were dislocated from the Amudarya to Lake Sarykamysh, and from the Syrdarya to the Turgai lake system 400 km to the north.

The realization of the possibility of losing ornithocomplexes in tugai forest and wetland ecosystems coincided with the rapid drying out of the delta in the 1970s. The Sudochinsky and Nurumtyubek game reserves and Badaitugai reserve were established at that time. In 1985 the protected area status was reconsidered and the Sudochinsky reserve was renamed the Delta reserve.

The measures taken, however, were obviously insufficient. Studies by Lukashenko (1990) determined that a satisfactory number of rare hydrophilic species were nesting at the Amudarya delta at its lowest level of water supply. Spoonbill (*Platalea leucordia*) was estimated at 500 pairs comprising 15–20% of the total population of that species nesting in the entire Soviet Union; egret, 300 000 pairs; mute swan (*Cygnus olor*), about 355 000 pairs.

A. B. Bakhiev, M. Ametov, V. S. Zaletaev and R. V. Lukashevich appealed to the Karakalpak Committee on Environment Protection, to the public, and to decisionmakers about the need to protect bird populations and water bodies by combining traditional methods (e.g., creating nature and game reserves) with alternative methods (e.g., regulating the hydrologic regime in the delta basins and land reclamation). Thanks to the continuous efforts of these scientists, numerous lakes were protected as nature reserves.

The formation of natural complexes on the dry Aral seabed

Dried land that had previously been wet is a unique, extremely destabilized setting for biological activity. Environmental changes here are so rapid that the processes of biological formation of complexes terminate before they reach their peak, and they must start anew. Zaletaev (1989) referred to them as 'short-lived' processes.

Kurochkina *et al.* (1991) proposed differentiating among the following:

1. short-term ephemeral ecosystems – on the newly exposed drying seabed formed near the annually shrinking coastline;
2. wasteland ecosystems – ecotopes with no vegetation;
3. residual ecosystems – tugai forest-type ecosystems formed in bays and shallow waters and on the newly exposed seabed in the first years as the sea level receded; and
4. neoecosystems – relatively stable formations at marine terraces within the 1960s and 1970s dry zones with renewal of the dominant species.

These ecosystems are typical of practically all coastal types in the six isolated regions. Ephemers of the *Chenopodiaceae* family dominated in the first decade followed by a delay (e.g., wasteland ecosystems) with perennial species constantly expanding from the main sea coast. From this moment on, the impact of regions adjacent to the dry sites grows. The trend of biocomplex formation is related primarily to the soil substrate on which it develops and on the characteristics of the adjacent dry land. Psammophylic complexes are related to sands; halophylic complexes are related to salinized sandy-clay and clay sites.

About 25 species of mammals, 15 species of birds, 10 of reptiles, and a relatively minor number of invertebrate species have populated the newly dried lands in the Aral Sea region (Zaletaev *et al.*, 1991). 147 plant species have been recorded on the dry seabed strip on the eastern coast (Dimejeva, 1990).

Desertified, newly dried lands have a negative economic impact, since they cannot be used for pasture or for irrigated farming now or in the near future. Today, most of the dried seabed remains unfixed by vegetation and open to severe wind erosion and to salt accumulation on the surface. Hence, the area has become a major source of salt and dust that can be carried by wind to adjacent areas out to a distance of about 400 km.

The problem cannot easily be solved in view of the vastness of the region. Isolated dry-land sites along the coast in the 1960s were fixed by bush species (Koksharova and Isakov, 1985), but phyto-reclamation on most of the exposed dried-out seabed is limited by the high salt content, as well as by the hardness of the ground.

Zonal ecosystem changes along the main coast

The ecosystems of the autonomous alluvial landscapes on the elevated pebble-gypsum plains of Ustyurt and the clay plains of the northern and northeastern Priaralye (Table 6.4) have undergone minor changes, as a result of the impacts of the drying out of the sea. Salt from the dried seabed is a key factor in the transformation of Ustyurt biocomplexes. Some researchers estimate that 15–20% of all the salt transported by powerful wind storms from the eastern Aral Sea area is deposited here. Observations by Ptichnikov (1994) and Kabulov (1990) have shown that in 1977–79 the groundwater table was deep (45 to 54 m), remaining stable and inaccessible to plants; the salt composition of soils remained practically unchanged; and the phytocenosis type, its initial structure and plant protective cover, were preserved.

The viability of plant communities decreased, and moisture-loving and less salt-resistant species disappeared (e.g., *Salsola arbusculiformis, Artemisia terrae-albae, Agropyron cristatum*). The above changes exposed the onset of a succession process. The ongoing changes have not yet crossed any environmental threshold. Yet these changes are found at a 30-km distance into the plateau, making the total area quite considerable.

In September 1996 we carried out investigations of the eastern margin of the Plateau. We discovered that, along its southern extension facing the dried-out seabed from the Amudarya delta to Cape Aktumsyk, wormwood shrubs (*Artemisia terrae-albae*) had no leaves because, during the previous two years, there had been no summer precipitation. On some bank areas extending out to the remaining pools of sea water, within a strip 5 to 10 m wide, the wormwood had leaves and blossoms.

Profound changes have been reported in mesophylic complexes at chinks (scarps) in the Ustyurt plateau inclined toward the Aral Sea. Degradation of tugai vegetation fragments occurred as a result of decreased water supply to biotopes and a decline in the groundwater table; i.e., the stability threshold of mesophyllic complexes was exceeded at local sites. Although this process was observed over a very small area, it should be viewed as negative, in the view of the disappearance of unique varieties of flora in this region.

The diversity of landscapes and biotic complexes is typical of the northern Priaralye region. Minor changes on a component level have been observed in semi-bush communities in clay deserts. Degradation processes have been reported in the Maliy (Small) and Bolshoi (Large) Barsuki. Some researchers (e.g., Ptichnikov, 1994: Basova, 1993) believe that these processes resulted from overgrazing and are not related to the declining level of the Aral Sea.

The most 'dangerous' change in this region is the salt and dust that is carried from the exposed seabed. Semenov *et al.* (1991) reported that the total annual amount of material carried by the wind from the northern part of the sea reached 1.1 million metric tons per year. The normal content of salts in the atmosphere near the earth is estimated at approximately 10000 to 20000 metric tons per year over an area with a radius of 300 to 500 km. It is clear that the natural ecosystems of the Priaralye have been subjected to exceptionally heavy salt loads.

The response of Priaralye ecosystems subjected to such impacts should be exhibited through deteriorating soil fertility and diminished plant productivity that, in general, results in a lower productivity of the rangelands. In this case, vegetation can be used as an indicator of the impact on natural ecosystems of the drying out of the Aral Sea.

Analysis of data on changed productivity of vegetation (obtained at meteorological stations in the northern Priaralye at different distances from the sea coast) has indicated that at the Ayakkum station (Figure 6.1), located to the west of the main source of salt drift, the effect of salt and dust drift has not yet reached a level that causes change in vegetation productivity. However, the effect of several years of salt drift on the dynamics of vegetation productivity at the Taup and Monsyr stations became most pronounced after 1975.

The eastern Priaralye region is a sandy-clay solonchak complex with diverse flora and vegetation (Batgalova, 1993). Similar to the northern Priaralye region, changes in this region are most obvious in the sandy desert. The degraded vegetative cover has been closely related to overgrazing.

Drastic changes have been reported on the Kyzylkum coast of the Aral Sea, including the Akpetkinski archipelago and the ancient clay deltas of the Amudarya and Syrdarya. Marked changes have also been observed in the groundwater's salt content resulting, according to Kabulov (1990), in the disappearance of *Haloxylon persicum* forest in the Janadarya delta and wind erosion of Akpetka sand mounds. The latter resulted in the replacement of stable perennial communities with annual species. A specific form of desertification – a mossy desert – lies to the east of Belitau highland. Drinking-water shortages due to the retreating sea coast and high salt content of sea water has resulted in the absence of wild and domestic animals. This has led to an expansion of desert moss *(Tortula desertorum)* and the disappearance of grass species.

Priaralye ecosystems, particularly those in sandy deserts, are markedly disturbed and are close to crossing a crisis threshold. Many researchers consider overgrazing to be the main cause of ecosystems degradation. Its impacts will aggravate an already destabilized environmental situation in the Priaralye region.

The lowering of the Aral Sea level also affects the population of mammals. Thus, in 1965–70, when the sea had receded from its former coastline by 3 to 5 km, the coastal vegetation started to thin out, animal habitats became degraded, and the population of animals of prey in the coastal strip of the Ustyurt and the Kyzylkum started to diminish. By 1975–77, wild cats disappeared, caused by their migration from the coastal strip. Their population density increased in some locations of the Ustyurt, the Amudarya delta and the Kyzylkum desert at distances from the sea of 25 to 120 km. But from 1978, due to the enhanced negative effect of the lowering of the sea level on vegetation, the population of game mammals started to decrease noticeably. Thus, in 1962–65 in the northwest of the Kyzylkum where a protective cover by plant communities was 25–50% on an area of 100 km², an average of 4 djeirans could be found. Djeiran and wolf, which had large populations before 1970, are now rarely found. Similar changes were observed in Ustyurt.

Conclusion

A major environmental crisis has put non-salt-tolerant hydrophylic ecosystems in the Priaralye region on the edge of extinction, resulting in the disappearance of more than 60 plant and 20 animal species, 4 types and over 50 biocomplexes. Understanding and public awareness of the situation only emerged 30 years ago, at the same time that initial signs of environmental change were noticed. Attempts to ameliorate the crisis have not been sufficient. A problem of ecosystem variation as a problem of maintaining biological diversity has only recently become the subject of serious discussion. Stabilizing environmental degradation and change in the Priaralye region cannot be successful as long as the continuing sharp sea level drop and the drying out of the deltas remain the main reasons for desertification processes in the region.

References

Abdreimov, T., 1981: *Birds of Tugai Forests and Adjacent Deserts in the Lower Amudarya*. Tashkent: FAN, 100 pp.

Ametov, M., 1981: *Birds of Karakalpakia and their Protection*. Nukus: Karakalpakstan Academy of Sciences, 138 pp.

Anon., 1970: *Report on the Aral Sea*. Materials for the session of Bureau and ad-hoc Committee of the Scientific Council, 'Integrated use and protection of water resources'. Moscow: USSR Academy of Sciences, 49 pp.

Bakhiev, A. B., 1981: Vegetation in the lower Amudarya (within Karakalpak ASSR boundaries), its ecology and dynamics. Dr. of Sci. (Biol) thesis. Tashkent: AN UzSSR, Institute of Botany, 44 pp.

Bakhiev, A. B., 1985: *Ecology and Vegetation Communities' Replacement in the Lower Amudarya*. Tashkent: FAN, 192 pp.

Bakhiev, A. B., K. N. Butov and M. T. Tadjitinov, 1977: *Dynamics of Vegetation Communities in the Southern Priaralye Region in Relation to Changes in the Hydroregime of the Aral Basin*. Tashkent: FAN, 84 pp.

Bakhiev, A. B., N. M. Novikova and M. E. Shenkareva, 1987: Changes in the economic value of plant resources under reduced water supply to floodplains. *Water Resources*, 2, 167–9.

Barykina, V. V., D. V. Panfilov and V. A. Timoshkina, 1979: Modern trends of changes in Priaralye biocomplexes. *Problems of Desert Development*, 2, 34–41.

Basova, T. A., 1993: Alterations in vegetation of northern Priaralye region under anthropogenic desertification and their mapping. Cand. of Sci. (Biol) thesis, Almaty: Institute of Botany, NANRK, 25 pp.

Batgalova, G. S., 1993: Assessment of anthropogenic damage to vegetation in the eastern Priaralye region. Candidate of Sci., (Biol) thesis. Almaty: Institute of Botany NANRK, 26 pp.

Dimejeva, L., 1990: *Flora and Vegetation of the Aral Seashores*. Alma-Ata: Institute of Botany, Kazakh Academy of Sciences. 27 pp.

Glazovsky, N. F., 1991: *Proceedings of the USSR Academy of Sciences*. Geographical Series 1(123).

Granitov, A. I., 1970: Uzbekistan plant communities. Cand. of Sci. Thesis, Tashkent: Institute of Botany, AN UzSSR, 23 pp.

Kabulov, S. K., 1990: *Desert Phytocoenosis Alterations under Aridization*. Tashkent: FAN, 238 pp.

Koksharova, N. E. and G. I. Isakov, 1985: Forest reclamation of the dry Aral seabed. *Problems of Desert Development*, 5, 48–55.

Kurochkina, L. Ya., V. V. Vukhrer, G. B. Makulbekova and L. A. Dimejeva, 1991: State of vegetation in the dry seabed and Aral Sea coast. *Transactions of the USSR Academy of Sciences, Geographical Series*, 4, 76–81.

Kust, G. S., 1994: Desertification and soil evolution in arid lands. Dr. of Sci. (Biol) thesis, Moscow: MGU, 50 pp.

Kuznetsov, N. T., R. V. Nikolaeva and I. D. Ryabova, 1980: *State of Art of the Aral Sea*. Moscow: VINITI, 53 pp.

Lukashevich, R. V., 1990: Water impacts on the structure and function of hydrophylic ornithocomplexes in the Amudarya Delta. Cand. of Sci. (Biol) thesis. Moscow: AN SSSR, IVP Russian Academy of Sciences, 15 pp.

Mamutov, N. K., 1991: Meadow vegetation transformations in the Amudarya Delta under aridization. Cand. of Sci. (Biol) thesis, Moscow: MGPI, 18 pp.

Medtullayev, J., 1974: Karakalpak SSR land resources. In: M. T. Tadjitdinov *et al.* (eds.), *Natural Resources in the Lower Amudarya*. Tashkent: FAN, 51–78.

Novikova, N., 1995: Water management for biodiversity conservation in the desertified delta plains of the Aral Sea basin. *Intergovernmental Oceanographic Commission Workshop Report No. 105*, E. Duunsma (ed.), Proceedings of Conference on Coastal Changes, Bordeaux, France, 991–4 (in English).

Novikova, N., 1996: Current changes in the vegetation of the Amudarya delta. In: P. D. Micklin and W. D. Williams (eds.), *The Aral Sea Basin*. Berlin: Springer, 69–78.

Popov, V. A., 1990: *The Aral Sea Problem and Amudarya Delta Landscapes*. Tashkent: FAN, 112 pp.

Ptichnikov, A. V., 1990: Priaralye landscape desertification dynamics in the last 15 years. Cand. of Sci. (Geog.) thesis. Moscow: MGU, 27 pp.

Ptichnikov, A. V., 1994: Regional features of Priaralye desert landscape dynamics under desertification. *Problems of Desert Development*, 2, 3–9.

Ptichnikov, A. V. and M. Saburov, 1990: Tugai forest landscape transformations in the Amudarya Delta under desertification (on the example of Badai-tugai reserve). *Bulletin of Karakalpak Branch of Academy of Sciences*, 2, 45–8.

Rafikov, A. A. and G. F. Tetyukhin, 1981: *The Aral Sea Level Fall and Environmental Changes in the Lower Amudarya*. Tashkent: FAN, 198 pp.

Reimov, R., 1985: *Mammals of the South Priarlye: Ecology and Conservation*. Tashkent: FAN, 96 pp.

Semenov, O. E., L. P. Tulina and G. N. Chicasov, 1991: Changes of the climate and ecological parameters in Priarlye. In: I. Yu. A. Izrael and Yu. A. Anochin (eds.), *Monitoring in the Priaralye*. St. Petersburg: Gidrometeoizdat, 150–76.

Sokolov, E. V. (ed.), 1984: *Red Data Book of the USSR* (Lesnaya promyshlennost). Moscow: T. I., second edition, 392 pp.

Treshkin, S. E., 1990: Structure and dynamics of tugai bush and forest communities in the lower Amudarya under anthropogenic impacts. Cand. of Sci. (Biol) thesis. Moscow: Research Institute of Environmental Protection and Natural Reserves, 23 pp.

Zaletaev, V. S., 1989: *Destabilized Environment: Arid Zone Ecosystems under a Changing Hydrological Regime*. Moscow: Nauka, 148 pp.

Zaletaev, V. S. and N. M. Novikova, 1995: Changes in the biota of the Aral region as a result of anthropogenic impacts between 1950 and 1990. *GeoJournal* **35**(1), 23–7 (in English).

Zaletaev, V. S., N. M. Novikova, and A. B. Bakhiev, 1991: Problems of conservation and rehabilitation of the biota within the Priaralye region in conditions of desertification. *Vestnik of Karakalpak Academy of Sciences of Uzbekistan*, **4**, 3–13.

Zhollybekov, B., 1995: Soil Cover Alterations in the Coastal Part of the Amudarya River under Aridization. Nukus: Bilim, 132 pp.

7 Public health in the Aral Sea coastal region and the dynamics of changes in the ecological situation

LEONID I. ELPINER

Introduction

The challenging problems of the Aral Sea coastal region, which have arisen following striking changes in ecological conditions in the area, now seem to have been the logical result of the ill-considered planning of anthropogenic impacts on natural resources and human habitat.

A comprehensive analysis of the process of development of current ecological conditions (and its impacts on the way of life and on public health) is of interest from both a scientific and a practical point of view. In this regard the Aral Sea coastal region is an appropriate subject of research, because the pronounced and ecologically important natural processes and phenomena observed here were accompanied by sharp, adverse changes in the region's medical and demographic conditions.

The existence of 'green' and 'brown' ecological problems (Vogel, 1994), most often the result of creeping environmental changes (i.e., slowly changing, incremental, cumulative processes), is fairly obvious:

> Green issues refer to those problems at the top of environmental agendas throughout the world, including ozone, greenhouse, and related global climatic change. Brown environmental issues … include waste removal, safe water provision, urban health problems, inadequate sanitation, and local air pollution.
> (Vogel, 1994, p. 233)

It is unclear, however, on what time scales the initially unmonitored processes are altered. To what extent and with what degree of degradation (or transformation) of the environment are they connected? Do these processes reach identifiable threshold values? Is it possible to change negative patterns and trends once we begin to speak of crisis? Perhaps we can establish thresholds (i.e., levels) of ecological change by assessing the character and importance of related processes. By doing so, one can determine the period of initiation of the various stages of creeping environmental change.

The attempt to answer these questions, using an analysis of medical and ecological conditions in the Aral Sea crisis zone, seems feasible. In the Aral Sea situation, these are interrelated processes. In fact, human health has become the major factor used to identify the Aral region as an ecological disaster zone.

Public health status of the Aral Sea coastal region

A comprehensive analysis in 1991 of the public health status of the 1970s and 1980s was carried out for the Aral Sea coastal region. The analysis was performed by researchers at the Water Problems Institute (1991) in cooperation with a number of specialized public health research institutes of the USSR. Their findings pointed to the fact that the medical and biological conditions that developed in the coastal region of the Aral Sea were characterized by a high level of morbidity from intestinal infections and an increase in non-infectious pathology, e.g., oncological, cardio-vascular, organs of secretion, digestion, and hemogenesis and respiratory system problems. The pathology related to pregnancy was also identified as was a high rate of infant morbidity and mortality. Innate malformation and other genetic problems were observed more frequently than had been the case in the past (Ministry of Public Health of the USSR, 1990; K ASSR, 1987).

Thus, according to the data of the Public Health Service in the Kyzyl-Orda region (Kazakstan) for the 1973–88 period, the incidence of typhoid fever increased up to 29 times in some years, viral hepatitis increased up to 7 times, and paratyphoid up to 4 times. More than 60 000 people were stricken with viral hepatitis in the region in this period and more than 70 000 suffered from acute enteric (intestinal) diseases. Water-borne outbreaks of typhoid and paratyphoid infections and viral hepatitis were registered annually in 30 settlements in the region. Between 1976 and 1986 several water-borne outbreaks of viral hepatitis took place, infecting several tens of thousands of people. From the total number of registered cases of viral hepatitis in the former Kazakstan SSR, nearly 40% fell within the Kyzyl-Orda region and about 45% of the total were infected by typhoid fever.

A trend can be found in the increase in enteric infection morbidity (Figure 7.1). Thus, if within 1976–80 the annual index of typhoid fever morbidity made up 58.9 cases per 100 000 people, then in 1981–85 it reached 75.4 cases per 100

Figure 7.1 Enteric infections in the Kyzyl-Orda region of Kazakstan.

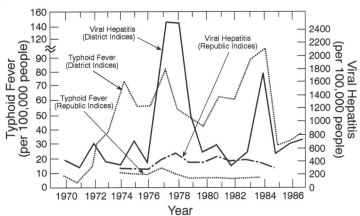

ooo people. It should be noted that nearly 75% of the people with acute enteric infections were children and 30% of the infant mortality came from this number (Kyzyl-Orda Regional Sanitary and Epidemiological Station, 1987).

No decline was recorded in morbidity by tuberculosis, nor by the natural outbreak and epidemics of infections (e.g., Crimean hemorrhagic fever, Q-fever, visceral Leishmaniasis).

The indices of somatic (noninfectious) diseases also changed for the worse: For example, the number increased of sick people with a pathology of the gastroenteric tract; liver and kidney problems increased threefold and cardiovascular system problems increased twofold. The level of morbidity with cancer of the esophagus in the region was highest in Kazakstan (Kyzyl-Orda Regional Sanitary and Epidemiological Station, 1988).

About 70% of women who are mothers suffer from anemia; more than half of pregnant women suffer from extragenital pathology, especially diseases of the kidneys (5%) and of the cardiovascular system (4.5%). All this results in high morbidity and mortality of the newborn. Diseases such as rickets, hypotrophy and anemia are widespread among children (Ministry of Public Health of Kaz SSR, 1988).

The population with registered parasitic diseases in the Kyzyl-Orda region (Kazakstan) increased threefold from the early 1980s to the early 1990s. As shown in Table 7.1, the main form of helminthiasis (i.e., infestation with or disease caused by parasitic worms) harmful to the health of the population, especially children, within the Kyzyl-Orda region is hymenolepiasis. Analysis of the long-term dynamics of affliction by this form of helminthiasis points to a stabilization of the rates at 0.7% of the number of examined, the equivalent to 150–220 cases per 100 000 people. Hymenolepiasis is a contact helminthiasis, and the intensity of affliction of the population in many respects depends upon the conditions of the water supply and the observance of sani-

Table 7.1 Parasitic disease incidence in Kyzyl-Orda region

Disease	Morbidity rate (per 100 000 people)					
	1980	1986	1987	1988	1989	1990
Visceral leischmaniasis	1.2	1.4	0.6	0.5	0.5	0.6
Dermal leischmaniasis	–	–	4.0	45.0	223.9	255.4
Malaria	–	0.3	–	0.5	0.1	0.1
Echinococcosis	–	–	–	0.6	0.8	0.1
Taeniorhynchus	3.1	1.7	2.8	1.2	1.4	1.2
Enterobiasis	283.4	700.8	844.9	966.0	1223.0	1053.0
Hymenolepiasis	188.2	195.4	544.1	157.1	137.6	121.7
Ascariasis	4.4	4.6	2.8	4.7	3.3	1.8
Scabies	–	31.1	11.7	15.4	12.2	3.5
Crimean hemorrhagic fever	0.3	1.4	0.9	0.9	1.2	0.5

tary and hygienic practices (*Medico-Ecological Problems of the Aral Sea Crisis*, 1993).

Similar severe, adverse medical and ecological conditions are also typical for other areas in the Aral Sea coastal region. The differences are characterized only by the level of intensity of specific types of the above-mentioned diseases. There were excesses in all cases at the republic and national levels. And clearly there was a marked trend in the growing number of cases of observed pathology.

Large-scale outbreaks of typhoid fever and paratyphoid A and B were reported in Karakalpakstan in 1981–85 with the greatest number of sick in the northern zone averaging 67.4 cases per 100 000 people (and 30.3 and 26.0 in the southern and middle zones, respectively). The northern group of districts became a large focal point of typhoid and paratyphoid diseases. Taking extraordinary measures and ensuring medical treatment of bacilli carriers enabled the reduction in the number of cases of typhoid fever and paratyphoid A and B by 62.7% and 60.6%, respectively (Table 7.2) (*Medico-Ecological Problems of the Aral Sea Crisis*, 1993).

In 1986–90, the incidence of salmonellosis doubled with the highest affliction rate appearing in the northern and middle zones. In some of these districts there was a manifold increase in morbidity. The average rate of morbidity with acute enteric infection (mainly young children) increased in Karakalpakstan by 9.4% and in the northern zones adjacent to the Aral Sea up to 51.5%. There was also an increase in the morbidity rate of NAG-infection up to 123.5% in the northern part and up to 62.5% in the southern part of Karakalpakstan; the number of NAG-carriers increased twofold. Five cases of cholera and vibrio-carriers were reported in 1981–85 and 1986–90 (K ASSR, 1987). In 1986–90 the incidence of viral hepatitis increased by 22%, with a sharp growth of up to 25.2% of hepatitis B morbidity within this group.

Table 7.2 The incidence of infectious diseases in the northern zone of Karakalpakstan (adjacent to the Aral Sea), 1981–90

Disease	10-year morbidity rate per 100 000 people		
	1981–85	1986–90	Increase + decrease −
Typhoid fever and paratyphoids A, B, C	86.8	17.6	−
Salmonellosis	7.9	13.2	+
Acute enteric infections	479.5	726.8	+
Dysentery	60.9	49.5	−
Viral hepatitis	623.9	947.2	+
Hepatitis B	5.1	40.5	+
Poliomyelitis	0.2	1.1	+
Measles	36.2	64.8	+
Diphtheria	2.7	1.7	−

Table 7.3 The incidence of parasitic disease in the Republic of Karakalpakstan

Disease	Morbidity rate (per 100 000 people)					
	1981	1986	1987	1988	1989	1990
Ascariasis	6.9	4.7	5.0	3.0	4.4	3.2
Taeniorhynchus	153.7	138.01	120.3	117.2	87.1	87.2
Echinococcosis	–	–	–	–	2.9	8.0
Strongyloidosis	–	–	–	–	1.5	0.7
Hymenolepidosis	503.1	686.8	855.0	604.4	685.1	649.4
Enterobiasis	899.9	1591.8	1787.2	2121.8	1769.4	1802.3
Trichocephaliasis	–	0.3	0.3	0.2	0.2	0.5

Of the group of airborne infections, the rate of measles morbidity increased by 25%, with the largest increase in the northern region (by 74%). The total number of poliomyelitis cases increased by 80% (by 45% in the northern zone)(Elpiner and Delitsyn, 1991).

Tuberculosis morbidity, which increased almost twofold within the past decade, should be distinguished from other infectious diseases. It was three-fold higher in comparison to the Soviet Union and to Uzbekistan levels. Inhabitants of districts subjected to a greater extent to the impacts of ecological disaster have been more frequently affected by tuberculosis (*Medico-Ecological Problems of the Aral Sea Crisis*, 1993).

As indicated in the official statistics and in the results of special studies carried out in 1991 by the Martsinovskii Institute of Medical Parasitology and Tropical Medicine, the rate of helminthiasis morbidity (per 100 000 people) in the Republic of Karakalpakstan increased within the 1980s from 615.6 (in 1981) to 887.1 cases (in 1990). As in the other parts of the Aral Sea coastal region, the most commonly encountered helminthiases are enterobiasis and hymenolepiasis (Table 7.3).

Particular emphasis should be placed on echynococcosis. Although the data on people affected with this disease are inadequate, it is known that the rate among animals affected by this severest form of helminthiasis is high (about 50%). Consideration must therefore be given to this high level, which may indicate the potentially high level of hazard faced by people (*Medico-Ecological Problems of the Aral Sea Crisis*, 1993).

Analysis of the total health dispensary system in Karakalpakstan in 1988–89 has shown that more than 66% of the adult population and 61% of the children suffer from various pathologies (*Medico-Ecological Problems of the Aral Sea Crisis*, 1993; Elpiner, 1995). The steady worsening of these indices has also been noted. For example, between 1981–87 deaths from acute infections of respiratory organs increased almost threefold, the number of liver cancer cases doubled, the incidence of esophageal cancer increased by 25%; digestive system cancer morbidity also increased (during certain years by up to 1.5–2 times). Of those with esophageal cancer, 43% were able-bodied and relatively young. Of the total number of ill, 70% were dying within one year of the date that the disease was discovered. In the 1980–85 period the rate of morbidity among young age groups increased more than twofold. Clearly, cancer victims have become younger.

From 1981 to 1987 the number doubled of persons infected with hypertonia, cardioischemia, and peptic ulcer, cases of cholecystolithiasis increased five times, and the rate of incidence of chronic nephritis doubled. Between 1985 and 1990, maternal mortality increased threefold compared with the previous five years. Infant mortality increased from 34.6 (per 1000 live births) in 1965 to 52 in 1989 (Elpiner and Delitsyn, 1991).

The steady growth in infant mortality cases has also been noted; in some regions it exceeded 100 cases per 1000 newborn. From 1980 to 1989 the number of child deaths per 1000 of newborn increased by about 20% (Elpiner, 1991). Among all of the causes of infant mortality, the share of premature births, congenital hypotrophy (diseases passed to fetus by mother), and uterine immaturity, among others, increased to about 30%.

An analysis of age structure of infant mortality in Nukus and Takhiatash demonstrated that the share of neonatal mortality in the total structure of infant mortality increased over time (1977, 22%; 1980, 25.8%; 1982, 30%). The share of infant mortality also increased during the first week of life in these cities from 15.6% in 1977 to 43.0% in 1982 (*Medico-Ecological Problems of the Aral Sea Crisis*, 1993).

According to official statistics, 60% of the deceased children up to one year of age in these Karakalpakstan cities died as a result of diseases of the respiratory organs, 97% of which died from acute pneumonia. Other causes were perinatal mortality (20.1%), infectious and parasitic diseases (14.5%) and congenital abnormalities (4.4%). However, the experts' assessment of the histories of the children who died within the first year of life revealed a significant hyperdiagnostics of pneumonia as the cause of death, hyperdiagnostics of infectious and parasitic diseases and an inadequate account of congenital anomalies along with other causes of perinatal mortality.

After the experts analyzed each case of mortality, the actual distribution of causes of infant mortality was as follows: diseases of respiratory organs, 26%; infectious and parasitic diseases, 30%; other causes, 36%; congenital abnormalities, 7.6% (*Medico-Ecological Problems of the Aral Sea Crisis*, 1993).

The 1988 analysis has shown that more than 80% of the women of childbearing age in Karakalpakstan suffered from anemia (the average All-Union rate was at 25–30%). This situation was aggravated by the fact that half of the pregnant women suffered from other extragenital diseases. One-third of pregnant women had a premature interruption of their pregnancy. There has also been a growing trend in the number of children with congenital developmental abnormalities, which, along with high maternal and infant mortality, is the sign of a weakening of the health of the human population, its degradation, and its ultimate extinction (*Medico-Ecological Problems of the Aral Sea Crisis*, 1993; Karakalpak SSR Academy of Sciences, 1989).

The data also revealed that 45.6% of the children in Karakalpakstan were ill. At the same time, according to the official data of the Ministry of Public Health of Uzbekistan in 1988, the at-risk group taken from the entire infant population made up only 8.9%, and the groups of sick children, only 1.2%. Therefore, only 10% were in poor health. This is just one example of the official distortion of the actual state of public health.

As for the Tashauz region of Turkmenistan, the situation has been no less acute. Within the 1980s, the rate of morbidity of the population in the region with acute enteric infections maintained a steady high level (from 55 to 528 per 100 000 people).

Viral hepatitis morbidity in 1988 in Turkmenistan's Tashauz region (547 per 100 000 people) was double that in Turkmenistan as a whole (264 per 100 000 people) and was one-third greater than for the Soviet Union (305.4 per 100 000 people). A similar trend was maintained during the whole subsequent period. Both the absolute figures and rates of morbidity within these years had a constant growth trend from 378 in 1980 to 706 in 1989 (316 per 100 000 people for the USSR) with a pronounced increase in the morbidity rate in certain years (Turkmen SSR, 1990*a,b*).

By the end of the 1980s a persistently unfavorable morbidity of the population remained independent of the enteric infections. Thus, the rates of population morbidity with tuberculosis in the Tashauz region (264–367 per 100 000 people) exceeded the average for all of Turkmenistan (240 per 100 000) and the USSR (210 per 100 000) (Turkmen SSR, 1990*a,b*).

The morbidity resulting from malignant neoplasms in the region (295–334 per 100 000 people) and in the whole republic (91–176 per 100 000 people) remained rather high (Turkmen SSR, 1990*a,b*).

Official statistical reports of the USSR Ministry of Public Health, presented in Table 7.4, point to the acuteness of social problems in the Aral Sea coastal region, among which the problems of public health are most important (Ministry of Public Health of the USSR, 1990).

As shown, the rate of infant mortality varied across the USSR. For each 1000 newborn children, the following numbers died during the first year of their life: 49 in Tajikistan, 46.1 in Uzbekistan, 24.7 in the USSR. The rates of infant mortality in the then-Karakalpak ASSR totaled 59.9; in Turkmenistan as a whole 56.4, and in the Tashauz region 75.2. At the same time in 1988, according to the data of the Ministry of Public Health of the Turkmen SSR, the Tashauz region experienced an acute shortage of physicians (22 per 10 000 people) in comparison with the Soviet Union data (43.8), and in-patient beds (93.9 per 10 000 people with a Soviet Union rate of 116.9).

Generalized data, relative to the state of public health in the Aral Sea coastal region, are also characteristic. By 1989, total morbidity in the region had increased about 4 times when compared with 1985, gastritis 3.3 times,

Table 7.4 Comparative data on public health status (1988)

Diseases	USSR	Turkmen SSR	Tashauz region (TSSR)	Uzb. SSR	KASSR[a]
Average life expectancy (years)	70	64.7	64.1	68.6	64.8
Maternal mortality (deceased mothers per 100 000 live births)	47.7	77.1	93.0	–	150.0
Infant mortality (deceased under one year of age per 1000 born)	24.7	56.4	75.2	46.1	59.9
Viral hepatitis (per 100 000 people)	305.4	264.3	547.8	–	253.2
Malignant neoplasms	–	–	295 (1985)	–	–
(per 100 000 people)	–	–	334 (1986)	–	–
Prenatal malformations	–	–	301 (1985)	–	–
(per 100 000 people)	–	–	437 (1988)	–	–

Note: [a] KASSR, Karakalpak Autonomous Soviet Socialist Republic.

cholecystolithiasis 10 times, nephrolithiasis 6 times. Anemia in child-bearing women in 1985 made up 40% and in 1988, 56%. According to the results of the review of the dispensary system, 74% of the population were suffering from some form of disease (*Medico-Ecological Problems of the Aral Sea Crisis*, 1993).

Morbidity of population in the Aral Sea coastal region

The pattern of population morbidity reveals the involvement of microbial and chemical factors. Water-borne, food, and routine contact modes of transmission are typical for the microbial factor, suggesting high morbidity from enteric infections. Water-borne, air, and food pathways of impact on human organisms are a conceivable reason for the second (i.e., chemical) factor, capable of causing noninfectious somatic diseases, oncopathology, genetic diseases, negative influence on the immune system, etc. The observed pathology has been progressing against a background of serious degradation of the human habitat, characterized by high levels of microbic pollution of drinking waters, by high levels of mineralization and of pesticide pollution, pesticide and salt particulates in the air, and by the pesticide pollution of food products (Figure 7.2). The aggravation of these negative factors because of unfavorable social and domestic conditions, e.g., undernourishment of the populace, has been quite evident.

The adverse impacts on human habitat in the Aral Sea coastal region have beyond question been directly connected to the processes sparked by the onset of the destabilization of the natural environment of the Aral Sea coastal region.

The significant increases in water withdrawal in the middle and upper

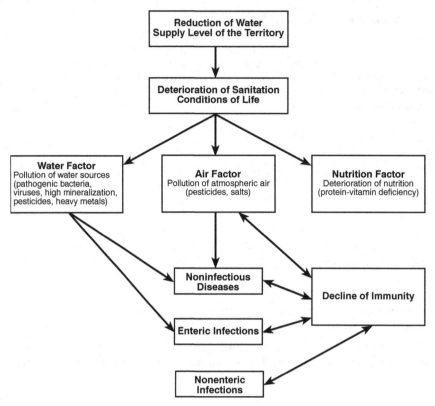

Figure 7.2 Cause-and-effect relationships of population morbidity in the Aral Sea and environmental factors.

reaches of the Amudarya and Syrdarya are connected to the following: the growth of areas under irrigation, resulting in the reduction of the volume of the river flow (within a number of years during the 1980s the inflow of river water into the Aral Sea was negligible), considerable extension of the hydrographic (main drain) network, increasing volumes of returned agricultural waste waters, increasing mineralization of the main water arteries of the Aral Sea coastal region, the drying out of the Aral Sea, changes in groundwater level and the intense secondary salinization of soils, changing climate with the reduction of relative air humidity, changes in the annual variation of precipitation, a growing number of days with dust storms, an increase in salt removal from the dried sea bottom (estimated between 40 to 150 million t/ha) accompanied by an increase in the mineralization of atmospheric precipitation (Glazovsky, 1990; Resolution of the State Economic Commission of the State Planning Committee of the USSR, 1989; Rubinova, 1979). All these changes had a direct effect on the state of public health in the region.

The microbic pollution of surface water is also connected with the sharp decline of dilution and self-purification of waters in locations where the sewerage system is inadequate. At the same time, the polluted waters of the

Amudarya and Syrdarya are the main water sources for the population of the Aral Sea coastal region. As a result, the population is often forced to use water from canals and even from the aryks for domestic use, including drinking purposes (Water Problems Institute, 1991). Thus, the fresh water situation in the Aral Sea coastal region can be characterized by its acute deficiency of water quality and quantity.

The water quality and human health problem is associated with the widespread use of water from the main water arteries (Syrdarya and Amudarya) by the population, including irrigation canals, brackish and desalted groundwaters. The water quantity problem is connected with the underdeveloped, low-power system of centralized water supply and the periodically low-water level of the sources (Kyzyl-Orda Regional Department of Municipal Engineering, 1981).

The discharge of highly mineralized drainage waters, containing pesticides from agricultural lands, into natural watercourses, along with a significant reduction of the flow due to excessive withdrawals of water for irrigation purposes, constitute the basic causes of the deterioration of water quality of the region's water resources (Melnikov *et al.*, 1977).

This process can be traced in time and space to within the most recent 10–15 years. According to the data of the water control services (Kyzyl-Orda Regional Sanitary and Epidemiological Station, 1987), the mineralization of water sharply increased in the lower reaches of the Syrdarya and Amudarya, reaching 2–3 g/l and more. The total 'hardness' of river water has reached 15–25 mg-equiv./l, the level of chlorides 450–750 mg/l, and the level of sulfates 700–1000 mg/l (Figure 7.3). Furthermore, residual amounts of toxic chemicals, synthetic surfactants, phenols, oil products and other ingredients of anthropogenic origin can be found in the water. For example, in Amudarya water one can find primarily pesticides and phenols. In the middle reaches of the river the concentration of both is higher and in the lower reaches still higher.

The rates of bacterial pollution of water, especially during the spring–summer period, is also an important consideration. The content of enteric bacillus has been recorded at tens and hundreds of times the level of pollution allowed for the sources of household-drinking and domestic water use. Pathogenic microorganisms are separated from the river water through cleansing by boiling or filtering.

The ecological situation in the lower reaches of the Amudarya, regulated by the reservoirs, is relatively less intense than in the lower reaches of the Syrdarya River, which apparently is associated with the less pronounced impact of waste and return waters (USSR State Committee for Science and Technology, 1984). However, the steadily reduced flow of Amudarya water into its lower reaches led to severe degradation of water quality (Tverdokhlebov *et al.*, 1988).

Studies carried out by the Sysin Research Institute of Human Ecology and Environmental Sanitation in 1991 provided a comprehensive picture of the

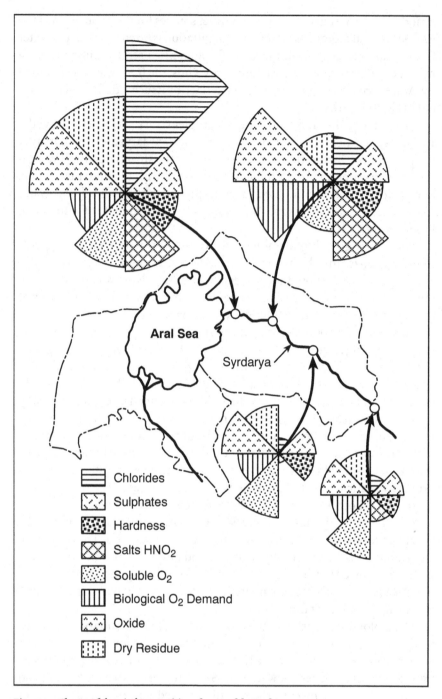

Figure 7.3 Change of chemical composition of water of the Syrdarya.

Legend:
- Chlorides
- Sulphates
- Hardness
- Salts HNO_2
- Soluble O_2
- Biological O_2 Demand
- Oxide
- Dry Residue

quality of water resources and drinking water at the peak of the Aral ecological crisis (*Medico-Ecological Problems of the Aral Sea Crisis*, 1993).

As a result of studies of water from the Amudarya and Syrdarya in the Tashauz, Khorezm and Kyzyl-Orda regions, 20 indices revealed excesses of standard rates of chemical oxygen demand (COD), iron, magnesium, and organochlorine compounds. The leading indicators of water pollution of the Syrdarya and Amudarya are water mineralization, hardness, biological oxygen demand (BOD), chemical oxygen demand, and concentrations of hydrocarbonaceous isomers.

Based on the composition of microelements and the content of heavy metals in Syrdarya and Amudarya water and in drinking water, sanitation regulations were exceeded, respectively, for bromine (1.4–4 times the maximum allowable concentration), barium (1.1–1.9), and manganese (1.1–2.6). The waters of the Amudarya and outflowing canals contain aluminum that exceeds maximum allowable concentrations by 2–3 times. The content of fluorine in river water in the Aral Sea coastal region is low (0.25–0.5 mg/l) and inconsistent with sanitation requirements for this geographic zone (Tverdokhlebov *et al.*, 1988).

Salinization of groundwater was also noted with mineralization reaching 1.4–2.9 g/l (Kyzyl-Orda region), 1.2–2.1 g/l (Tashauz region), 1.3–2.3 g/l (Khorezm region), and 1.9–2.4 g/l (Karakalpakstan).

Bacterial pollution of surface water bodies in all regions (*E. coli* index of 10 000 and 50 000) demands special methods of treatment. As a rule, groundwater resources are shallow and unprotected and, to a significant extent, their quality is a function of the levels of bacterial pollution of surface waters (rivers, irrigation canals), on the almost complete absence of sewage in the settlements, and on storm runoff.

The low efficiency of treatment facilities and desalting plants creates unfavorable conditions for public health. The drinking water of four areas in the Aral Sea coastal region can be characterized by high levels of bacterial and viral pollution. Also found are agents of enteric infections, which points to its high risk of epidemic problems (CNIIE, 1988; USSR State Committee for Science and Technology, 1984).

The low quality of water resources will inevitably have a negative effect on the water quality of centralized water pipelines. Standard treatment facilities here (settling, coagulation, filtration, chlorination) are not able to provide for the adequate treatment to make the water free of bacteria. Furthermore, disinfection of drinking water by chlorine causes the formation of highly hazardous halogen compounds (chloroform bromdichlormethane, dibromchlormethane, bromform) in concentrations that exceed health standards by 2–4 times. This is connected to the high content of organic matter in the water. High corrosivity of water is the cause of frequent deterioration of the airtightness of water pipeline systems, which results in secondary infections. As a result, the water quality of tap water in the Kyzyl-Orda region has

Table 7.5 Dynamics of the change in bacterial composition of the Syrdarya water (near Kyzyl-Orda, Kazakstan)

Years	Microbial number		Coliform index	
	min.	max.	min.	max.
1956	136	2400	17	90
1963	214	4730	23	2500
1976	1800	175 000	280	23 800
1978	2900	212 000	900	2 300 000
1986	3700	250 000	1200	2 700 000

Table 7.6 Water quality of municipal water pipelines of Karakalpakstan (percent rate of water quality nonconformity with State Standard requirements)

Water quality indices	Year						
	1981	1982	1983	1984	1985	1986	1987
Bacteriological	21	94	24	31	32	17	7
Chemical	16	20	20	26	24	24	26

reached 25–35% of bacterial indices of samples and did not meet the USSR standard (see Table 7.5)(GOST, 1988). In rural localities this index increases to 30–46% (Kyzyl-Orda Regional Department of Municipal Engineering, 1981; Ministry of Public Health of Kaz. SSR, 1988); in Karakalpakstan as a whole (for details see Table 7.6); in Dashkovuz (formerly Tashauz) region 46.8%; in certain areas in 1991 this index reached 70%. This picture worsens when one takes into consideration the data on mutagenic activity in a number of drinking water samples (*Medico-Ecological Problems of the Aral Sea Crisis*, 1993). For example, genotoxicity of mother's milk, revealed in the urban areas of Nukus and Turkul where the mutagenesis of drinking water was found, is probably connected with their use of this water.

Centralized water pipelines using Syrdarya water provide for 65% of the population of the Kyzyl-Orda region, mainly those living in the largest settlements. Water consumption here is 3–4 times below the rate of 20–80 l/day per person. The lack of drinking water in the lower reaches of the river has been about 10 l/day per person (Kyzyl-Orda Regional Sanitary and Epidemiological Station, 1987).

The rural population has been forced to import desalted water from brackish groundwater resources. This sharply inhibits water consumption. The problems of desalted water on human physiology remains unclear (Elpiner, 1963). Wells continue to be the sources of water supply for 4–10% of the rural population, depending on the region.

Only 33% of the population in the lower reaches of the Amudarya, i.e., in Karakalpakstan) are served by a centralized water supply (59.8% in the cities and 8% in the settlements). Water quality is low (Table 7.6). More than 90% of the rural population in Karakalpakstan use water of the irrigation network during the spring–summer period along with water from wells dug in the dry channels.

Soil and geological conditions, the high groundwater level, and the great penetrability of soils are conducive to water pollution in wells. Such pollution is confirmed by the high *E. coli* index and pathogenic enterobacteria (*Medico-Ecological Problems of the Aral Sea Crisis*, 1993).

Inter-republic canals and subcanal water lenses are used essentially as drinking water sources in the Tashauz region. Mineralization of these waters has exceeded standard rates by 2–3 times, and the content of phenols by 6–14 times. The content of nitrates and pesticides have also been in excess. Nearly 98% of the region has water with mineralization levels of more than 1 g/l, which, as a matter of principle, makes it impossible to categorize it as drinking water. Other water alternatives have not been available (Turkmen SSR, 1990*b*).

Along with serious defects of water supply, the low level of sanitation in settlements is typical for the population of the Aral Sea coastal region. For example, only three cities are partially equipped with a sewerage system: Nukus 30%, Takhiatash 20%, and Kungrad 10%. Thus, there are very serious problems with water treatment in the settlements.

Of special importance for human habitat in the Aral Sea coastal region has been the pollution of agricultural produce by chemical substances, in particular pesticides and fertilizers, in quantities that significantly exceed permissible levels. The long-term extensive development of monocrop agriculture was accompanied by the widespread use of both organophosphorus and organochlorine chemicals, including such highly toxic and stable ones as DDT, butyphos, hexachloran, and Lindane. Their annual application was on the order of thousands of metric tons (Glazovsky, 1990). Up to 54 kg/ha of various herbicides, defoliants, etc. have been used in Uzbekistan in the recent past (Glazovsky, 1990; Kyzyl-Orda Regional Sanitary and Epidemiological Station, 1988).

The USSR level of fertilizer application in this region has been significantly exceeded by 10–15 times (Elpiner and Delitsyn, 1991). Today, pesticides are found in excess in water resources, drinking water, the air, food products, and even in mother's milk (Table 7.7). The percentage of cases where pesticides were revealed in water samples taken from open-water bodies of Karakalpakstan from 1981 to 1988 varied annually from 1.3% to 13.5%. The percentage of cases where maximum permissible concentrations were exceeded was over 90%. According to the available data, the percentage of food product samples containing pesticides in 1979–85 in Karakalpakstan was in excess by 1.3% to 37% (by 2.8% to 32% in milk). In a lesser number of cases, pesticides

Sample No.	γ-hexachloro-cyclohexane	BY-58	γ-hexachloro-cyclohexane	Malathion insecticide	Metabolites, DDT	DDE, ODD, OPD, PPD	Others
Aral region							
1.	–	–	–	0.06	–	0.008	–
4.	–	0.0013	–	0.008	0.016	–	–
6.	–	–	–	0.036	0.012	–	–
7.	–	–	–	0.288	0.012	–	–
8.	–	–	–	0.02	0.024	–	propanide, 0.004
9.	–	–	–	0.008	0.008	–	–
10.	–	–	–	–	0.02	–	–
Kazalinsk region							
32.	–	–	–	0.008	0.016	–	–
33.	–	–	0.008	0.004	0.016	–	celtan, 0.00004
35.	–	–	–	0.012	0.004	–	–
36.	0.0002	0.0176	–	0.02	0.008	–	–
38.	0.000008	0.0184	–	0.02	0.004	–	–
45.	–	–	0.0008	0.016	0.004	–	propanide, 0.0016
47.	–	0.00002	–	0.008	0.0008	–	–

Source: Findings of the Research Center of Nutrition Problems, Almaty

were found in grain products, fruits and vegetables (Kyzyl-Orda Regional Sanitary and Epidemiological Station, 1988).

According to the data of SANIIRI (Central Asian Scientific Research Institute for Irrigation) based in Tashkent, by the end of the 1980s in Karakalpakstan, α and γ hexachlorocyclohexane, DDE, and DDT were found in the bodies of commercial fish (pike perch, carp).

In the Kyzyl-Orda region pesticides were found in each second sample, and each seventh sample was in excess of permissible concentrations. Such an excess was noted in each 50th sample of fruit and in 30–50th sample of fodder. The appearance of pesticides in the atmosphere in rural areas is connected with the treatment of agricultural lands and soil followed by natural weathering, and in the cities with the activity of mills that contaminate the air with dust containing pesticides. Thus, when processing 1000 metric tons of raw cotton with a 10–15% impurity level because of pesticides, the dust with pesticides is about 31 metric tons. This has led to dust in the air that exceeded permissible rates up to 0.7 km from the industrial point sources (Resolution of the State Economic Commission of the State Planning Committee of the USSR, 1989).

The aerial application of highly toxic chemicals to fields results in air pollution for 4 to 8 days by pesticides, at a distance of one kilometer from the treated fields and in concentrations often in excess of maximum permissible concentrations (Reference Book on Pesticides, 1977). Pesticides were found in every other air sample (Rubinova, 1979).

Dust–salt load on the environment

The removal of salts from the dried bottom of the Aral Sea has been assessed in various ways by different research organizations. There is agreement among reviews of the data on this subject (IGAN, 1985). It makes up 12.0–38.6 t/ha of the 1960–80 average. The annual average amount of salts removed from the entire dried seabed has been estimated at 43 million metric tons between 1960 and 1984. Also, in this period data suggest that large-scale dust storms, which can occur as frequently as 10 times a year, account for the transport of 15 to 75 million metric tons of dust annually. The amount of dust which can fall out in the region of the Amudarya delta during dust storms has been estimated at about 1.5 million metric tons annually. Dust fallout during a single large-scale dust storm contained up to 10% of salts in the Amudarya delta and was estimated to be 36–58 kg/ha.

Although there are studies by Uzbek researchers that disagree with the degree of severity of the impact of the removal of salts by winds from the dried sea bottoms in the eastern part of Central Asia, the possibility of the atmospheric transfer of salts has not been disputed (Glazovsky, 1990). Clearly, there is a relationship between the increase of chemical pollutants (pesticide, salts) in human habitats and changing conditions of land and water in the region. Assessments of the microbial pollution of the aquatic environment failed to consider its impacts on society.

The pathogenic (hazardous) trend of the impact of biological and chemical substances in the environment makes it imperative to consider the changing morbidity of the population along with the state of the environment. The studies carried out in Karakalpakstan by the Research Institute of Epidemiology of the Ministry of Public Health of the USSR (within the framework of integrated studies headed by the Water Problems Institute of the USSR Academy of Sciences) made it possible to confirm the importance of the water factor in the spread of viral hepatitis, typhoid fever and dysentery. This has been confirmed by our data (Figure 7.4).

Thus, the importance of the water factor became accepted. For the time being, it has not been established that there is direct epidemiological proof of the relationship between noninfectious morbidity and the water factor, or, for that matter, any other environmental factor in the Aral Sea coastal region. Nevertheless, the present state of scientific knowledge on the importance of environmental factors to public health allows us to seek out and identify those relationships (*Handbook of Forecasting the Medical and Biological Consequences of Hydraulic Engineering*, 1990; Elpiner, 1963).

The influence of the high level of mineralization of drinking water on the

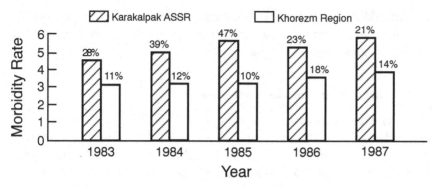

Figure 7.4 Dynamics of the rate of acute enteric morbidity and quality of drinking water in the Aral Sea coastal region of Uzbekistan.

rate of morbidity of the organs of digestion, cardiovascular, and genito-urinary systems, and the development of a pathology of a woman's reproductive system (and pregnancy) have been well established (Levin *et al.*, 1982; Livshitz *et al.*, 1975; Novikov *et al.*, 1980; Elpiner and Vasil'ev, 1983).

Of obvious interest are the following data (*Medico-Ecological Problems of the Aral Sea Crisis*, 1993): the increased content of lead in the hair of 30% of the newborn and cadmium for 60% of older infants in Muynak, and molybdenum in the hair of parents and their children. In the opinion of researchers, due to the suppression of hemapoiesis, the development of the central nervous system is retarded, and lesions occur in kidneys and osseous tissue due to the excessive accumulation of heavy metals in the organism. Their opinion has been supported by the high frequency of pathology of such systems in the population of the Aral Sea coastal region. The same data revealed a relatively low content of copper in the hair of the women examined in comparison with an assumed standard rate. Copper deficiency can also develop, however, with an excessive intake of molybdenum, manganese, zinc, lead, cadmium and strontium by an organism. These metals to varying degrees are functional protagonists. As a result, the study authors concluded that there was a relationship between anemia and the increased content of the above-mentioned metals in the water. The deficiency in an organism of such an important element as copper adversely affects normal hemapoiesis.

The assessment of the causes of the medico-ecological situation that emerged in the Aral Sea coastal region forced us also to consider the food production situation. Official statistics (IGAN, 1985) showed the existing disproportion of crop land distribution between industrial and food crops (Table 7.8). As a result, food production in the Aral Sea coastal region is far below the rates required for good nutrition, especially animal products. Actual production does not comply with the requirements for the growth and development of humans.

Special studies carried out in the Aral Sea coastal region by the Research Center of Regional Problems of Nutrition (centered in Almaty) have shown

Table 7.8 Crop areas in the Republics of Central Asia in 1986, occupying the largest part in the Aral Sea coastal region (× 1000 ha)

Republic	Total crop area	Grain crops	Industrial crops	Potatoes, vegetable and melons	Fodder crops
				Including	
Uzbekistan	3953	700	2091	205	957
Tajikistan	770	151	321	40	258
Turkmenistan	1161	164	651	69	277

Source: IGAN (1985).

Table 7.9 Dynamics of basic food product consumption per capita in the Kyzyl-Orda region (Kazakstan) in 1970–88 (as percent of republic indices)

Food product	1970	1975	1980	1985	1990
			Year		
Meat and meat products (on conversion to meat)	58.0	53.0	57.0	68.9	85.2
Milk and milk products (on conversion to milk)	52.0	52.5	58.1	86.0	97.0
Fish and fish products	93.1	83.3	60.7	33.0	40.3
Eggs (pieces)	58.0	49.4	47.0	52.5	75.6
Bread and bakery products (on conversion to bread)	100.0	100.0	96.0	95.2	95.6
Potatoes	30.0	28.7	29.7	26.9	52.3
Melons and vegetables	43.0	41.2	35.7	47.7	55.4
Fruits and berries	50.0	69.6	63.3	60.0	79.4
Vegetable oil	81.0	83.6	83.3	90.8	65.1
Sugar and confectionery (on conversion to sugar)	97.0	91.3	99.0	90.8	65.1

that the actual nutrition of the population in the areas adjacent to the Aral Sea (the Muynak region of Karakalpakstan, Kazalinsk, and Aral districts in the Kyzyl-Orda region) and in more distant areas do not meet the physiological requirements of humans (*Medico-Ecological Problems of the Aral Sea Crisis*, 1993).

The deficiency in the consumption of meat and meat products in this locale ranged between 17% and 61% when compared with the whole region. Their intake of fish and fish products, eggs, berries, fruits, vegetables, and melons has been extremely low. The total deficiency of daily caloric intake has been compensated for by the excessive consumption of bread and other bakery products. The data illustrating this situation, as shown for the Kyzyl-Orda region, are presented in Table 7.9.

Table 7.10	Fish catches in the Aral Sea basin ($\times$ 1000 centners)				
Location	1961	1965	1970	1976	1980
Northern part of the sea	162.0	122.0	59.1	51.1	–
Southern part of the sea	224.2	160.9	70.6	68.6	24.0
The whole sea	386.2	282.9	129.7	119.7	24.0
Lakes of the Syrdarya	51.1	27.6	45.2	15.2	13.1

A sharp decline in the consumption of fish products by the end of the 1970s demands an explanation. According to official data (Resolution of the State Economic Commission of the State Planning Committee of the USSR, 1989), the decline was connected primarily to the degradation of the Aral Sea as a water body that could support a healthy fish population (Table 7.10).

The commercial fishing industry in the northern part of the sea ended in 1976. The area of lacustrine fishery systems (more than 200 000 ha in 1960), which was located only in the northern part of the Sea and provided for more than 80 000 centners of high-quality fish, began to dry out gradually. In 1981, 30 000 ha of water area remained in the whole Kyzyl-Orda region. The catch of fish from these waters in 1980 was only 12 000 centners.

The decline in the share of meat and milk products in human diets in the region was a consequence of the reduction of pasture and hayland areas because of a monoculture policy (e.g., cotton) and because of desertification processes, an increase in the frequency and intensity of dust and salt storms, and a regional change in climate (e.g., reduction of humidity, precipitation, etc.). All these changes have been directly related to the consequences of the decline in sea level.

A medico-ecological interpretation of this situation (*Handbook of Forecasting the Medical and Biological Consequences of Hyraulic Engineering*, 1990) suggests a connection between widespread malnutrition-related diseases among infants under one year old and the health condition of their mothers. Diseases included hypovitaminosis, anemia, hypo- and paratrophy, rickets, lagging in physical development.

Poor nutrition had its effect on the changes of protein, fat, vitamin and microelement provision to the organism of those examined as shown in Table 7.11. The level of protein and its fractions in blood serum were below standard for all women examined in the Aral Sea coastal region. An imbalance was reported in the amino-acid spectrum with an alteration in the ratio between replaceable and irreplaceable amino acids. A direct correlation was identified between the level of vegetable and fruit consumption and the content of vitamin C (ascorbic acid) in blood serum; the value was 2 to 2.5 times below the standard. Protein and vitamin deficiency of the population has been one of the primary causes of the reduction in the immunity level of infants and women.

Table 7.11 Average per capita provision of energy and nutrients in the zone of the most unfavorable effects of sociological factors in the Aral Sea coastal region (as percent of standard levels)

Nutrients	Physiological daily ration	Muynak district, Karakalpakstan (Uzbekistan)	Kyzyl-Orda region (Kazakstan) Kazalinsk district	Aral district
Proteins (total), g	90	87.2	70.7	72.2
(including animal)	50	79.7	50.2	53.2
Fats (total), g	90	77.2	79.6	74.5
(including vegetable)	27	139.7	–	–
Carbohydrates, g	450	90.1	73.0	72.4
Caloricity, kcal	3000	86.4	73.9	72.4
Trace minerals, mg:				
Calcium	900	93.0	76.0	76.5
Phosphorus	1250	109.5	86.5	87.2
Iron	15.0	120.6	126.7	126.6
Vitamins, mg:				
A	1.0	18.5	17.0	30.0
B_1	1.7	50.3	47.0	47.0
B_2	2.2	55.0	45.4	50.0
E	15.0	–	86.0	72.0
PP	20.0	60.0	–	–
C	70.0	15.1	20.7	14.0

The schematic presentation in Figures 7.5, 7.6, and 7.7 of the cause–effect relationships between the dominant human pathology in the Aral Sea coastal region and adverse changes in the environment sums up these considerations.

On the realization of changes of medico-ecological situation in the Aral Sea coastal region

Concerns about the state of public health and the attempts to establish the highest quality of life were frequently used for propaganda purposes by leaders of the former Soviet Union.

The appearance of selected official documents mentioning medico-ecological problems of interest to the public (and on the basis of such publications alone it is possible to know the level and the extent of these problems) does not mean that the necessary information for decision-making had not been available much earlier.

To analyze the peculiarities and sequence of the process of the actual ecological degradation in the Aral Sea coastal region, it is important to know, in

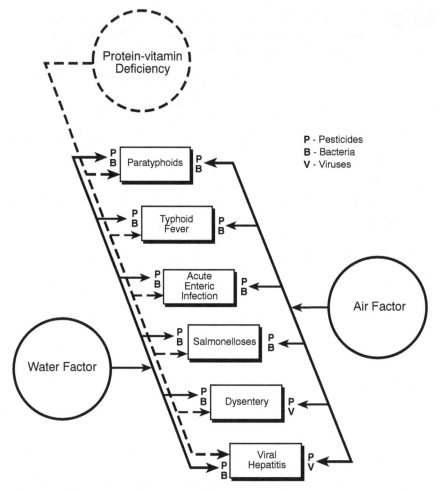

Figure 7.5 Relation between infectious enteric diseases of the population and the state of the environment in the Aral Sea coastal region.

general, the pattern of development of ecological awareness at the legislative level in the Soviet Union.

The concentration of attention of Soviet legislators on ecological aspects of the use of natural resources was typical for the end of the 1960s and the 1970s. Within that period, the 'fundamentals' of land (1968) and water (1970) legislation of the USSR and the Union Republics appeared. The paramount importance of providing the population with clean drinking water and water for household needs was emphasized. The following laws were approved: the Fundamentals of the USSR and the Union Republics on mineral resource legislation (1975) and in 1977 the Fundamentals of forestry and public health legislation. The latter established the responsibility for pollution of the human environment, prohibiting the undertaking of any economic activity without appropriate conservation measures. That legisla-

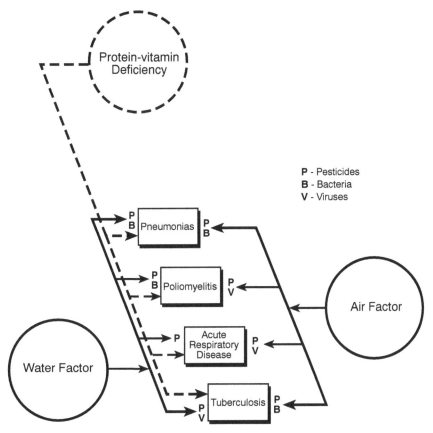

Figure 7.6 The relationship between infectious nonenteric diseases and the state of the environment in the Aral Sea coastal region.

tion was further supported by the passage of an act for atmospheric protection (1981).

Several other fields of legislation in that period – economic, civil, labor, administrative, criminal – were also written to include ecological considerations. The standards, directly or indirectly oriented at environmental conservation, were raised in those fields of legislation.

Rather stringent resolutions of the Central Committee of CPSU and the Council of Ministers of the USSR were also typical for that period. However, their resolutions reflected an irresponsible and neglectful attitude toward environmental conservation of the authorities at all levels. So, too, did the resolutions related to the problems of Lake Baikal (1977), of the Ladozhskoe and Onezhskoe and Il'men' Lakes (1984), the Baltic Sea (1976) and other problems of general importance, e.g., 'On the ecological situation in a number of regions and industrial centers in the country' (1987), 'On the radical reorganization of the conservation of nature in the country' (1988).

Those resolutions can be referred to for serious criticism of the ministries

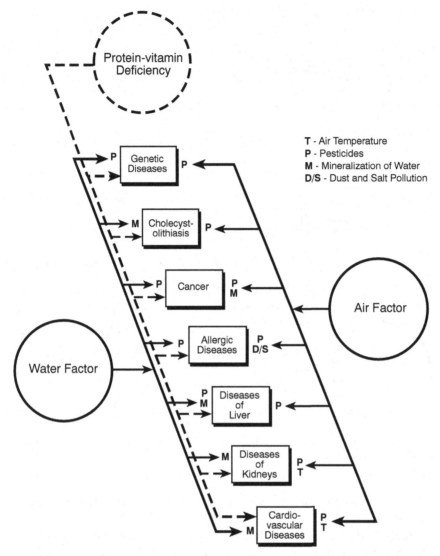

Figure 7.7 Relationship between noninfectious diseases of the population and state of the environment in the Aral Sea coastal region.

and departments, mainly for the low level of concern for conservation, e.g., the low level of 'restraint on the intensification of production under conditions of the growing interrelationship between the environment and economic development'. This phrase lifts up the proverbial curtain on the stage where the authorities played the role of conservationists. None of the noted resolutions include any mention about the population, its health, environmental diseases, etc. The necessity for more detailed research on the societal and, thus, on the medico-ecological aspects of the whole problem, was noted only casually in the resolutions of the Central Committee of CPSU and the

Council of Ministers of the USSR 'On the cessation of the works of the partial diversion of northern flowing Siberian rivers' (1986).

However, if one refers to legislation of the former USSR, one will find that it had been declared in the constitution of 1977 that 'The interests of present and future generations of people constitute the main aim of the ecological policy of the USSR. All measures needed for the conservation and scientifically justified rational use of land and its mineral, water vegetative and animal resources, purity of air and water, securing reproduction of natural wealth and improvement of the human environment, are taken for this purpose.'

As can be seen, ten additional years were needed for resolutions assessing the activities declared by the constitution a decade earlier. As for official reactions of the central Soviet authorities to the Aral crisis, they emerged even later: in 1990–91, when the crisis became so pronounced (and 'globalized') that it took on the characteristics of a national disaster.

The history of awareness of the medico-ecological situation appears somewhat different in research studies directly or indirectly related to the Aral problem.

If we address the scientific reports of the Water Problems Institute of the USSR Academy of Sciences, on the medico-ecological aspects of the diversion of the flow of Siberian rivers into Kazakstan and Central Asia in 1978–1980, we can find rather alarming data on population morbidity in the Aral Sea coastal region. The data were also presented in a memorandum of the Institute of Geography of the USSR Academy of Sciences on the Aral Sea problem (1983) and in the scientific report of the same institute (1993) which had been commissioned by special instruction of the State Committee for Science and Technology (GKNT) in 1985, and was linked to the social and economic problems of development in the Aral region.

However, expanded integrated medico-ecological research on the consequences of the Aral crisis was begun only in 1989 by the Water Problems Institute. The results of those studies have been reported in a monograph in 1993 (*Medico-Ecological Problems of the Aral Sea Crisis*, 1993).

Foreign organizations and scientists have shown great interest in this problem, especially during 1988–90. In 1990 the UNEP launched an international project at the request of Soviet authorities to prepare an action plan for the conservation of the Aral Sea. In 1992, UNEP produced a diagnostic study which was the first adequate international statement of the medico-ecological situation in the Aral coastal region.

On thresholds

To identify thresholds of change in the medico-ecological situation in the Aral Sea coastal region, we used official data, data of research studies, and the levels of reaction by various decision-makers. Comparably reliable and opportune is the knowledge of local situations provided by documents including statistical reports at the district level. It would be wrong to rely only on the official documents of the republic and the central government (e.g., resolutions of the Central Committee and governments or documents of the various Soviet Union ministries). Moreover, it is possible to get an idea of the development of events over time and space, having identified the feasible thresholds change. When analyzing actual data, it is also possible to identify patterns in observed ecological changes and to determine the extent of their importance from a medico-ecological stance. To achieve reliable conclusions, it is correct to undertake an analysis of data of any single administrative territory. However, this would require further research on the correlation of trends in the data from adjacent areas.

Thus, an analysis of the officially reported data from local authorities and the Sanitary-Epidemiological Service of the Kyzyl-Orda region points to the growth of such enteric infections as typhoid fever in the beginning of the 1970s. During that period a steadily increasing trend was revealed in the level of mineralization and bacterial pollution of surface water sources and of drinking water. The continuation from 1972 to 1984 of a high level of morbidity (exceeding Kazakstan's levels 5–10 times) suggests the increasing impact of these factors in reducing the immunological status of humans.

In fact, after a noticeable reduction of typhoid fever morbidity in 1980 there was an increase with a peak in 1984. The period 1980–84 is especially distinguished by the growing impacts of pesticides, judging by the frequency and levels of pesticide detection in the environment and in food products. In addition, that period can be characterized as a period of the creation of a protein-vitamin deficiency in the population.

The development of officially recorded pathological noninfectious processes after 1980 took place against a background of pesticide, mineral and toxic impacts, directly or indirectly connected to water.

However, the acknowledgment of the whole situation was fully reflected in local official documents only by 1981 (Kyzyl-Orda Regional Department of Municipal Engineering, 1981; Kyzyl-Orda Regional Soviet of People's Deputies, 1981). Beginning from that time, at least in the Kyzyl-Orda region of Kazakstan, thresholds of awareness occurred. The author of this chapter worked to reveal such adverse changes. The analysis of similar data in Karakalpakstan yielded distinctive results. Nevertheless, the official realization of thresholds of changes in the health situation were lagging here. Perhaps, apart from political directives to withhold negative information about the health of the population in the Aral region, one finds in this region a lower level of medical facilities and of trained personnel than in the nearby Kyzyl-Orda region.

By the end of the 1980s and in the early 1990s, the shortcomings in the activities of local specialists were compensated for by medical teams from central public health research institutions. However, their contribution was

Peculiarities in the state of public health	Warning of change	Threshold 1 Realization of changes	Threshold 2 Realization of the problem	Threshold 3 Realization of a crisis	Threshold 4 Realization of the need for action	Threshold 5 Action (beginning)
Increasing rate of infectious enteric morbidity	1970–72	1974–76	1976–78	1981–83	1981–84	1988–91
Increasing rate of infectious nonenteric morbidity	1978–79	1981–82	1986	1988	1988	1988–91
Increasing rate of noninfectious morbidity	1980	1985	1987	1988	1988	1988–91

Table 7.12 Sequence of the realization of changes of medico-ecological situation in the Aral Sea coastal region

the collection of reliable data (i.e., data undistorted by political considerations). Thus, the answer to the question about the causes of rather belated assessments centers on political constraints. The results of the analysis of official data is presented in Table 7.12.

Research findings allow us to affirm that designated thresholds of changes in the medico-ecological situation can delay official reactions by 2–3 years. Undoubtedly, this applies to thresholds 1–4. In any case, the first recorded important change in the state of public health was related to infectious bacillus E. coli morbidity and can be dated at the beginning of the 1970s, i.e., a decade after the process of the sea level decline and the reduction in water quality became obvious. The thresholds of acknowledgment of these adverse changes are noted in the various tables.

Therefore, already a decade after noting the first serious signals of the unfavorable water ecological situation in the Aral Sea coastal region, changes in the Republic of the rates of population morbidity (at least that of E. coli) became known. However, judging from official documents, the activities of the decision-makers aimed at intensifying the application of preventive measures were delayed for yet another 8 to 10 years. And their actions were not major measures designed to eliminate the ecological causes of the outbreaks of water-borne enteric infections. Instead, they focused on the intensification of medical therapy and palliative remedies for the known defects in the water supply system.

The development of the medico-ecological situation in the Aral zone of ecological disaster is the result of the activity of persons characterized as having a conservative approach to the resolution of problems, and also by an indifference with regard to public health. Responsibility rests with such authorities at the Republic level. The more so because, in retrospect, there are identifiable thresholds of realization of the medico-ecological situation outlined in the process of development of the gradually increasing ecological changes in the Aral Sea basin.

Solution to the medico-ecological problems associated with creeping environmental problems in the Aral crisis zone demands intense international financial and economic assistance. This has been especially necessary and obvious with the disintegration of the USSR and the acquisition by the Central Asian Republics of the status of independent states.

References

CNIIE, 1988: *Report on Scientific and Research Work: Epidemiological Justification of Preventive Measures Against Enteric Infections in Connection with Territorial Redistribution of Water Resources.* Moscow: CNIIE, 89 pp.

Elpiner, L. I., 1963: Theoretical prerequisites of the assessment of biological full value of drinking waters. In: *Trudy NII gigieny vodnogo transporta*, Moscow: Gidrometeoizdat, 273 pp.

Elpiner, L. I., 1991: Agony. *Radical*, 1, p. 1.

Elpiner, L. I., 1995: Medical-ecological problems of the eastern Aral land. *GeoJournal*, 35(1), p. 43.

Elpiner, L. I. and V. M. Delitsyn, 1991: The Aral ecological disaster and problems of public health. *Melioratsiya i vodnoe khozyaistvo* (Land Reclamation and Water Management), 4, p. 7.

Elpiner, L. I. and V. S. Vasil'ev, 1983: *Problems of Drinking Water Supply in USA,* Moscow: Nauka, 168 pp.

Glazovsky, N. F., 1990: *Aral Crisis.* Moscow: Nauka, 135 pp.

GOST, 1988: *Sources of Centralized Domestic and Drinking Water Supply.* Moscow: GOST, pp. 2761–84.

Handbook of Forecasting the Medical and Biological Consequences of Hydraulic Engineering, 1990: Moscow, 172 pp.

IGAN, 1985: *Final Report: Investigation of the Impact of Water Management Activities on the Aral Sea Regime, Social and Economic Processes of the Development of the Aral Sea Coastal Region, Connected with the Decline of its Level. Development of Scientific Fundamentals and Measures on the Efficient Use and Conservation of Natural Resources under Conditions of Anthropogenic Desertification of the Aral Sea Coastal Region.* Moscow: IGAN, 384 pp.

K ASSR, 1987: *Data of State Sanitary Supervision of Centralized Water Supply in K ASSR 1981–87.* Ministry of Public Health, K ASSR.

Karakalpak SSR Academy of Sciences, 1989: *Medical and Genetic Studies of the Population of Karakalpak ASSR.* Karakalpak Branch of the Uzb. SSR Academy of Sciences, 83 pp.

Kyzyl-Orda Regional Department of Municipal Engineering, 1981: *On the State of Water Supply in Cities and District Centers of Kyzyl-Orda Region with Drinking Water.* 23 October 1981.

Kyzyl-Orda Regional Sanitary and Epidemiological Station, 1987: *Natural Climatic and Social and Economic Factors Affecting Public Health in Kyzyl-Orda Region.* 8 May 1987.

Kyzyl-Orda Regional Sanitary and Epidemiological Station, 1988: *On the State of Public Health in the Kyzyl-Orda Region in Connection with the Disturbance of Ecological Equilibrium in the Region.* 21 October 1988.

Kyzyl-Orda Regional Soviet of People's Deputies, 1981: *On Certain Negative Effects Connected with the Decline of Sea Level, Influencing the Development of the National Economy of the Region.* 28 October 1981.

Levin, A. I., S. I. Plitman, Yu. V. Novikova *et al.*, 1982: *Methodical Problems of Investigation of the Influence of Mineral Content of Drinking Water on the Health of Population.* Collection of Scientific Works, MNIIG, 141 pp.

Livshitz, L. A., A. T. Kokina and A. E. Okeanov, 1975: On the problem of water hardness and morbidity of population with malignant diseases. *Gigiena i sanitariya* (Hygiene and Sanitation), **1**, p. 106.

Medico-Ecological Problems of the Aral Sea Crisis, 1993: Moscow: VINITI.

Melnikov, N. N., A. I. Volkov and S. A. Korotkova, 1977: *Pesticides and the Environment.* Moscow: Khimiya.

Ministry of Public Health of Kaz. SSR, 1988: *Information on Sanitation and Epidemiological State of Kyzyl-Orda Region.* 30 March 1988.

Ministry of Public Health of the USSR, 1990: *Information on the Indices of Public Health of the Tashauz Distict of Turkmen SSR, Karakalpak ASSR, and Khorezm Region of Uzb. SSR in 1980, 1986-88,* No. 04-4/13-4, 21 March.

Novikov, Yu. V., Yu. A. Noarov and S. I. Plitman, 1980: Importance of hard waters in prevention of cardio-vascular diseases. *Gigiena i sanitariya* (Hygiene and Sanitation), **9**, p. 69.

Reference Book on Pesticides, 1977: *Hygiene of Application and Toxicology.* Second ed. Kiev: Urozhai, 374 pp.

Resolution of the State Economic Commission of the State Planning Committee of the USSR, 1989: *On Experts' Assessment of a Complex of Measures for Regulation of Water Regime of the Aral Sea and Prevention of Desertification of the Amydarya and Syrdarya Deltas.* 27 May.

Rubinova, F. E., 1979: *Change of the Syrdarya River Flow under the Impact of Water Management Construction in its Basin, Proc. SARNIGMI.* Moscow: Gidrometeoizdat, **58**(139).

Turkmen SSR, 1990a: *Data on the State of Health of Adult and Infant Populations Residing in the Tashauz District,* No. 8/298-34, 21 March. Ministry of Public Health.

Turkmen SSR, 1990b: *Information of the Executive Committee of Tashauz District Soviet People's Deputies on Quality of Drinking Water in the Tashauz District,* No. 379, 3 May.

Tverdokhlebov, E. N., O. I. Sklyarov and O. N. Lesnik, 1988: Hydrochemical regime of the Syrdarya River and its impact on adjacent lands. *Melioratsiya i vodnoe hozyaistvo* (Land Reclamation and Water Management), **9**, p. 38.

USSR State Committee for Science and Technology, 1984: *Report on Section of Assignment to Carry out Studies and Develop Hygienic Requirements on Improvement of Conditions for Water Use of the Populations in the Regions of Water Diversion and Partial Use of Flow of Siberian Rivers within the Uzbek SSR.* Tashkent: NIISGIPZ, 25 pp.

Vogel, C., 1994: Creeping environmental problems in South Africa. In: M. H. Glantz (ed.), *Workshop Report on Creeping Environmental Phenomena and Societal Responses*

to Them, pp. 233–9, Boulder, Colorado: National Center for Atmospheric
Research.
Water Problems Institute, 1991: *Scientific Justification and Elaboration of Hygienic
Recommendations on Improvement of Sanitation Conditions of Water Use for the
Population of the Aral Sea Coastal Region*. Moscow: Academy of Sciences of the
USSR, 80 pp.

8 The impact of political ideology on creeping environmental changes in the Aral Sea basin

IGOR S. ZONN

The Aral Sea is a part of a self-regulating hydrological system. It receives water from the two largest rivers in Central Asia, the Amudarya and Syrdarya, and the sea water evaporates into the atmosphere. Changes in this balance have a primary effect on the water level of the sea. These rivers flowing to the Aral have exhibited extremely uneven flow from one year to the next. For example, during a 60-year period, the Syrdarya's annual flow ranged from 22 to 57 km³, with a mean value of 34 km³; the Amudarya's flow ranged from 48 to 101 km³, with a mean value of 63 km³. A 'periodicity' of low-flow and high-flow periods has been observed to last 10–12 years. Clearly, the decrease of river flow to the sea results in a drop in sea level, in a reduction of sea surface area and volume, and in changes in other characteristics of the sea.

Irrigation farming in the Aral Sea basin began at least as early as 4000 BC. The local population of this region, like those in the valleys of the Tigris, Euphrates and Nile rivers, shared the experiences and knowledge accumulated over generations about the use of regional water resources, and about how to carry out irrigation farming in river floodplains and deltas in arid areas without disturbing the balance of nature.

The major disturbance of the long-lasting natural balance between ecological change in the Aral basin and the sea began early in the twentieth century as a result of human activities. Before then, the sea was abundant in water and even had a tendency to rise, despite the onset of intensive colonization of the region by Tsarist Russia.

A.I. Voeikov, an outstanding Russian geographer and climatologist around the turn of the twentieth century, put forward the 'constructive' idea of intercepting for irrigation purposes water flowing 'uselessly' to the Aral Sea (Pokshishevskii, 1949).[1] As early as 1882, he wrote in his report, *The Rivers of Russia*:

> The existence of the Aral Sea within its present limits is evidence of our backwardness and our inability to make use of such amounts of flowing water and fertile silt, which the Amu and Syr rivers carry. In our country, which is able to use the gifts of nature, the Aral Sea would serve to receive water in winter

---

1. In 1901, the Russian-American Commercial Treaty was not renewed, and in connection with the 'cotton hunger' threat, Voeikov used all available scientific means to show that Russia had every possibility to ensure its self-sufficiency in cotton production.

[when it is not needed for irrigation] and release it in summer during high
flow.   (quoted in Bostandzhoglo and Epshtein, 1988).

In 1908, Voeikov wrote:

> Meanwhile, there was a reason that could cause the water level drop in the
> Aral Sea. In the last 40 years Russian power gave peace to Turkestan. Farming
> the land and the need for artificial irrigation were growing in importance.
> Hence, great quantities of water evaporate from the surface of the fields and
> from the orchards and do not reach the Aral Sea.

Thus, the natural changes of the Aral Sea and the maintenance of its functional role as a sea have clearly been identified and acknowledged. Human
impacts on the sea became apparent only after 1960.

In fact, from the early 1960s (the anthropogenic period), Aral Sea desiccation and the stressed water management situation were also clearly identified
and not in question. As now emphasized by nearly all representatives of
science and the public, recent problems of the Aral Sea basin were caused by
the command-and-control system of the authoritarian Soviet government.
The established structure of agriculture and the mono-agricultural policies
(i.e., cotton and rice) were typical of Central Asia both in the center and in out-
lying areas. Various political aspects of the Aral Sea problem have already
been reviewed: Micklin (1991), Zonn (1992), Feshbach and Friendly, Jr. (1992),
Glantz *et al.* (1994), and Agarwal (1996).

Ideology of total irrigation and cotton self-sufficiency in the period of the Soviet Union

THE LENIN PERIOD (1917–1929)

The destiny of the ancient region of Sogd, Maverannakr and the
former Bukhara and Khiva emirates was determined by the takeover of those
places by the Soviet regime. By that time 'cotton fever' had turned into a
rather permanent 'disease' spreading across the natural 'body' of Turkestan.
Economic development in Central Asia in those days had been predetermined by the Decree signed by V. I. Lenin in 1918 'On the Allocation of 50
Million Rubles for Irrigation Development Works in Turkestan and on the
Organization of these Works' and by the Resolution of the Council of People's
Commissars of 20 December 1920 which sought to restore the cotton culture
in the Turkestan and Azerbaijan Socialist Republics.[2] By doing so, the Lenin
path of irrigation and related land resources development in Uzbekistan and
other republics was aimed at solving the high-priority strategic problem of
achieving cotton self-sufficiency. As of 1926, the production of cotton fiber in
the whole country was 250 000 metric tons.

---

2. Cotton was cultivated not only in Central Asia and Azerbaijan but in Georgia, Armenia, the
   Ukraine, as well as on the rain-fed lands of the Crimea, in Northern Caucasia, and in the
   Stalingrad Region.

In 1931 cotton production had increased to 388 000 metric tons, in 1937 it was 845 000 metric tons and by 1940 it had reached 893 000 metric tons. By that time:

> Lenin's concept of 'obligatory crop rotation' has disappeared forever from the vocabulary of the obedient Uzbek agrarian ... and nobody thinks of water: how much it is needed, how it can and must be used, how to irrigate and for what purpose. Canals and reservoirs are being designed.   (Zuev, 1991)[3]

Only cotton! The distortion of irrigated production began. Planning authorities from the Center and the local administration acting for the benefit of the Center expanded the plans of cotton purchases with no regard to the actual conditions. The command and control mechanism functioning throughout the country was also put into practice for cotton production. These methods of economic management were aimed at maintaining increasing trends.

The accelerated development of cotton production in Central Asia was associated with Stalin's five-year plans. It was impossible to maintain increases in cotton production indices without the continual extension of land under irrigation.

The Resolution of the Central Committee of the All-Russian Communist Party (of Bolsheviks) and the USSR Council of People's Commissars of 12 December 1939 'On Measures Concerning Further Increase of Cotton-Growing in Uzbekistan' outlined the tremendous program of irrigation construction meant for the development of irrigated farming. Usman Yusupov, the First Secretary of the Central Committee of the Communist Party of Uzbekistan stated in his report 'On Irrigation Construction in Uzbekistan' that

> We cannot resign ourselves to the fact that the water-abundant Amudarya River carries its waters to the Aral Sea without any use, while our lands in the Samarkand and Bukhara regions are insufficiently irrigated.   (Yusupov, 1939)

The following task was put forward at the same meeting:

> to bridle the Syrdarya and Amudarya rivers, to control them and to make their water serve the cause of socialism, for the purpose of raising the living standards of population and developing the country.   (Yusupov, 1939, p. 11)

During World War II, the work related to the implementation of this program (the construction of irrigation systems), although somewhat slowed down, went on. When the war ended, irrigation development associated with cotton growing assumed priority again. Already by the middle of 1945, the Party and the Government adopted the Resolution 'On Measures Concerning the Rehabilitation and Further Development of Cotton-Growing in Uzbekistan', which was enacted in February 1946 by the special Decree of the USSR Council of Peoples' Commissars 'On the Plan and Measures Concerning

---

3. The above-mentioned Resolution of the Council of People's Commissars of 20 December 1920 made it mandatory to 'put into practice the obligatory crop rotation with the dominating crop of cotton in all state and collective farms of cotton-growing areas.'

the Rehabilitation and Further Increase of Cotton-Growing in Uzbekistan for the period of 1946–53' (*Pravda*, 3 February 1946). It became the basis for pledging an oath to Stalin at the Tashkent Kurultai, the text of which was announced by Usman Yusupov. On behalf of the Uzbek people he promised to raise the production of raw cotton in Uzbekistan to 2 400 000 metric tons by 1953. The policy was in direct conflict with the limits of the natural environment. The attempt to increase cotton yields and production was carried out quickly without due regard for the agricultural conditions (i.e., limitations) of the republic's land and water resources. Regardless of such limits, the oath pledged to Stalin, the 'father of all people', had to be kept.

Orchards and vineyards were uprooted to provide the best soils for cotton production. The area under fodder crops and pastures were also decreased. Land development in the middle reaches of the Amudarya and Syrdarya basins was initiated; at that time the need for drainage of contaminated irrigation water was ignored. Provision was made only for the quantity aspects of water delivery and excluded consideration of seepage and other irrigation water losses. No steps were taken with regard to technology development or implementation to ensure the proper conditions for such irrigation. Moreover, in 1950 the erroneous Resolution was adopted 'On the Transition to the New System of Irrigation for More Comprehensive Use of Irrigated Lands and Improvement of Agricultural Operations'. For mechanized farming, the Resolution called for a major reduction in the number of irrigated plots and an increase of their areas up to 20, 40, 60 and even 100 ha. The proposed measures provided for the transition within 3–4 years to the new system of irrigation. This restructuring was completed by the year 1960.

Conceptually, the transition to the new system of irrigation should have caused the groundwater level to drop. The result was to the contrary of what might have been expected (Sirozhidinov, 1991). The groundwater rose, as did the level of salinization.[4] In spite of its desperate efforts, the Republic could not reach the planned (arbitrarily selected) targets.

The designs of numerous hydraulic structures were included in the 'Great Stalin Plan for the Transformation of Nature in the USSR', along with the projects of planting massive windbreaks and shelterbelts of trees (much like US President Roosevelt's shelterbelts in response to the US droughts in the Midwest in the 1930s) and putting into practice what Russians refer to as *travopolnaya,* alternating wide strips of grass and crops. To implement this Plan, the 'great construction projects of communism' were proposed. A Main Turkmen Canal was one of these proposed projects.

This proposed project was later replaced by the one to construct the

4. This refers to the erroneous anti-drainage concept of V. A. Shaumyan; it was a guiding concept for many scientists, planners and builders from 1948 to the 1960s. According to this concept, it was possible to irrigate without drainage, providing that water application rates were drastically reduced and the system of farming with alternating wide strips of grass and crops was put into practice.

Karakum Canal in the 1950s. Its construction would have contributed to the development of irrigation of and water supply to large areas. At that time, however, no one took into account the damage that would be caused by seepage from the unlined canal, by salinization of hundreds of thousands of hectares of lands, and by the resettlement of tens of thousands of villages on new, less fertile, lands. Attention was paid only to the engineering aspects of the construction of the canal, reservoirs, etc. The irrigation operations in the fields and water application in general were to be implemented by collective farms and by peasants who did not have the required experience and were exhausted by the war effort. Later, the afforestation efforts to protect soils were abandoned. Meanwhile, hydraulic construction went on supported by state funds. The ideological basis for all construction projects was captured by the following slogan: 'We cannot wait for favors from nature, our goal is to take them from it!'

THE KHRUSHCHEV PERIOD (1954–1964)

With the accession of Nikita Khrushchev to power in the Kremlin, agrarian policy was oriented toward the comprehensive development of virgin and long-fallow lands and irrigation farming (e.g., see Zonn *et al.*, 1994). Wide-scale development of irrigated cotton cultivation was proposed in the Resolution of the CPSU Central Committee and the USSR Council of Ministers on 9 February 1954 'On the Future Development of Cotton-Growing in the Uzbek SSR and the Turkmen SSR in 1954–1958'. As usual, the top priority program of developing irrigation in the southern part of the USSR was fully supported:

> The great deed of the Soviet people performed during the development of the virgin lands in Siberia and Kazakhstan was repeated in the irrigation of millions hectares of lands. (Selyunin, 1990)

It was cotton that made these activities worthwhile. The relatively low cost of cotton fiber from Central Asia was very attractive, in addition to the strong will to surpass America in cotton production. The Communist Party was quite aware of the importance of the assigned task to provide cotton not only to Soviet textile enterprises but also to provide it to light industries in other socialist countries. And there was one more important reason, the steadily growing militarization of the national economy which required more and more cotton for use by armaments.

The expected quotas or targets for cotton delivery from the republics of Central Asia grew from year to year, and to meet those increasing quotas there was a need for the extension of land devoted to cotton. Khrushchev, then the First Secretary of the Central Committee of the Communist Party of the Ukraine, at that time had wanted to cultivate cotton even in the Ukraine. That idea was soon rejected and attention was shifted to Central Asia.

In February 1956, the 20th Congress of the CPSU was held, where the guidelines were adopted for the 6th five-year plan of the USSR (for 1956–60). In accordance with this document agricultural development included regions located within the river basins of the Amudarya and Syrdarya. Such development also focused on irrigated cotton production.

Proceeding from the decisions of the 20th Party Congress, a special Resolution 'On the Irrigation of Virgin Lands of the Golodnaya Steppe in the Uzbek SSR and the Kazakh SSR for Increasing Cotton Production' was adopted on 6 August 1956. The Resolution led to the increasing of the cultivated areas under cotton by 200 000 ha in Uzbekistan and by 100 000 ha in Kazakstan.

In continued pursuit of 'white gold' (e.g., cotton) another Resolution of the CPSU Central Committee and the USSR Council of Ministers was adopted on 14 July 1958 'On the Further Expansion and Acceleration of the Works on Land Irrigation and Development in the Uzbek SSR, Kazakh SSR and Tajik SSR'.

The implementation of all these resolutions was accompanied by the large-scale construction of irrigation networks and the growing use of water resources of the Amudarya and Syrdarya. The first stage of the Karakum Canal, the Kairakumskoye, Farkhadskoye, Kattasaiskoye, Sariyazinskoye, Iolotanskoye, Kattakurganskoye reservoirs, etc., as well as many main irrigation canals, were constructed in this period.

Khrushchev visited Uzbekistan many times and familiarized himself with the activities of collective and state farms. He did his best to accelerate the development of cotton growing. He put forward the new goal of producing 3 million metric tons of raw cotton by the year 1960. Once again, orchards and vineyards had to be destroyed in order to develop new cotton lands. Central Asia became a central region for cotton-growing. It was a major agronomic mistake. Nobody was concerned about the limits imposed by nature. Water was free of charge.

According to the data of the Uz SSR Academy of Sciences, the areal extent of potentially irrigable lands in Central Asia and southern Kazakhstan was estimated to be about 20 million ha. However, by 1960 only 4 million ha had been irrigated. The distribution of irrigated lands in the Amudarya and Syrdarya basins at that time is shown in Table 8.1.

The formation of the recent structure of the water budget in the Aral Sea basin began in 1959, when the first stage of the Karakum Canal was put into operation (Budagovsky, 1992). At that time the Aral Sea represented a water body of about 66 000 km² in surface area, with its level at 53.3 m above the world ocean level. This was ensured by the perennial inflow of the Amudarya and Syrdarya.

The mean annual water budget of the sea in the period 1911–1960 can be characterized as follows:

	mm/year	km³/year
Input		
Surface and subsurface water inflow	850	56.2
Precipitation	150	9.9
Output		
Evaporation	1000	66.1

Table 8.1 Lands available for irrigation in the Syrdarya and Amudarya basins

Region	Total area available for irrigation (× 1000 ha)	Irrigated lands in 1960 (× 1000 ha)	Potential irrigated lands (× 1000 ha)
Syrdarya Basin			
Naryn upstream	170.0	131.0	39.0
Ferghana Valley	1478.5	1064.0	414.5
Golodnaya Steppe, Djizak and Dal'versa areas	1127.0	290.9	837.0
Chirchik-Angren basin	500.0	315.0	185.0
Arys'-Turkestan basin	1183.0	150.0	1033.0
Total	5258.5	2040.0	3218.5
Amudarya Basin			
Tajikistan Southwestern oases	710.9	255.0	455.9
Syrkhandarya basin	536.5	250.0	286.5
Kashkadarya basin	943.5	80.0	863.5
Zeravshan Valley	2559.4	557.0	2002.4
Amudarya middle reach and Obruchev steppe	356.0	110.0	246.0
Amudarya downstream	1925.0	479.0	1446.0
Turkmenia South Regions	1831.6	194.0	1637.6
Total	8862.9	1925.0	6937.9
Total for the two basins	14 121.4	3965.0	10 156.4

Source: Petrov (1964).

At that time the Aral Sea was rich in fish. In addition, about 2 million muskrat pelts from the sea coast were purchased annually. In other words, marine and animal wildlife concentrated in and around the Aral Sea area was abundant.

Irrigation construction and the use of the newly irrigated lands in the Syrdarya and Amudarya basins, underway in the 1950s, did not actually result in noticeable changes of the Aral Sea's water regime. At that time, the increased water withdrawals for irrigation and other purposes were compensated for by the reduction of considerable evaporation losses from numerous lakes previously located in the rivers' floodplains. In 1960 the surface area of these lakes was about 400 000 ha, including 100 000 ha in the Syrdarya delta and 300 000 ha in the Amudarya delta (Zaletaev *et al.*, 1991). In 1960, 40.4 cubic km were withdrawn for the irrigation of over 5 million ha; the drainage water flow was 5–6 cubic km/year (Reshetkina, 1991). In addition to this, a rather favorable hydrological situation occurred in Central Asia. The Aral Sea level benefited from the fact that water availability during the 1951–1960 period in

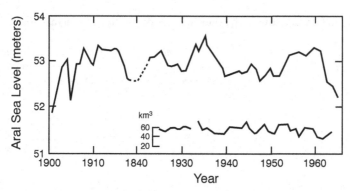

Figure 8.1 Aral Sea regime, according to published data (Kunin, 1967; figure drawn by R.V. Nikolaeva). Upper curve: level of the Aral Sea (m above Baltic Sea level). Lower curve: inflow to the Aral Sea (km³/year).

the Amudarya basin exceeded the long-term average value by 5% and in the Syrdarya basin by about 18% (Figure 8.1).

As Voropaev (1992) noted,

> by the 1960s ... all factors responsible for the compensation of the reduction of water inflow to the sea were exhausted: these factors involved the substitution by cultivated crops of wild water-loving, non-productive vegetation in floodplains, river flow regulation in reservoirs, reduction of flooding accompanied by considerable water losses in lakes, oxbows, deltas, overgrown tugai lands, etc. The growth of water withdrawal from rivers at that stage was already manifested by the reduction of water flow into the sea.

At that time, the discrepancy between the demands of the national economy and the available water supply had become evident. The deficiency of water resources in the region resulted in a slow-down of development in the majority of industrial sectors. The growth of water withdrawals began to surpass the increment of newly added irrigated lands because of (a) the construction of collector drainage systems, (b) the propagation of the so-called leaching regime of irrigation, and (c) the increase of seepage and other major water losses. The gap between available water resources and the demands of irrigable lands resulted in the need to redistribute water and to transfer it across distances. The gap was also worsened by the need to construct reservoirs for the interannual flow regulation of the rivers of the region.

The early 1960s were marked by a new desire for wide-scale and comprehensive development of desert and semi-desert regions. A new Party Program was adopted at the 22nd Congress of the CPSU, held in October 1961. The following was written into the Program:

> To ensure stable high, ever-increasing yields, to protect agriculture against the adverse effects of the elemental forces of nature, particularly against drought, and to sharply increase soil fertility and to rapidly increase animal husbandry, it is necessary ... to implement a wide-scale program of irrigation construction meant for the irrigation and water supply of millions of hectares

of new lands in arid regions and for the improvement of the existing irrigation farming.

To this end, it was proposed:

- to organize a new, large, cotton-growing area in the Syrdarya basin where, according to preliminary estimates, it would be possible to irrigate 800–850 000 ha of lands of the Golodnaya Steppe on the territories of the Uzbek, the Kazakh and Tadjik Republics;
- to construct the Nurek hydropower plant with a view to irrigate up to 1.2 million ha of lands under cotton, paddy and other crops in the Uzbek and Tadjik Republics;
- to organize irrigation and land development for cotton-growing in Turkmenistan on the basis of the Karakum Canal for the land area covering 600 000 ha;
- to organize new rice-growing areas in the lower reaches of the Amudarya and Syrdarya, an area of irrigated lands covering about 900 000 ha.

The implementation of the resolutions of the Party and Government adopted in the 1950s with regard to irrigated lands contributed to the gradual changes in the water budget of the Amudarya and Syrdarya. Starting in 1961, these changes affected the terminal sea, i.e., the Aral. Its level started to drop by 18 cm a year; the process continued slowly but steadily and was governed by the growth of irreversible water consumption for irrigation needs and was further aggravated by the *reduced natural flow availability*. In that period, water flow into the sea dropped by about 10 cubic km/year and the natural water availability decreased from 105 to 100 cubic km/year (Ratkovich, 1979a). The mechanism for the destruction of natural processes was put into motion at that time.

Uzbekistan produced the desired 4 million metric tons of raw cotton in the autumn of 1964, when Khrushchev was relieved of his post. Leonid Brezhnev succeeded him.

THE BREZHNEV PERIOD (1966–1985)

Widespread work on land reclamation was initiated in compliance with the resolutions of the 23rd CPSU Congress, the May 1966 Plenary Meeting of the CPSU Central Committee, and the 24th CPSU Congress.

The 23rd CPSU Congress paid much attention to the new trends in agricultural development in the country, to the intensification of production, and to the increases in crop yields and in the productivity of animal husbandry. It had been recognized that one of the important conditions for agricultural intensification was the implementation of a wide-scale program for improving soil fertility on the basis of land reclamation both within the borders of the whole country as well as within the boundaries of every state farm and collective farm.

The May 1966 Plenary Meeting of the CPSU Central Committee defined concrete measures and ways of implementing the program adopted by the 23rd Party Congress. The General Secretary of the CPSU Central Committee,

Brezhnev, said at the Plenary Meeting:

> In our opinion, today the matter of great importance is land reclamation in the broad sense, as it has been discussed at the present Plenary Meeting of the CPSU Central Committee. We must realize and provide the answer to ourselves, to the Party and the people that it is not a current campaign but an agricultural long-term program, whose implementation will require concentrated efforts and significant capital investments as well as material and technical resources. The program is based on the findings of science and practical experience, on real possibilities available now for the Soviet economy.   (CPSU Central Committee, 1966, p. 14)

The main feature of this statement was that land reclamation was to assume a broad scope and was to be practiced in large regions. Great importance was attached to further land reclamation in Central Asia with a view to increase cotton production. The goal was also to meet the country's needs in rice as soon as possible, mainly as a result of the accelerated construction of irrigation systems for rice-growing in the lower reaches of the Amudarya and Syrdarya.

Government authorities supported this decision based on their belief that the increment in the output of agriculture and animal husbandry in the Aral Sea basin as a whole would be much higher, in terms of value, than the total damage caused by the desiccation of the Aral Sea. Direct losses of such sectors as fisheries and the fur trade were identified, while indirect losses and secondary effects were not properly taken into account. Together with Rashidov, the then-First Secretary of the Central Committee of the Communist Party of Uzbekistan, Brezhnev put forward the task to produce by any means available 5–6 million metric tons of raw cotton in Uzbekistan. He was responsible for the popular slogan: 'golden hands create white gold'.

The period of 1966–86 featured the large-scale construction of water management projects and the commissioning of large irrigation canals in Central Asia, such as the Karakum Canal (stages II–III), the Karshi and Amu-Bukhara canals, etc., in which the headwater intake ranged from 200 to more than 500 cubic m/s. Large reservoirs, in particular the Andizhan, Charvak, Chardarin, Tuyamuyun, and Nurek reservoirs, among others, were put into operation. To a great extent, they permitted the regulation of river flow in the region. The land surface devoted to irrigation was permanently expanded (Table 8.2).

The implementation of the decisions passed by the May 1966 Plenary Meeting of the CPSU Central Committee resulted in putting into cotton cultivation 794000 ha of irrigated lands in Uzbekistan during 1966–75; in Kazakstan the sown areas under paddy along the Syrdarya and in its lower reaches increased from 42000 to 105000 ha during the same period.

The increase in the irrigated area contributed to the growth of the overall production of raw cotton. For the 1966–70 period, the gross yield was 3982000 metric tons, on average, and during 1971–75 it was 4894000 metric

Table 8.2 Irrigated lands in the Aral Sea basin (× 1000 ha)

Political unit	1960	1965	1970	1975	1980	1985	1990
Uzbekistan	2570	2639	2750	2995	3527	3908	4171
Karakalpakstan	196	217	250	254	350	455	490
Tajikistan	427	442	524	566	627	660	703
Turkmenistan	496	509	670	855	960	1160	1350
Kyrgyzstan	313	323	338	364	378	410	420
Kazakstan (Chimkent and Kyzyl-Orda Oblast)	305	328	373	570	635	700	760
Basin total	4111	4458	4655	5350	6127	6930	7403

tons. The highest production of raw cotton was reached in 1977: 5 660 000 metric tons. In Kazakstan, where cotton was grown only in the Chimkent region, production of raw cotton in 1975 was 283 000 metric tons.

It was impossible to produce more cotton. In essence, all lands that could be withdrawn from agriculture and allocated to cotton were already in use. In the 1970s irrigation was being practiced in areas which obviously could not produce economically feasible yields (drainless zones, gypsum-bearing and pebbly soils, areas of ancient accumulation of salts, areas with significant surface slopes, etc.). Additional new lands for irrigation were found, but required the sacrifice of rural settlements. A special resolution provided for the eviction of rural dwellers from small kishlaks (villages) to one large central farmstead. Kishlaks were leveled by bulldozers, and the inhabitants of 15–20 kishlaks had to move to a central farm. The destinies of these local populations were forever changed and their way of life was destroyed forever. However, in spite of this campaign and other efforts of the republic's administration, Uzbekistan could not reach the planned target of 6 million metric tons of raw cotton as promised to Brezhnev.

It should be remembered that in Kazakstan beginning in 1961, large-scale measures were initiated for paddy irrigation in the lower reaches of the Syrdarya, in the Kyzyl-Orda region. In 1961 the area under paddy made up 13 100 ha; by 1965 it had reached 40 600 ha (Zhapbasbaev, 1969). At that time the lower reaches of the Syrdarya were called the 'rice granary', providing over three-quarters of the total rice production in the republic. Rice, much like cotton in Uzbekistan, acquired monopoly status. Its share of rice production was 85%. It became impossible to practice crop rotation and, as a result, soil fertility became completely exhausted. In the course of time, rice-growing turned into a privileged economic sector at the great sacrifice of other sectors. It consumed the major portion of capital investments for agriculture and, more importantly, it consumed practically all the water resources of the Syrdarya. During the 1955–85 period, the following irrigated rice-growing areas were developed: Kyzyl-Ordynsky (75 000 ha), Chiilisky (34 000 ha), Kazalinsky (16 000 ha), Toguskentsky (10 000 ha).

The policy aimed at cotton self-sufficiency and approved over 70 years ago by the government of the former Soviet Union led to development of and dependence on the irrigation of cotton monoculture. The share of cotton in crop rotation systems reached 75–78% and in some regions and farms it was 100%. This was accompanied by a drastic reduction in fodder production for animal husbandry and in land devoted to orchards, vineyards, vegetables, and melons. Any chances for intensification of agricultural production on the basis of technocratic and engineering approaches to land reclamation, with little regard for environmental protection, were completely exhausted (Minashina, 1990).

This resulted in the fact that Central Asia with its rapidly growing population (from 18 million in 1965 to 30.7 million in 1985) could no longer meet its requirements in food production. In addition, the regional environmental situation had worsened.

By the 1970s, the reduced availability of water resources limited further development of irrigated farming in the basins of the Amudarya and Syrdarya. At the same time, the pressure to expand the irrigated areas was dictated by seemingly unlimited land reserves and by high population growth rates, along with a return to favorable climatic conditions (Table 8.3).

As noted by Ratkovich (1979b), the problems of water management in the Aral Sea basin at that time attracted the attention of government authorities

Table 8.3 Dynamics of irrigation in the Aral Sea basin (× 1000 ha)

	Kazakh SSR	Kirgiz SSR	Tajik SSR	Turkmen SSR	Uzbek SSR	Total for basin
Irrigated lands in 1930	317	228	317	279	1392	3071
Increase of irrigated lands:						
1930–35	82	92	34	53	339	600
1935–40	99	54	26	80	407	664
1940–45	6	42	3	−9	−96	−54
1945–50	13	−17	−19	51	234	262
1950–55	41	0	79	14	94	228
1955–60	1	−8	−13	28	201	209
1960–65	88	−68	41	18	68	147
1965–70	70	54	63	155	35	377
1970–75	108	−13	35	186	314	630
Total increase of irrigated lands 1930–75	508	136	249	576	1596	3065
Irrigated lands in 1975	825	902	566	855	2988	6136

Source: Based on the data of 'The Aral Sea Scheme'.

and was studied by numerous engineering and scientific research institutions.

In July 1970 the CPSU Central Committee and the USSR Council of Ministers approved the report on 'The Prospect of Land Reclamation Development in 1971-85, River Flow Regulation and Redistribution'. The report called for the development (in 1971–75) of main guidelines for the transfer of Siberian river flow to the Aral Sea basin and for feasibility studies of the first stage of this proposed diversion. At that time, an important role in taking action in support of the Aral Sea basin was provided by the Report of the Special Commission of the USSR Council of Ministers 'On the Complex of Measures Concerning Rational and Economic Water Resources Use and Improvement of Water Supply to Branches of the National Economy up to the Year 1990'. The first part of this Report, dedicated to the developments in Central Asia and Kazakstan, was examined and approved by the State Expertise in April 1976. It was stated in a resolution that the area of irrigated lands in the Amudarya and Syrdarya basins might reach 8.1 million ha by the year 1990 ('the Main Principles' provided for the extension of irrigated areas up to 8.9 million ha by the year 1990 and up to 12.1 million ha by the year 2000). From that time on, Central Asians lived in expectation of and waiting for the transfer of water from Siberia. Uzbekistan was interested most of all in the river transfer. Rashidov, then First Secretary of the Central Committee of the Communist Party of Uzbekistan, supported the idea of river transfer in every possible way. This is evident from the open letter published not long ago in connection with the 75th anniversary of Rashidov:

> We are witnesses of the great interest attributed by Rashidov to the fate of the Aral Sea. But it was impossible to change anything at that time. Nevertheless, under the condition of monopoly power of Moscow, Rashidov had enough courage to put forward and uphold at the highest level the historical project of Siberian river flow transfer to the Aral area; this project was called 'the project of the century!' After Sh. Rashidov's death, the attained positions were given up and the opportunity was missed.   (*Pravda Vostoka*, 12 March 1992)

Glazovsky (1990) pointed out that 'wide-scale development of irrigation in the republics of Central Asia ... in the minds [of Central Asians] was closely related to the rapid implementation of a partial Siberian river flow transfer to the Aral Sea basin.' The intent to carry out the Siberian diversion scheme was proven by the construction of the Tuyamuyun off-channel reservoir on the Amudarya in the Khorezm sands. Filling the reservoir with water started in the late 1970s and early 1980s:

> This reservoir was constructed with the aim of using Siberian water in the future; it had to be one of the largest reservoirs in the region – 150 km long and 15 km wide. The reservoir basin had to intercept millions of cubic meters of Amudarya water which could reach the Aral Sea very soon.   (Yaroshenko, 1989)

In 1982, the Food Program of the USSR, proposed for the period ending in the year 1990, was adopted. Emphasis was put on the production of animal husbandry, vegetables, and melons. The further development of irrigation

Table 8.4 Dynamics of cotton yield and output of cotton fiber in Uzbekistan
(× metric tons/ha)

Produce	1966–70	1971–75	1976–80	1981–85	1986–88
Raw cotton	2.49	2.85	2.94	2.67	2.47
Cotton fiber	0.82	0.92	0.89	0.78	0.79

Source: Ataniyazov (1991).

suggested by the Program was, as a rule, associated with the river basins, which had tight water budgets. It was a fictitious program, which 'died' quietly with time. Nevertheless, it served as a basis in 1984 for the adoption of the 'Long-Term Program of Land Improvement and an Increase of Reclaimed Land Use Efficiency for the Stable Increase of Foodstock of the Country'.

Some drawbacks were revealed when the Program was adopted at the Plenary Meeting of the CPSU Central Committee. As a result of these drawbacks, the desired production was obtained on only one-third of the irrigated lands. This was enough to analyze and decide whether it was worthwhile to accelerate the development of new lands in the future. Environmental problems were touched upon only in passing.

By that time, extensive farming had resulted in the deterioration of the soil and the hydrogeological conditions of the irrigated lands. Nearly half of the irrigated soils in Central Asia became saline. According to Khakimov (1986), as a result of soil salinization, the harvest of agricultural produce was reduced in Uzbekistan by 30%; in Turkmenistan, 40%; in Kazakstan, 30–33%; in Tajikistan, 18%; and in Kyrgyzstan, 20%. Agricultural production slowed down and crop yields failed to increase (they often became lower) (Table 8.4).

Crop rotation systems were not used and in some regions (e.g., Karakalpakstan) it became impossible to put them into practice. Year after year, cotton planting followed cotton planting, and rice followed rice. Lack of crop rotation could not help but have an adverse effect on the quality of cotton and on the nutritional value of rice.

In 1986, of 6.9 million ha of irrigated lands in the Aral Sea basin, only 1.8 million ha were provided with engineering systems meeting the recent standards. In 1990, this became 7.4 million ha and 2 million ha, respectively (Khosrovyants, 1991). On rice-growing farms the increase of sown area was higher than the increase of engineering systems, because in order to fulfill the plans, state farms had to expand rice-growing onto lands developed using primitive methods, and as a result, water application rates had to be increased 3 to 4 times in order to improve the soil conditions.

Cotton is a high water-consuming crop. To ensure its successful growth in the northern areas, such as Karakalpakstan, the irrigation rate should in theory be about 7000 m³/ha; in southern areas the irrigation rate should be

Table 8.5 Amount of water delivered for irrigation and the required amount regarding the rates in the republics located in the Aral Sea basin (1984)

Index	Uzbekistan	Kyrgyzstan	Tajikistan	Turkmenistan	Kazakhstan	Total for the Aral Sea basin
Irrigated area (thousand ha)	3577.9	336.8	588.4	1039.9	634.2	6177.2
Water consumption (billion m^3)	37.8	2.3	5.5	11.0	7.4	64.1
Actual amount of water delivered for irrigation (billion m^3)	42.3	3.6	11.4	17.9	8.2	83.4

Source: Putyato (1991).

about 9000 m^3/ha. In fact, in the Samarkand and Navoi regions water was applied at a rate of 20–25 000 m^3/ha, in the Khorezm and Tashauz (now spelled Dashkhovuz) regions of Turkmenistan, up to 30–40 000 m^3/ha and in the Chimbai region of Karakalpakstan the water allocation rate sometimes reached as high as 50–55 000 m^3/ha (Yanshin and Melua, 1991). The former Minister of Land Reclamation and Water Management of Kazakhstan, Sarykulov (1994), wrote that in 1965 seepage losses in canals amounted to 50–60% of the total water withdrawal from irrigation sources. In 1967–68 the rice-growing areas of the Kyzyl-Orda region featured excessive water withdrawal for rice irrigation, which had no justification. At the designed irrigation rate of 22–26 000 m^3/ha only, water conveyance and excessive discharge required up to an additional 31 500 m^3/ha. Thus, the total amount of water consumed by a rice field was, on average, 53 500 m^3/ha. This circumstance led not only to the excessive use of irrigation water (that in itself was particularly harmful), but also to overloading of the drainage network, which had been put into operation before it was completed.

Excessive water application to the land could be observed everywhere. The tentative calculation of water requirements for the main irrigated crops in the Aral Sea basin revealed that the optimal amount of water application was 64.1 m^3, a figure that was about 30% less than the amount actually delivered (83.3 m^3) (Table 8.5).

An attempt was made to overcome the backwardness of irrigation technology in the region by introducing advanced technologies, which required higher doses of mineral fertilizers and chemicals. Nobody could foresee the pitiful result of such 'therapy'. As a consequence, soils were poisoned for a long time, and the water bodies were poisoned as a result of drainage water discharges. According to Glazovsky et al. (1992), not long ago 300–400 kg/ha of fertilizers and 55 kg/ha of chemicals (including persistent pesticides) were applied to irrigated lands in the Aral Sea basin. These figures exceeded by 10 times the amounts used in other parts of the former Soviet Union.

Widespread irrigation resulted in the elimination of the natural drainage system. The newly constructed drainage systems were designed without regard for existing natural drainage and, as a consequence, they often were unable to fulfill their natural functions. Century-old methods of irrigation (e.g., furrow irrigation and overland flow irrigation) were practised, and such outdated methods caused considerable water losses and uneven moistening of the soil. The only improvements in methods were for water withdrawal (e.g., reservoirs, pumping stations) and for water conveyance (canal lining, automatic water outlet systems). Finally, a considerable imbalance was created between the geographic extent of new lands opened up to irrigation and the water-saving modifications made to existing irrigation systems.

Great damage was caused by the methods called for in the 'directives' of local authorities, by the unreasonable increase in water withdrawals, by land preparation and by unsanctioned (i.e., unofficial) plantings. There was no effective inter-republic, inter-regional, or inter-sectoral water management and distribution system. Decision-making depended only on the authorities of the USSR Ministry for Land Reclamation and Water Management in Moscow. Since there was no charge for water, there was no economic incentive in place to encourage users to conserve water. The greatest problem was the irresponsibility of decision-makers. Nobody was held responsible for the damages caused.

From the late 1960s onward, the consequences of the unprecedented construction of water management facilities and the expansion of irrigation farming became evident as a result of the deteriorating environment of the Aral Sea and its basin. Increases in water withdrawal in the upper and middle reaches of the Amudarya affected the water supply in the lower reaches to a great extent from 1986. It was in the lower reaches where the problems of rice-growing, cotton-growing and other agricultural sectors had to be addressed.

The drastic drop of the Aral Sea level in the 1970s was caused by a coincidence of excessive water withdrawals under unfavorable (i.e., drier) climatic conditions. The natural water supply to the sea was about 75–80 km^3/year, or 20–25 cubic km/year less than it had been in the 1960s. Water consumption in the basin grew and reached about 70 cubic km^3/year. By 1970 the water flow into the Amudarya and Syrdarya deltas fell to 33 km^3 and continued to decline following the completion of flow regulation in the lower reaches and in connection with the commissioning of the Kazalinsk headworks in 1970 and the Takhiatash headworks in 1974 (Sarybaev, 1991; Akramov and Rafikov, 1990).

From 1971 to 1980 the water level in the Aral Sea dropped 53 cm/year and from 1981 to 1985 it dropped about 78 cm/year. A long low-flow period was observed in both river basins during 1974–86. The Amudarya flow decreased by 8%, on average, while the Syrdarya River flow decreased by 13%.

Against this background flow, withdrawal from the Syrdarya and Amudarya for domestic needs, but primarily for irrigation, steadily increased during 1961–86. Water withdrawal during 1981–85 increased by 45%, as compared with the period of 1966–70, while the irrigated area had increased by 22% (Table 8.6).

Table 8.6 Dynamics of the growth of water withdrawal from irrigation sources and irrigated land expansion in Uzbekistan

Index	1966–70	1971–75	1976–80	1981–85
Water withdrawal (billion m^3)	40.6	42.1	56.6	58.7
Irrigated lands ($\times$ 1000 ha)	2650.0	2834.4	3238.9	3670.6
Irrigation rate ($\times$ 1000 cm/ha)	15.3	14.9	17.5	16.0

Source: Ataniyazov (1991).

Thus, each 1% increment of irrigated land area was accompanied by a 2% increase of water consumption. It should be noted that during 1975–90 the increment of irrigated lands in the Syrdarya and Amudarya basins was rather high, amounting to 800 000 and 1 500 000 ha, respectively.

In addition, an increase in unproductive water losses from the hydrographic network was noticed. This increase was caused by the filling up of new reservoirs up to the dead storage level, by the evaporation of water from the surface of these reservoirs, and by the partial diversion of 'used' water to natural depressions. According to Minaeva (1980), about 215 cubic km of water (14 cubic km/year) were extracted from the hydrological cycle. Out of this amount, 15% was used for filling reservoirs and enclosed depressions, 29% for accumulation in the soil and subsoil, 15% was lost to evaporation from the water surface, 36% for evaporation from crops, and 5% for evaporation from natural vegetation and solonchaks. The drainage runoff in the Aral Sea basin amounted to 35 ± 5 km^3. Of this amount, over 13 km^3 of drainage water were diverted beyond the boundaries of irrigated areas and discharged either into lakes, which served as evaporation pans, or to drainless depressions in the desert. In this way there were over 40 artificial water bodies formed with a water surface area exceeding 7000 km^2 and a storage capacity of more than 50 km^3 (Altunin *et al.*, 1991). In all, from 1960 to 1987 the Aral Sea failed to receive about 600 km^3 of water.

Changes in the input and output components of the water and salt budgets of the Aral Sea immediately affected its regime. In this connection, restrictions were introduced on water delivery for irrigation water distribution in the river basins during low-flow years. In addition, water resources in some stretches of the river were re-used several times, thus causing considerable deterioration of their quality, particularly in the middle and lower reaches. The depletion of water resources in rivers of the Aral Sea basin resulted in a drastic reduction of flow into the sea. Beginning in 1982, the inflow to the Aral Sea became irregular and dropped three times to zero by 1987. In 1985 and 1986, when the surface water inflow to the sea was practically nil, the water level drop was 81 and 84 cm a year, respectively. In 1988 the sea's elevation was 40.1 m above sea level. In 1987, the separation of the Small Sea from the Large Sea occurred for the first time. The second, and most probably final, separation occurred in 1989.

THE PERIOD OF RESTRUCTURING AND PUBLIC ATTENTION (1985–91)

Many facts cited above became known to the public only after the onset of Gorbachev's policies of *perestroika* and *glasnost*. *Perestroika* (restructuring) entailed an unprecedented surge of public consciousness and a concern for environmental quality, a peculiar response of the population to the destruction of nature induced by the totalitarian regime. Changes in the public's relations with the environment were much more rapid than changes in the environment itself.

The debate over the 'project of the century' – the Siberian River diversions to Central Asia – was settled by public pressure; in 1986 the Political Bureau of the CPSU Central Committee adopted the Resolution 'On the Cessation of the Work on the Partial Flow Transfer of Northern and Siberian Rivers'.

The problem of the Aral Sea was not so acute when consideration of the river flow transfer project was taken into account, because the water of these Siberian rivers could more than compensate for the decline in sea level and could even be used to maintain the sea, though maintaining it did not fully depend on this water diversion.

As soon as the Resolution of the Political Bureau of the CPSU Central Committee on the cessation of the Siberian river diversion project was adopted, public opinion and science focused on the Aral Sea problem, elevating it to the rank of a global-scale environmental disaster.

From late 1988 and early 1989, the Aral Sea problem acquired an international character. UNEP, the World Bank, UNDP, UNESCO, and foreign representatives and scientists became involved in the study of this problem.

Government officials, scientists, engineers, writers, poets, and the public formed a united front in an effort to preserve and rehabilitate the Aral Sea and, following this, to save the Aral Sea coastal area. The aforementioned participants had their own interests in the Aral region. The apparent repentance of those in governmental circles for their mistakes in land reclamation in the Aral region also came under public scrutiny.

By 1988 the Aral Sea level had dropped by about 14 m, and the CPSU Central Committee and the USSR Council of Ministers had to admit that serious mistakes were made in the use of water and land resources of the Aral Sea, in supplying the population with drinking water of good quality, and in construction of an up-to-date sewage and waste-water treatment system in urban and rural areas. Over a long period of time, primary attention had been paid to putting new irrigated areas into use with little regard for environmental and social impacts.

No efforts were made to develop and implement effective measures to prevent negative changes in the environment or in the health situation in the

Aral Sea area. With further growth in the irrevocable water use for irrigation, there was a drastic drop of water level and reduction of the surface area. The lack of a comprehensive approach and poor-quality engineering work were exposed during and after the construction of the irrigation system.

In spite of this, the USSR Ministry for Land Reclamation and Water Management[5] submitted in 1989 a memorandum to the Government 'On Further Measures Concerning the Development of Land Reclamation to Radically Solve the Food Program and to Ensure the Steady Development of the Agro-Industrial Complex of the Country'. Uzbekistan's President Karimov once again raised the idea about water supply from outside the Aral region and asked Moscow to render assistance (*Pravda Vostoka*, 23 September 1989).

Being aware of the cutback in land reclamation activities, a large group of agricultural scientists appealed to USSR President Gorbachev. The appeal, made in October 1990, sparked consideration of the problem 'On the Future Development of Land Reclamation in the Country' by the Committee for Agrarian Problems and Food of the USSR Supreme Soviet. A resolution was passed in 1991 elaborating a nationwide program for land reclamation. At that time the extension of irrigated lands and the construction of new hydraulic structures were planned for the sake of 'saving' the Aral Sea, which had been ruined by such activities. However, this program was not carried out.

Table 8.7 summarizes Communist Party and Soviet Government resolutions concerning land reclamation, water management and agriculture in the Central Asian Republics of the former USSR (resolutions cited above). They are correlated with the growth of irrigated areas and the drop in Aral Sea level.

The retrospective review of the stages of 'political' development of irrigated lands in Central Asia revealed that agricultural and water management activities depended on political and socio-economic factors, such as the socialization of private property, the ideology of 'cotton self-sufficiency', the overall extensive development of irrigation and so forth. The concept of irrigation (i.e., land reclamation) was based on 'development' playing the main role. Development was primarily understood to be the intense growth of irrigated land areas and use of water and other natural resources. The goals of land reclamation were formulated in terms of: the need for the extension of agricultural areas, the involvement of new lands into crop rotation systems, the increase of the volume of production, and the reduction of the cost of production. Under these circumstances, the problems of land reclamation and environmental conditions were relegated to the background, and in the 1960s–70s they were not taken into consideration at all. In other words, the technocratic, engineering approach dominated considerations of land recla-

5. This Ministry was replaced in 1989 by the Vodstroi Concern (Company).

Table 8.7 Party (CPSU CC) and Government (CM) Resolutions on land improvement in Central Asia and their consequences for the Amudarya and Syrdarya flows and the Aral Sea

Year	Irrigated lands in the Aral Sea basin (×1000 ha)	The Aral Sea level drop (m)	Average input of the Amudarya and Syrdarya to the Sea (km³)	Resolutions of the Governmental authorities
1955–60	3965	–	58.3	February 1954. 'On the future development of cotton growing in the Uzbek SSR in 1954–1958' April 1954. 'On the future development of cotton growing in the Turkmen SSR in 1954–1958' June 1954. 'On the future development of cotton growing in the Tajik SSR in 1954–1960' August 1956. 'On the irrigation of Virgin Lands of the Golodnaya Steppe in the Uzbek SSR and the Kazakh SSR for increasing cotton production' July 1958. 'On the further expansion and acceleration of the works on land irrigation and development in the Uzbek SSR, Kazakh SSR and Tajik SSR'
1960–65	4241	0.91	42.8	October 1961. XXII Congress of the CPSU. Adoption of the CPSU Program which included land reclamation
1965–70	4698	1.56	43.7	May 1966. 'On the wide-scale development of land reclamation to obtain high and stable yields of cereals and other agricultural crops'
1970–75	5350	3.91	26.9	July 1970. 'On the perspective land development in 1971–1985, regulation and redistribution of river flow'
1975–80	6125	7.14	12.3	
1980–85	6895	11.03	4.8	May 1982. USSR Food Program until 1990, which included the problems of land reclamation development October 1984. 'On longterm Program of land improvement, increase of reclaimed land use efficiency for the stable increment of foodstock of the country'
1985–90	7403	14.9	9.5	August 1986. CPSU Politboro Resolution 'On the cessation of the work on the partial flow transfer of Northern and Siberian rivers' September 1988. 'On measures for fundamental improvement of the environmental and sanitary conditions in the Aral Sea region, and for increased effectiveness in the utilization, strengthening and safeguarding of water and land resources in this basin'

Table 8.7 *(cont.)*

Year	Irrigated lands in the Aral Sea basin (×1000 ha)	The Aral Sea level drop (m)	Average input of the Amudarya and Syrdarya to the Sea (km³)	Resolutions of the Governmental authorities
				November 1989. USSR SC Resolution 'Concerning urgent measures for improving the environmental conditions of the country'
1990–91	–	16.6	–	March 1991. USSR SC Resolution 'Concerning the implementation of the Resolution of the Supreme Soviet of the USSR concerning urgent measures for improving the environmental condition of the country related to the problem of the Aral Sea'

Notes: All Resolutions are limited to the period of the existence of the USSR.
CPSU CC, the Central Committee of the Communist Party of Soviet Union; CM, the Council of Ministers.

mation (Table 8.8). The latter was easy, because the approach was promoted by an army of hydraulic engineers of the Central Asian school. The short-term profit was always estimated to be higher than long-term well-being, hence the panacea, a reliance on technological decisions, came into being: 'We shall turn deserts into flourishing orchards'. This concept matched very well with the notorious notion of 'cotton self-sufficiency', and the notion of 'Everything for the sake of the man' proved, in essence, to have been only slogans.

The Aral Sea disaster is the inevitable result of the profound crisis in the regional economy. Dukhovny and Razakov (1988) wrote that there was 'no need to be afraid of the truth: the Aral Sea's destiny was predetermined by the trends of economic development in the region.'

In the 'Main Principles of the Concept of the Preservation and Rehabilitation of the Aral Sea' (Anon, 1991), it was noted that

> the fundamental reason for the Aral Sea crisis centers on the profound discrepancy between the established production structure of the national economy and the potential limitation of the region's ecosystem. The crisis resulted from the fact that, formerly, the commanding administrative system ignored the laws of nature and economic laws and made unlimited use of water and other resources.

The opinion of American researchers Feshbach and Friendly (1992) about the Aral situation was scathing: 'the disaster of the Aral Sea was the most frightful and recent example of official Soviet policy toward villages and villagers with respect to agriculture and to Central Asia. The transformation of the sea into a desert became a symbol of the ecocide that took place there during the last sixty years. In reality, the Aral Sea was sacrificed in order to develop a 'glittering southern showcase of socialism'.'

Table 8.8 Changes in the character of land development within the Aral Sea basin

Years	Character of land development	Response to environmental changes
1960s	Expansion of irrigated areas without proper drainage resulted in a permanent rise in groundwater level and in soil salinization. The transition from drainless irrigation to comprehensive development of arid lands (e.g., land levelling, drainage, leaching, fertilizer application) was in practice limited by water supply through hydraulic construction. Water engineering dominated over land reclamation development, because of the lack of a sound scientific and technological foundation. Erroneous decisions concerning the structure of the economy and the strategy of land development (e.g., extension of irrigated lands, irrigation of saline soils, monocultures of cotton and rice). Almost completely destroyed the natural regime of water and salt exchange which was substituted for by an artificial (manmade) regime.	Speeding up of the process of the degradation of nature. Passive contemplation of the threatening scope of human interference in the regime of the regional rivers and sea. Serious environmental disturbance and considerable damage to the national economy. The scientific and design data did not consider an objective socio-economic assessment of the living conditions of the population in the region. Historically, these conditions were extremely difficult. Vague awareness of scientific, engineering and water management organizations about the desiccation of the Aral Sea and degradation of deltaic areas.
1970s	The push for the maximum expansion of the irrigated area at all costs. Realization of the social significance of irrigation in combination with the intense use of chemicals which resulted in the chemical pollution of soils, water and air. New land development involved less fertile soils with unfavorable hydro-geological conditions. The development of these soils required considerable investment and time. Constantly increasing water consumption, particularly for consumptive use, was aimed at achieving national economy development (e.g., agriculture, industry, power production, fisheries, transportation) and at meeting the requirements of domestic water supply. Feasibility studies were undertaken on the partial transfer of Siberian river flow to Central Asia.	There was a disregard of scientific forecasts about Aral Sea level drop. Land reclamation scientists and water management organizations oriented the planning authorities toward achieving a balance between the development of new lands with the rehabilitation of old lands. However, the actual plans and capital investments did not match the desired balance. The justification of the need for irrigation development was based on rapid demographic growth and food production needs and also on erroneous calculations suggesting a high level of economic efficiency in the use of irrigated lands.
1980s	Low quality design, construction and operation of irrigation systems was an inevitable consequence of rapid rates of expansion of irrigated lands; the introduction of intensive technologies, requiring the application of large quantities of fertilizers and plant-protecting chemicals, resulted in pollution exceeding the	A call was proposed for the urgent salvation of the Aral Sea, as the main factor in improving the living conditions of populations in the region. Supporters of this position led the decisionmakers and public away from understanding the real problem of the region. An increase in the general concern for the Aral Sea's destiny. Numerous

Table 8.8 *(cont.)*

Years	Character of land development	Response to environmental changes
	maximum permissible concentrations by factors of ten. Complete regulation of the Amudarya and Syrdarya flows, the accumulation of large quantities of water in the upper and middle parts of their basins resulted in drastic changes of environmental factors in their deltas because of the cessation of floods, reduction of water availability, decrease of the input of silt, organic and mineral nutrients, an increase in toxic salts in river water; the environmental changes caused the development of unfavorable processes and phenomena and, hence, the degradation of land and water ecosystems and the deterioration of living conditions in the region.	decisions, sometimes for shortterm advantage, made at the highest levels under pressure of national and international public opinion, turned out to be ineffective, because the priority values of the socialist system still remained. The programs adopted proved ineffective.
to the mid-1990s	The existing structure of irrigation farming established for large-scale state agricultural production was maintained. Insignificant reduction in the amount of irrigated land in order to make allocations of land for the development of individual household plots and for changes in cropping patterns. Statements of high level government leaders in the region about the possibility of using water resources delivered from outside of Central Asia.	Transition from the Soviet Union's propagandistic alarmist policy with respect to the Aral Sea to a policy of concrete regional cooperation. Decisionmaking was undertaken at the governmental level concerning the cooperation of the Central Asian Republics in the distribution of water resources in the Aral basin in order to solve environmental and social problems in areas adjacent to the Aral Sea. Establishment of regional organizations (the Inter-State Council, Executive Committee, and the Aral Sea Foundation) to implement the plan of action aimed at controlling the Aral Sea crisis. International assistance of the UNEP, UNDP, the World Bank and other organizations. Potential increases for conflicts over water allocation and water quality in the region.

Political decisions to extend the irrigated area under cotton and rice (e.g., monocultures) caused a series of adverse environmental changes that affected the Aral Sea basin. This supported the belief that there was an absence of a scientifically based program of agricultural development in Central Asia.

The fact that profound environmental changes in the Aral Sea basin were inevitable had been known long before the changes began.

Now reproaches are heard concerning the necessity of thinking the problem out, forecasting the consequences and stopping water withdrawal from the Syrdarya and Amudarya rivers at the level observed in the 1960s. In such a way it would be possible to preserve the sea in its previous state. Those who made decisions in the field of development of irrigated areas knew about it.

(Bostandzhoglo and Epshtein, 1988)

However, as noted by Glantz (1998), the identification of incremental environmental change is not an easy task:

> Societies (individuals as well as government bureaucracies) are for the most part not able to see any changes that would prompt them to interact with their environments any differently than they had on previous days.

Gradual changes in environmental conditions were accumulating over time. Since the extension of irrigated areas went on gradually and the final result of the extension was an increase in cotton production, politicians of the highest rank were reluctant to reveal adverse impacts of creeping environmental changes that had taken place alongside these economic achievements. The Aral Sea level dropped and

> there was no adverse impact on the decision-makers in Moscow making decisions about adverse environmental changes resulting from a declining Aral Sea level. Their non-intervention in this environmental change was, in fact, a form of intervention on behalf of the adverse environmental change.

(Glantz, 1998)

There is no doubt that the adverse effects of the Aral Sea level drop were realized by scientists too late and too many reasons were given. Glazovsky (1990) dedicated one of the chapters of his book *Crisis of the Aral Sea* to the awareness of the Aral Sea problem by scientists and the public at large. Alaev (1991) thought that the main reason was the scientific miscalculation made at each of the three organizational levels of science, i.e., regional geography, ministerial departments, and academic disciplines. On the one hand, this was the result of 'the confidentiality of some scientific findings of the 1970s' (Kuznetsov, 1992). On the other hand, in the 1970s, when scientists had raised the alarm about the Aral Sea, the 'project of the century', the Siberian river transfer to the Aral basin, was introduced into scientific thinking. Many Soviet scientists took part in the feasibility studies of this project without a murmur. Moreover,

> It is important to note that all published forecasts were too optimistic. The sea level drop turned out to be much more rapid than had been supposed by all analysts.
> (Murzayev, 1991)

The main distinguishing feature of decision-makers was their neglect of scientific recommendations and warnings. This was typical of the socio-economic policy of the past and may not have as yet been eliminated by Central Asian policy-makers. That is why the *political realization* of the problem occurred only at the end of the 1980s, when the Governmental Commission for the Aral Sea was established under pressure from scientists, intellectuals, and the public. The commission's chairman, Yuri Izrael, noted:

> The Aral Sea problem is the most dramatic and delicate example of the fact, how an environmental problem, *which was not realized in due time and remained unsolved for a long time*, turned into an urgent social problem.

(*Pravda*, 12 September 1988)

At the same time the *realization of the necessity of actions* appeared aimed at improving the environmental and health situations in the Aral Sea region and a special resolution of the CPSU Central Committee and the USSR Council of Ministers was adopted. Unfortunately, the resolution was never put into effect.

In the meantime, environmental changes affected the whole area of the Aral Sea basin. However, their manifestations were different within the region. Table 8.9 identifies threshold situations appearing during the processes of creeping environmental changes.

Creeping environmental changes, which have reached one of several critical thresholds, continue to destroy the environment as a result of political and decision-making inertia. Selyunin (1990) was right when he wrote:

> The Aral Sea death is not a large-scale disaster but only a small part of the whole problem. As everybody can see, the sea existed and now it disappears. However, a more menacing process is not so noticeable, i.e., the degradation of the living conditions of the 30 million people inhabiting Central Asia.

In his book, *Earth in the Balance*, Gore (1992) cited a proverb often used by people in the State of Tennessee: 'If you've got yourself into a hole, stop digging.' This saying perfectly suits the situation of the Aral Sea basin. It is necessary to stop developing new irrigated areas and aggravating existing damages until studies of the situation have been undertaken.

Post-Soviet period: independence of the republics of Central Asia and the heritage of cotton self-sufficiency

The disintegration of the Soviet Union in December 1991 and the formation of the five newly independent states of Central Asia radically changed the region's geopolitical situation. This could not help but affect the Aral Sea problem (e.g., Zonn, 1992; Glantz *et al.*, 1994). The Aral Sea basin and its main rivers, the Amudarya and Syrdarya, have acquired international status. As mentioned earlier, the economies of the republics of Central Asia depend primarily on water resources which are relatively scarce. The leader of each of these republics is aware of the fact that in seeking national economic development, scarce water resources could become a source of regional competition. In attempts to head off such potential conflicts, these leaders have adopted international agreements on water use and water allocation. Russia took on the role of observer, while appearing to be an active participant in the search for a solution to Aral Sea problems. However, it was unable to finance its own studies on the implementation of its projects aimed at improving the situation in the area around the sea. At the same time, the national interests of newly independent Central Asian Republics arose, including problems related to shared water resources.

The Aral Sea problem reminds us about Lewis Carroll's Cheshire cat; even if the sea disappears, its problems will remain and will long remind us about the Aral Sea problem.

Table 8.9 Creeping environmental changes in the Aral Sea basin

Creeping environmental changes	Early warnings about problem	Threshold 1 Awareness of a change	Threshold 2 Awareness of problem	Threshold 3 Awareness of crisis	Threshold 4 Awareness of the need for action	Threshold 5 Actions
Expansion of cotton production; development of irrigation farming	A.I. Voeikov (1908), S.Yu. Geller (1969) promoted the development of natural resources in the Aral Sea basin and of irrigation farming. They were aware of some of the problems that might arise in connection with river water diversions for irrigation use.	Awareness of the problem of water resources deficiency began with the start of intensive development (rapid expansion) of irrigated lands in the Aral Sea basin in the 1960s.	(a) Recognition of the deficiency of water resources in the lower reaches of the river; (b) realization of the necessity to solve the problem of water allocation among irrigation areas and republics.	Awareness of a crisis came with the wide-spread salinization of irrigated lands, and the increased necessity for leaching irrigated soils was identified at a time of growing water deficiency.	Awareness of the need for action aimed at reducing the areas under cotton and other water-consuming crops dated back to the 1970s; but there was the all-union economic production plan, which had to be fulfilled by the republics	The area under cotton has noticeably decreased (by 30%) to give way to grain crops (grain self-sufficiency) and other crops. A farmers' movement is developing.
Drop in Aral Sea level	A scientific and production-oriented Conference dedicated to the present state and future reproduction of fish and methods of rational use of the Aral Sea fish resource (held on March 22–23, 1962 in Nukus) warned (written into its resolution), that the drying up of the Aral Sea would result in the	The seriousness of the problem was realized with the increase of the rate of Aral Sea level drop, after the high river flow year of 1969.	The sea level drop hampered navigation and fishing. Problems arose relating to the outlet to the sea, the development of artificial lagoons for ship mooring and the need to provide the recreation zone to the south of the Aral Sea with sea water.	The crisis set in after the cessation of navigation, fishing and the closing of the recreation zone in Muynak. Vessels were trapped by desert sands as level dropped. The dried up sea bottom became a source of salt drift.	The demand for partial flow transfer of Siberian rivers to the Aral Sea basin became urgent. The issues of Aral Sea level stabilization were considered to preserve the sea at least at a smaller size.	Effective measures concerning sea level stabilization, and moreover saving the sea, have not been taken so far. Water inflow now depends, to a greater extent, on nature's generosity, rather than on the will of people and active non-governmental organizations.

Reduction of water inflow to the sea from the Amudarya and Syrdarya (rivers)	The problem became evident, when for the first time the Amudarya and Syrdarya channels dried up in their lower reaches. Before that, all problems concerning the reduction of water inflow to the lower reaches and to the sea had been perceived as associated with natural low-flow periods in the Aral basin.	The reduction of water inflow to the sea was associated with the degradation of coastal areas, the drying up of river channels, and the water deficiency in lower reaches of rivers.	Awareness of the crisis caused by the reduction of river water inflow to the sea came in connection with water deficiency and the deterioration of water quality. People perceived the loss of the sea as something that was inevitable. Emigration of people from the coastal zones became more frequent.	Awareness of the need for urgent actions came in the mid-1970s. Academician I.P. Gerasimov, the great strategist in the field of geography, led the movement for 'saving the sea' and for solving these problems. Higher authorities, however, suspended the carrying out of solutions to the problems.	Actions connected with the reduction of river-water inflow to the sea were of opposite character, i.e., they were aimed at retaining as much water as possible in the delta for irrigation purposes and in the river channel, especially, in long-period storage reservoirs.

expansion of solonchaks and salt transport from the newly exposed seabed onto adjoining areas.

The specific features of creeping environmental changes in the Aral Sea basin did not approach invisibly. On the contrary, everybody knew about them and understood that they were inevitable. However, the Aral Sea was sacrificed for the sake of cotton self-sufficiency of the USSR. The reduction of water inflow to the rivers' lower reaches was predicted by all those connected with plans for irrigation development in the region. However, it is most likely that not everybody understood the seriousness of the environmental changes. The total ignorance of environmental protection in the 1960s–1970s was to blame.

Table 8.9 (*cont.*)

Creeping environmental changes	Early warnings about problem	Threshold 1 Awareness of a change	Threshold 2 Awareness of problem	Threshold 3 Awareness of crisis	Threshold 4 Awareness of the need for action	Threshold 5 Actions
Deterioration of river and sea water quality	Information concerning the deterioration of river-water quality had not been available for a long time. However, specialists in fishery were aware of the Aral Sea level drop and salinity increase, and had already in the 1960s become very much concerned about this problem.	In 1975–1979, river-water salinity noticeably increased, even at the house-hold level. For several months a year salinity exceeded the MPC by 1.5–2 times. The problems of supplying the population with drinking water in the lower river reaches arose.	River water in the lower reaches became unsuitable for drinking. Salt concentrations in the irrigation water became elevated. Indigenous fish species ceased to reproduce in the sea.	Awareness of the crisis with respect to river-water quality came as a result of the rate of growth of sickness among humans. As for the sea-water quality, the crisis was revealed through the loss of living sea resources.	Steps were taken to find ways to improve the quality of river flow, to remove drainage effluent from the rivers and to provide people living in the lower reaches of the rivers with drinking water.	Construction of drinking water pipelines; adoption of the 'Pure Water' Program; removal of drainage effluent from the river channels.
Drastic reduction of fish resources	Foreseeing the drop of Aral Sea fishery importance to Soviet leaders, in 1959 the governments of Uzbekistan and Kazakstan adopted resolutions concerning the organization of fish-rearing through the establishment of fish farms, fish hatcheries, etc. A conference held in Nukus in 1962 empha-	There was a more than 25% reduction of fish yields in 1965, compared with the year 1960 which contributed to the awareness of policymakers of the problem. Moreover, a reduction in the recruitment (replenishment) of fish resources in the sea was recorded, as a	There was a drastic reduction of fish yields. This led to a shortage of fish for fish-canning industry.	The curtailment of the sea fishery, the closing of fish hatcheries and fishery collective farms, and the resul-tant unemployment of fishermen. The abandonment of fishing settlements (e.g., Urga among others).	There was a search for new sources of water and way to acclimatize salt-tolerant fish species in the sea in addition to elaboration of the concept of the sea-saving solution of the employment problem for fishermen.	Provision of new water bodies, development of pond fisheries, establish-ment of lake fishery farms. Transfer of water bodies to farmers. Arrangement of water bodies in the bed areas of former lakes, sea bays of the work of forestry farms, basin irriga-tion. However, the

	sized the concern of scientists and practical fishery specialists of Central Asia about (a) the growth of river water withdrawal for irrigation and (b) the forecasts of sea level drop.	result of drying up of spawning pools.	Salinization of deltaic soils, desertification, loss of tugai, drying up of lakes. Continued reduction of pastures and haylands.		Were forced to search for new pastures for sheep flocks; forced to reduce cattle population.	efficiency of such work is low.
Degradation of deltaic ecosystems	The degradation of deltaic ecosystems was also expected in connection with river water withdrawal for irrigation, as was the development of the process of salt drifts from the drying sea bottom (due to wind erosion) and the salinization of soils and subsoils of the Priaralye.	These changes arose as soon as the natural river floods ceased and the haylands and pastures began to dry up.			The need for artificial water supply of deltaic zones to prevent further degradation was recognized as was the loss of hunting resources, e.g., the decrease of muskrat output (from 1 million pcs to 20 thousand pcs).	Establishment of the nature reserve 'Badai Tugai'; soil stabilization of the dried-up bottomland.
Groundwater table rise	The groundwater table rise was anticipated in planning the irrigation development but great importance was not attached to this problem.	Awareness of the problem occurred from the outset with the general groundwater table rise in Priaralye, and the depletion and salinization of freshwater resources.	Soil salinization and elevated groundwater in irrigated areas, in settlements, and in urban areas.	The salinization of 60–80% of irrigated lands, the reduction by one third of gross output from 1 ha of irrigated land. The drying up of orchards and vineyards in the Aral Sea region.	The need for an improvement of the drainage network was accepted.	A decrease in the share of cultivated water-consuming crops and saving of irrigation water. Construction of drainage collectors to remove drainage water to the sea. The latest construction project is the Right-Bank Collector.

Table 8.9 (*cont.*)

Creeping environmental changes	Early warnings about problem	Threshold 1 Awareness of a change	Threshold 2 Awareness of problem	Threshold 3 Awareness of crisis	Threshold 4 Awareness of the need for action	Threshold 5 Actions
Increase in the number and in the frequency of dust storms	The problem was slightly obscured by the forecasts of scientists. According to S.Yu. Geller (1969), the transport of salts from the drying Aral bottom by wind did not appear to be a serious hazard; on the contrary, the transport of calcium carbonate and calcium sulfate was believed to prevent soil alkalinization in irrigated areas of the Priaralye.	Awareness of changes occurred from the start of the formation of barkhans and barkhan ridges (dunes) out of the shifted sands of the dried up sea bottom. Awareness also of the removal from the seabed of salty dust.	The growth in the frequency of dust storms and salt-dust transport to the oases to the south of the Aral Sea.	The dried-up seabed (3 million ha) became the source of major salt and dust transport. In 1990 the frequency of dust storms was 15. A total of 40–150 million metric tons of salty dust were transported each year.	Recommendations of the All-Union Conference on the 'Processes in Desertification in the Aral Sea Region' (14–16 May 1985 in Moscow) concerning the phytoreclamation of the dried-up sea bottom and soil stabilization.	The work on soil stabilization in the dried-up Aral Sea bottom is ineffective.
Population morbidity and mortality	The situation concerning adverse changes in the population's health were not expected in the 1960s. Formerly, there had been no recorded environmental pollution problems in	Awareness of these changes occurred in connection with the growth of alimentary and kidney pathology, anemia, and cancerous diseases among population.	Increase in morbidity (illness) of the population. The index of esophagus cancer was seven times higher than the average index of the former Union.	A 1989 general health survey in Karakalpakstan revealed serious health problems in 66% of adult and 61% of children (Abdirov, 1990).	Consolidation of efforts of Central Asian states to solve the Aral Sea problems; conferences in Kzyl-Orda, Tashkent, Nukus, Dashkhovuz, and the International Conference in Nukus	The work of 'Health Trains' of Ecosan (an Uzbekistan non-governmental organization), free assistance from outside, construction of hospitals, system of local sanatoriums.

Central Asia with the exception of large and industrial cities. The use of chemicals in agriculture began later.

held in September 1995.

Nevertheless, continued increases are recorded of tuberculosis, blood diseases, cancer; shortage of medicines.

References

Agarwal, A., 1996: *The Curse of the White Gold: The Aral Sea Crisis*. New Delhi: Center for Science and Environment, 56 pp.

Akramov, Z. M. and A. A. Rafikov, 1990: *The Past, Present and Future of the Aral Sea*. Tashkent: Mekhnat. 143 pp.

Alaev, E. B., 1991: *Social-economic Problems of the Aral Sea Region (Priaralye Region)*. Report for 1991. Manuscript, p. 7.

Altunin, V. S., Ye. N. Kupriyanova and A. A. Tursunov, 1991: Internal water reserves for stabilizing the Aral Sea and restoring ecological equilibrium in its basin. *Izvestiya AN SSSR, seriya geografīcheskaya*, **4**, 118–24.

Anon., 1991: The main principles of the concept on the preservation and rehabilitation of the Aral Sea, and on normalization of the ecological, sanitary-hygienic, medico-biological and socio-economic situation in the Aral Region. *Izvestiya AN SSSR, seriya geographicheskaya*, **4**, 8–21.

Ataniyazov, B., 1991: Ecological and economical problems of irrigated agriculture in Uzbekistan. *Melioratsiya i vodnoe khoziyaistvo*, **3**, 16–19.

Bostandzhoglo, A. A. and L. V. Epshtein, 1988: Problem of the Aral Sea. In: *Teoriya i metody upravleniya vodnymi resursami sushi*, Part I. Moscow: VASKHNIL, 35–54.

Budagovsky, A. I., 1992: Hydro-ecological aspects of the problems in the Aral and PriAral Region. *Water Resources*, **19**(6), 110–21.

CPSU Central Committee, 1966: Proceedings of the May 1966 Plenary Meeting of the CPSU Central Committee, Politizdat, p. 14.

Dukhovny, V. A. and R. M. Razakov, 1988: Aral: looking the truth in the eye. *Melioratsiya i vodnoe khoziyaistvo*, **9**, 27–32.

Feshbach, M. and A. Friendly, Jr., 1992: *Ecocide in the USSR: Health and Nature Under Siege*. New York: Basic Books.

Geller, S. Yu., 1969: Certain aspects of the Aral Sea problem. In: S. Yu. Geller, *Problems of the Aral Sea*. Moscow: Nauka, 5–24.

Glantz, M. H., 1998: Creeping environmental phenomena in the Aral Sea basin. In: I. Kobori and M. H. Glantz (eds.), *Caspian, Aral and Dead Seas: Central Eurasian Water Crisis*. Tokyo: United Nations University, 25–52.

Glantz, M., A. Rubinstein and I. Zonn, 1994: Tragedy in the Aral Sea basin: Looking back to plan ahead? In: *Central Asia: Its Strategic Importance and Future Prospects*, Hafeez Malik, ed. New York: St. Martin's Press.

Glazovsky, N. F., 1990: *Aral Crisis*. Moscow: Nauka. 135 pp.

Glazovsky, N. F., R. Kasperson, G. V. Sdasjuk and B. L. Turner, 1992: The world's critical ecological regions: principles of the distinguish and study methodology (approaches of the Soviet and American geographers). In: *Global Changes and Regional Interconnections*. Moscow: AN SSSR, Institute of Geography. 58 pp.

Gore, A., 1992: *Earth in the Balance: Ecology and Human Spirit*. New York: Houghton Mifflin Company.

Khakimov, F., 1986: Desertification Processes and Soil Reclamation in the Downstream Area of the Amudarya. *Proceedings, V All-Union Conference 'Ecological Problems of Desert Development and Nature Conservation'*. Ashkhabad: Ilym Publishing House.

Khosrovyants, I. L., 1991: Problems of Water Resources Development in the Central Asian Region and Ways for Aral Sea Preservation. *Melioratsiya i vodnoe khoziyaistvo*, **12**, 2–10.

Kuznetsov, N. T., 1992: Geographical and ecological aspects of Aral Sea hydrological functions. *Post-Soviet Geography*, **33**,(5), 324–31.

Micklin, P., 1991: The Water Management Crisis in Soviet Central Asia. The Carl Beck Paper in Russian and East European Studies. No. 905, University of Pittsburgh. Center for Russian and East European Studies. 120 pp.

Minaeva, E. N., 1980: Discharge of river waters didn't enter the Aral Sea because of anthropogenic activity during 1961–1975. *Vodnye Resursy*, **5**, 82–8.

Minashina, N. G., 1990: Ecological consequences of water reclamation construction in the Aral Sea basin. In: *Ecologicheskaya Alternativa*. Moscow: Progress Publishing House, 350–70.

Murzayev, E. M., 1991: A short review of the study of the Aral Sea and its region. *Izvestiya AN SSSR, seriya geographicheskaya*, **4**, 22–35.

Petrov, M. P., 1964: *USSR Desert and its Development*. Moscow: Nauka, 146 pp.

Pokshishevskii, V.V. 1949: Aleksandr Ivanovich Voeikov and his works on man and nature. In: A. I. Voeikov (ed.), *Man's Impact on Nature*. Moscow: State Publ. Geograph. Literature, p. 12.

Putyato, N. S., 1991: Irrigation water use efficiency in the Aral Sea basin. *Melioratsiya i vodnoe Khoziaystvo*, **3**, 19–21.

Ratkovich, D. Ya., 1979*a*: Problems of water allocation in the seas and lakes of the southern USSR. In: Anon., *Problems of USSR Water Economy in the Future*. Moscow: AN SSSR, 396–467.

Ratkovich, D. Ya., 1979*b*: Perspectives of water supply in the Central Asia and Kazakhstan. In: *Problems of USSR Water Economy in the Future*. Moscow: AN SSSR, 362–94.

Reshetkina, N. M., 1991: The Aral Sea basin as a self-regulating natural system. *Melioratsiya i vodnoe khoziaystvo*, **9**, 3–7 (Part 1), **10**, 13–17 (Part 2).

Sarybaev, K., 1991: Development of irrigation farming in the Karakalpak ASSR (1960–1980). In: *Aral Crisis*. Moscow: AN SSSR,199–216.

Sarykulov, D., 1994: *Golden Water*. Kazakhstan: Almaty. 53 pp.

Selyunin, V. I., 1990: Action time. In: *Nature's Fate is Our Fate*. Moscow: Khudozhestvennaya literatura, 363–411.

Voeikov, A. I., 1908: Irrigation of the Zacaspian area from the geography and climatology view point. In: A. I. Voeikov, *Man's Impact on Nature*. Moscow: State Publ. Geograph. Literature. 1949, 158–79.

Voropaev, G. V., 1992: Can the Aral Sea be recovered today? *Water Resources*, **19**(6), 97–102.

Yanshin, A. L. and A. I. Melua, 1991: The Aral can be saved yet. In: *The Lessons of Ecological Miscalculations*. Moscow: Mysl, 28–53.

Yaroshenko, V. A., 1989: Amudarya oasis. In: *The Expedition 'Live Water'*. Moscow: Molodaya gvardija, 312 pp.

Yusupov, V., 1939: About irrigation construction in the Uzbekistan. *Report on the Meeting in the Time of Tsk Kp(b) Uzbekistan*. Tashkent: Party Publisher.

Zaletaev, V. S., V. I. Kuksa and N. M. Novikova, 1991: Some ecological aspects of the Aral problem. *Vodnye Resursy*, **5**, 143–54.

Zhapbasbaev, M., 1969: *The Agroclimatic Conditions for Growing Rice in the Continental Climate*. Hydrometeoizdat, 15 pp.

Zonn, I., 1992: Environmental stress and the search for sustainable development in arid and semiarid Central Asia: the case of the Aral Sea basin. In: *International*

Conference on Climatic Variations and Sustainable Development in Semiarid Regions.
    Brazil: Fortaleza. Conference Proceedings, Vol. III, 809–31.
Zonn, I., M. H. Glantz and A. Rubinstein, 1994: The Virgin Lands Scheme in the
    former Soviet Union. In: *Drought Follows the Plow: Cultivating Marginal Areas,*
    ed. M. H. Glantz, Cambridge, UK: Cambridge University Press, 135–50.
Zuev, V., 1991: *Aral Dead End* (Aral'skij tupik). Moscow: Prometey, 101 pp.

9   Change of the rivers' flow in the Aral Sea basin (in connection with the problem of quantitative assessment and consideration of environmental after-effects)

K.V. TSYTSENKO AND V.V. SUMAROKOVA

The environmental disaster in the Aral Sea basin is a direct result of societal overuse of the water resources of the basin's two major rivers, the Amudarya and the Syrdarya. Societal activities are responsible for serious reductions in both the quantity and the quality of river flow and for negative impacts on the major components of the region's ecosystems which are directly dependent on the availability of surface water. Thus, an assessment of the region's water resources, including changes in their use, is necessary to identify the impact of anthropogenic factors on river flow and to identify hydrological measures that could improve environmental conditions in the basin.

The Aral basin lies in the heart of the Eurasian landmass, far from the direct influences of the oceans. It therefore has a continental climate (i.e., marked seasonality) and an abundance of warm and sunny days. The major part of the basin is covered by the Turan Lowland deserts bordering on the Tien Shan and Pamir-Alai mountains. Glaciers and snowfields are the main sources of water feeding the region's two major rivers – the Amudarya with a catchment area of 1.1 million km², and the Syrdarya with a catchment area of 0.44 million km². Unlike the mountainous part of the basin, which is the zone of formation of riverflow, the lowlying flat land is a zone where water resources are dissipated as a result of evaporation from both irrigated and natural areas (Shults, 1965).

Irrigation farming ever since the distant past has been the main economic activity in the Aral Sea basin. Table 9.1 shows the expansion of irrigation in the twentieth century. By 1990, nearly 35% of all irrigated lands of the former Soviet Union was situated in the Aral basin. The continual development of irrigation was accompanied by increases in water withdrawals from the rivers en route to the sea. The maximum increases in both the amount of irrigated

Table 9.1 Area under irrigation in the Aral Sea basin (million ha)

Beginning of the century	1960	1990
2.4	3.5	6.7

Table 9.2 Volume of river water withdrawal for irrigation (km³/year)

Basin	1930	1961	1970	1980	1990
Syrdarya	–	27.8	41	48.3	43.5
Amudarya	–	28.8	44.5	60.8	53.9
Total	35–40	56.6	85.5	109.1	97.4

land and in water withdrawals occurred after the 1960s as shown in Table 9.2. These increases were directly connected to the implementation of the large-scale, water-dependent Soviet programs for the 'reclamation' of Central Asia's natural desert lands. Water use in other economic sectors such as for industry and municipal services increased from 7.5 to 18.0 km³/year, during the 1961–90 period.

When reviewing irrigation development in the Aral Sea basin, one must consider two important circumstances:

1. irrigation at the beginning of the twentieth century was based mainly on the flow of secondary streams – the tributaries of the Amudarya and Syrdarya – through the use of technologies and know-how available at that time.
2. At that time surface water from several rivers feeding the Syrdarya had long failed to reach the channels of the main rivers. In this respect, the Syrdarya river basin was quite typical; only 3% of all irrigated lands in this basin was irrigated from this river (Dingelstedt, 1893). For the Amudarya river basin that number was more than 40% (Tsinzerling, 1927).

The intensification of irrigation farming started after 1960 with the implementation of engineering measures to take water directly from the main channels of the rivers. At that time, construction of dams, reservoirs, and large headworks was in progress. This resulted in the main irrigation areas in the Aral Sea basin receiving water that was taken directly from the middle and lower reaches of the Amudarya and Syrdarya channels.

Along with the expansion of land under irrigation, major changes took place in cropping patterns. Whereas up to several decades ago grain crops had dominated (on the order of 60–80%), today industrial crops dominate (such as cotton and fodder grasses). In the Aral basin, the portion of the land given to grains gradually decreased to 10–15%, while the portion for the crops with high water demands, such as cotton and rice, increased in some regions from 25% to 50% or more.

Trends in the decrease in the amount of flow in the river channels were recorded. These trends resulted from irrigated land development and by changes in cropping patterns. These conditions were responsible for the worsening of the impacts of irrigation on the characteristics (e.g., intensity and variations) of the rivers' flow across their entire length, i.e., from their mountain sources to the deltas.

The degree of changes in the flow of the rivers toward the Aral Sea depends on variations in the volume of water coming from the zone of formation of the flow, the areal extent of the irrigated land, cropping patterns, the total volumes of water withdrawals, non-renewable water consumption, and other human-induced as well as natural losses of river water. The interaction between surface and subsurface water and changes in the water table are also of major importance. Thus, the process of anthropogenically induced changes of the river's hydrological regimes has been multifaceted and complicated.

An assessment of such changes was undertaken based on the annual values for the 1931–90 period. Because annual hydrological data have not been published after 1988, the estimates for 1989 and 1990 were based on limited data. (The method used to evaluate water resources is described in Volftsun *et al.*, 1988*a,b*.)

For many years surface water resources in the Aral Sea basin totalled almost 114 km³/year, with the water of the Amudarya being about twice that of the Syrdarya. Table 9.3 shows the values of surface water resources by basin, coming from the flow-formation zone.

The highest flow into the Aral Sea from the formation zones of the two rivers combined was recorded in 1969 at 193 km³/year and the lowest was 83 km³/year in 1974. The 1931–60 period taken as a whole can be characterized by relatively high water availability, when compared with the long-term mean.

Cumulative water consumption in the basin was extremely high after 1960. Between 1961 and 1970, about 70% of the flow (Table 9.4) was withdrawn from the rivers for societal purposes, including the delivery of water from the Amudarya to the Karakum Canal in Turkmenistan. Later, water withdrawal actually exceeded the water resources! This is accounted for by water that had returned from irrigated lands to the rivers and could be delivered again for the irrigation of the lands in the lower reaches of the rivers.

Table 9.3 Surface water resources (km³/year)

Basin	Water resources 1931–1990	Including	
		1931–1960	1961–1990
Syrdarya	37.9	38.2	37.7
Amudarya	75.9	78	73.9
Aral Sea	113.8	116.2	111.6

Table 9.4 Use of water resources (km³/year)

Description	1961–1970	1971–1980	1981–1990
Water resources	117	111	115
Total water withdrawal	82	111	127

Because water consumption for irrigation is about 80–90% of the total water withdrawal from the rivers (Table 9.2), an assessment of the irreversible losses of water and of the return flow in this economic sector is extremely important. We developed a method for calculating these factors, based on generalizations and on an analysis of multiyear water-budget studies undertaken by the State Hydrological Institute about the irrigated lands in the arid regions of the former Soviet Union (Levchenko *et al.*, 1990; GGI, 1984; Sumarokova and Degtyarev, 1985; Tsytsenko and Vonsovskaya, 1985).

Two categories of irreversible water consumption are net and gross (GGI, 1984). Net refers to water consumption on the lands that are actually irrigated. Gross refers to the total water flow intended for irrigation use. This includes net consumption and evaporation losses from fallow lands, open ditches and channels and from closed depressions. Return flow refers to water that returns to the rivers from the irrigated lands, reaching the river networks either by surface or subsurface flow.

During the period of 1931 to 1960, gross water consumption was on average 32 km^3/year (Volftsun and Sumarokova, 1985). From 1961 to 1990 it almost doubled, reaching 60 km^3/year. This large increase was proportional to the increase in irrigated lands (see Table 9.1). Irreversible water withdrawal losses (e.g., gross consumption) were estimated at 65–70% (Sumarokova and Degtyarev, 1985; Tsytsenko and Vonsovskaya, 1985). However, between 1961 and 1990, on average 42 km^3/year – the net loss – were lost on the irrigated lands. The difference between gross and net water consumption characterizes the inefficient use of river water and should really be viewed as a 'hidden reserve' in the event governments choose to redesign these irrigation systems Volftsun *et al.*, 1988*a,b*). Inefficient water use for irrigation purposes was approximately 18–20 km^3/year of which about 75% was lost to evaporation collection points for runoff from the fields located on the edges of the irrigated areas, the so-called 'irrigation-escape lakes'. The largest of such collection points are Lake Sarykamysh in the Amudarya's lower reaches (with a water surface area greater than 3000 km^2) and Lake Arnasai in the middle reaches of the Syrdarya).

Taking into account water diverted from the Amudarya into the Karakum Canal in Turkmenistan – an irreversible loss of the water for the Aral basin, the total consumption of river water devoted to irrigation between 1961 and 1990 was 70 km^3/year.

If irreversible water consumption is the direct cause of the decrease in river flow, then the return flow adds to some extent to the rivers' flow. However, accompanying the positive aspects of return flow (i.e., more water for irrigation use), there is a major negative effect. Return flow is contaminated by salts and man-made chemicals and is therefore a serious source of surface and groundwater pollution. On average, between 1981 and 1990 over 25 km^3/year of return flow was identified, or about 30% of the river water originally withdrawn for irrigation).

An annual increase in return flow had been observed before 1980. However, in the 1980s a decrease of up to 26% was noted in the annual volume of return flow, both in absolute values and relative to the amount of water for irrigation. This could, however, have been related at least in part to improvements in water use in the region.

The amounts of return flow at various points along the length of the Amudarya and the Syrdarya have been uneven (Levchenko *et al.*, 1990). In the upper part of the zone of river water use about 40–50% of the flow taken for irrigation returns to the rivers, because of these regions' hydrogeological features. In the middle and lower reaches, the volume of return flow decreases by up to 30% of water withdrawals, because of the relatively poor drainage conditions associated with the flattening out of the river valleys, the decrease in the number of channel cuts, and in their slopes. This is the region where most of the irrigated lands of the Aral Sea basin are concentrated. Thus, changes in the major aspects of irrigation farming, such as putting more land into production, could only add to the existing human-induced changes in the flow regimes of the Amudarya and Syrdarya.

This problem drew attention as early as the 1920s, when plans were developed for the reconstruction of irrigation facilities that had been destroyed during Russia's Civil War. Even then, some researchers believed that the proposed development of irrigation farming would result in a serious decrease of water of the Amudarya and Syrdarya, especially in their lower reaches (Dubovikov, 1922).

The lack, for a long time, of the occurrence of the anticipated decrease in flow, despite the considerable growth in the areal extent of irrigated land, provoked fiery discussions about the probable impacts of irrigation on streamflow (Tsytsenko and Sumarokova, 1990; Dunin-Barkovsky, 1956; Kharchenko, 1975; Shiklomanov, 1979; Yunusov, 1974). This debate forced a consideration of the matter, when planning began again for new land reclamation projects. Discussions of this problem in the 1960s and 1970s added stimulus for the organization and implementation of research. We participated in these activities (Kharchenko, 1975; Tsytsenko and Vonsovskaya, 1985; Sumarokova and Tsytsenko, 1978).

Human-induced flow changes (Δ_{YA}) were assessed according to the relationship:

$$(\Delta_{YA}) = Y_{VF} - Y_{OBS}$$

where Y_{OBS} = flow observed at the design site
Y_{VF} = natural restored flow at the same site

The value of Y_{VF} equals the water discharge which would have occurred in the absence of irrigation (e.g., virgin flow).

Serious difficulties arise when determining Y_{VF} values, because the length of hydrological observations in this region is very limited, even though it improved during the period of increased irrigation activities. This made it difficult to determine the true virgin streamflow of these two important rivers. Human-induced flow changes are assessed, taking into account the so-called conventional natural period when the impact of economic activity on streamflow volume was still negligible.

To assess river flow and identify anthropogenic impacts in the natural flow period a method of flow-series analysis was used, based on the comparison of discharge values for synchronous time intervals at sites found in the flow-use zone (Y_H) and the availability of water of the rivers in the zone of flow formation (W_{ZF}). We did this by plotting relationship curves of the type $Y_{NF}(W_{ZF})$. Several relationships are usually seen on such graphs (an example of one such relationship is given in Figure 9.1). The upper line permitted the evaluation of the length of the conventionally natural period and, correspondingly, the mean multiyear flow value (Y_N). The difference between the Y_N values of the upper line (conventional natural flow period) and the lower line (the period of serious impact of anthropogenic factors) represents the flow component characterizing changes as brought about by economic activities (Δ_{YA}).

Studies show that the conventional natural period ended about 1960. The lack of impact of irrigation on Amudarya and Syrdarya flows, until 1960, can be explained by some of the following compensating factors:

- The replacement by evaporation from the irrigated lands of natural flow losses resulting from evaporation from the natural unused lands.
- The construction of a collector and drainage network that, for many years, enabled the re-entry to the rivers of return and subsurface waters of natural origin.
- The peculiarities of the development and siting of new irrigation areas undertaken with regard to the rivers (irrigation sources) and return flow intakes.
- The relatively high amount of natural water availability in the flow-formation zone between 1932 and 1960 (Table 9.3).

The multi-purpose uses of river water resulted in a relatively stable Aral Sea level. However, interannual fluctuations were also influenced by the multiyear negative evaporation trend from the sea's water surface (Golubev and Zmeikova, 1991).

This equilibrium that existed between ever-increasing irreversible water consumption and the compensating effects of the factors cited above broke down with the construction of the Kairakkum Reservoir in the Syrdarya basin and with the sharp increase in water withdrawals from the Amudarya basin for the Karakum Canal (Sumarokova and Tsytsenko, 1978).

Annual flow variations of these rivers are given in Figure 9.2. Average values for some periods between 1961 and 1990 are given in Table 9.5. The gauge sites are located in the upper reaches of the deltas. By the end of this period of measurement, the human-induced decrease in flow on both rivers exceeded 80% of its value during the conventional natural period. The adverse

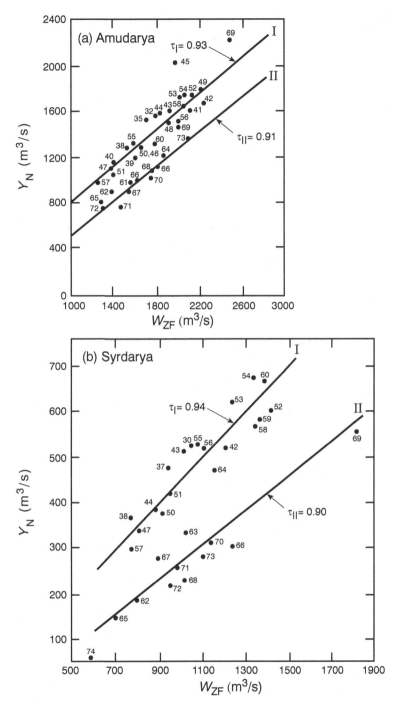

Figure 9.1 Annual flow curves $Y_N = f(W_{ZF})$ for the (a) Amudarya and (b) Syrdarya. I: Conventional natural period; II: Period of anthropogenic impact. Dots represent year of observation.

Table 9.5	Anthropogenic changes of river water availability in absolute km³/year) and relative (in % of conventional natural) flow values		

| Average annual flow at the top of deltas, 1931–60 (km³) | Decrease of flow due to anthropogenic factors | | |
	1961–70 (km³)	1971–80 (km³)	1981–90 (km³)
	Amudarya (at Samanbai [Chatly] village)		
47.0	10.9	25.7	38.1
(100%)	(23%)	(55%)	(81%)
	Syrdarya (at Kazalinsk town)		
15.0	6.24	11.1	12.9
(100%)	(40%)	(74%)	(86%)

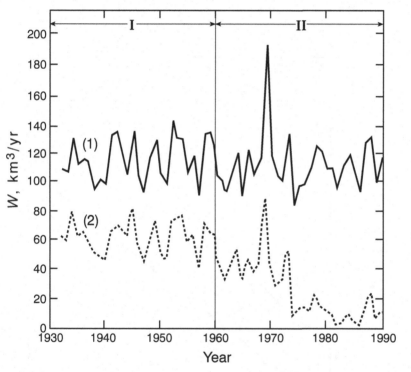

Figure 9.2 (1) Variation in volume of water resources and (2) total inflow to the upper reaches of the Amudarya and Syrdarya deltas. For other conventional signs, see Figure 9.1.

impacts of irrigation were relatively greater for the Syrdarya. The changes in water availability of the Amudarya and Syrdarya turned out to be 2 to 3 times lower in the upper reaches of the flow-use zone than in the lower reaches. The sharp decrease in the amount of river water reaching the deltas adversely affected environmental conditions in the deltas. When the hydrological regime was relatively undisturbed by human activities, flora and fauna developed under conditions of natural flooding with desert areas surrounding them. The presence of waterlogged lands, an exposed water surface resulting in high rates of evaporation, and water-loving vegetation needed to be taken into account when calculating the water balance of the Aral Sea. It was expected that, as irrigation was expanded and streamflow decreased, evaporation losses would also decrease because of the resultant drying out of the deltas. Planners hoped that this would compensate somewhat for the decrease of river inflow into the sea. Their assumptions proved to be unrealistic.

The impacts of human activities manifested themselves in two ways. The development of irrigation within the deltas was accompanied by the construction of a network of irrigation canals and an increase in water delivery as a result of it. The decrease in river inflow within the natural hydrographic network led to a decrease in the total area of open water surface and of reed communities and tugai vegetation. On the eve of the 1980s, the processes of human-induced desertification in the Amudarya and Syrdarya deltas appeared to have been reversed as a result of additional water supplies that resulted from artificial river-flow retention, e.g., the construction of dams, levees and temporary cofferdams, and the delivery of drainage flow directed to provide water to reed communities. The latter was aimed at improving conditions for fodder production, as well as for water storage.

Compared with 1982 (and based on aircraft and space survey data analyses) hydraulic and water management developments resulted in a 1.5 to 2.0 times increase in the area of open water and water-loving (hydrophylic) vegetation in the Amudarya delta. A modest increase in such conditions was also observed in parts of the Syrdarya delta (Sumarokova et al., 1991).

The diversion of streamflow for human-directed water supply of the deltas was so large that the river flow into the Aral Sea was appreciably reduced in those years and in some years, as the record shows, it was totally absent.

We have completed many years of research focused on assessing changes in the active surface of the Amudarya and Syrdarya deltas based on aircraft and remote sensing information. We also researched the water balance and river flow into the sea. Variations over time in flow losses in the deltas varied between 10 to 13 km³/year. The scope and dimension of the environmental disaster caused by human-induced flow changes and their accompanying impacts turned out to be major (Izrael et al., 1988).

A review of the history of the development of societal awareness of the environmental consequences of decreased flow of the Amudarya and Syrdarya can be very informative. As noted earlier, the decrease in the availability of river water had been the focus of attention of scientists and water-resource planners, long before the actual beginning of this process of environmental change. For example, studies in the 1920s and 1930s noted that the likely decrease of water availability in the lower reaches of the rivers would accompany irrigation development. The possibility of a decrease in flow, however, caused anxiety and concern primarily about the possible need for limitations on the scale of irrigation development in the region and *not* about the need to protect the environment in the region.

As early as the 1950s and 1960s, fishery specialists were among the first to draw attention to the adverse environmental impacts that would result from a decrease in river flow (Nikolsky and Fortunatov, 1959). They pointed out the real danger of the drying out of some parts of the deltas which served as spawning grounds for fish populations.

Long-term reclamation plans carried out between 1960 and 1980 were based on a belief in the need for the greatest possible expansion of irrigation land. Driven by such beliefs (or desires), government officials either ignored assessments that accounted for adverse river flow changes or used the calculations to 'prove' that there would be improved economic efficiency with a comprehensive expansion of irrigation farming.

Appropriate politically correct terminology was often used to support irrigation development. For example, government officials made an arbitrary distinction between 'productive' and 'unproductive' evaporation, with the latter referring to evaporation losses from natural unexploited lands. Statements made by Voeikov (1908) were widely cited to suggest that the Aral Sea was a useless evaporator of Amudarya and Syrdarya water. They argued that it would be much better if that water evaporated (i.e., was lost) from the 'useful' irrigated lands rather than from the exposed surface of the 'useless' Aral Sea.

It would be unfair to say that no consideration at all had been given to the need for environmental protection in the various proposed land reclamation projects. The way government officials intended to accomplish this, however, was quite peculiar, because sections of the plans treating environmental aspects, as a rule, followed the results of water-budget studies that generalized the requirements for water needs of all water users. The water requirements of ecosystems, therefore, had to be satisfied according to the so-called 'remaining' principle; that is, from the water resources that were left unused.

This situation lasted for a very long time and was not even influenced by the dramatic consequences of the severe water shortages in 1974–75. In this period, both the river-flow decrease and the Aral Sea water-level drop became very obvious, when serious difficulties arose in supplying sufficient amounts of water to existing irrigation activities. This period also witnessed the setting up for the first time of several commissions as the result of resolutions by Soviet authorities concerning the region's critical water situation and ways to improve it.

Expectations were raised at that time with a proposal to divert Siberian river water into Central Asia. Such diversions would have enabled the Soviet government to continue its proposed expansion of irrigation farming, while also stabilizing the Aral Sea level at a desired level. It is important to note that the possibility of Siberian diversions were extremely detrimental to an immediate resolution of the Aral Sea problem, because it delayed for a long

Table 9.6 Sequence of realization of environmental changes in the Aral Sea basin

Gradually accumulating environmental changes	Prevention of the problem initiated	Thresholds				
		Realization of changes	Realization of problem	Realization of crisis	Realization of need for action	Actions
1957–1973	1974–1975	1976–1980	1981–1985	1986	1987	1988–1991

time the search for ways to resolve the urgent need for a more efficient use of water resources and for an improvement of the entire water system in the Central Asia Republics (Sumarokova and Tsytsenko, 1991).

For the sake of fairness, it is important to note that neither scientists nor practitioners at that time realized or understood the magnitude of possible adverse effects of cumulative changes in regional environmental conditions, changes caused by tendencies toward excessive irrigation development. Awareness of the scope of the problem came much later, after the problem had passed through several thresholds (Table 9.6). The actual widespread recognition of the Aral crisis dates back only a decade to 1986–87. That recognition was the result of the mass media's coverage of the Aral Sea problem and of the setting up of a Soviet government commission on the Aral situation.

The commission's conclusions and recommendations formed the basis of the Resolution of the Government (1988) which dictated the approaches to resolving the Aral problem, as well as its methods and expected time frame for accomplishment.

'The Major Principles of the Scheme of Integrated Use and Conservation of the Water Resources in the Aral Sea Basin' were developed in 1989. For the first time during the entire period of designing water projects, problems associated with the stabilization of the region's environmental conditions received primary emphasis.

Immediately following the disintegration of the USSR in the early 1990s, the search for practical measures to deal with the Aral Sea crisis became the responsibility of the newly independent Central Asian Republics. The measures yet to be undertaken to address the Aral crisis notwithstanding, it had been the earlier studies on assessing changes in Amudarya and Syrdarya flow, until at least the mid-1990s, that served as the primary resources used to develop criteria for improving environmental conditions of this region in the future.

References

Dingelstedt, N., 1893: *Some Experience on Studying Irrigation in the Turkestan Region.* St. Petersburg: Gidrometeoizdat, 513 pp (in Russian).

Dubovikov, F.G., 1922: Impact of water withdrawal in the Syrdarya River upper reaches and on its lower reaches. *Khlopkovoye delo,* **9–10,** 67–87 (in Russian).

Dunin-Barkovsky, L.V., 1956: On the water balance of the irrigation area. *Izd. AN SSSR*, Ser. Geogr., **6**, 61–73 (in Russian).

GGI, 1984: *Recommendations on Determining Irrigation and Return Flow Volumes from Irrigated Lands.* Moscow: Soyuzvodproject, 25 pp (in Russian).

Golubev, V.S. and I.V. Zmeikova, 1991: Interannual change of evaporation conditions in the Aral Sea adjoining area. In: *Environmental Monitoring in the Aral Sea Basin*, eds. Yu. A. Izrael and Yu. A. Anochin. St. Petersburg: Gidrometeoizdat, 80–6 (in Russian).

Izrael, Yu. A., A. L. Yanshin, and P. A. Pola-zade, 1988: Present states and proposals on radical improvement of environmental and sanitary epidemiologic situation in the region of the Aral Sea and Amudarya and Syrdarya river lower reaches. *Meteoriologiya i Gidrologiya*, **9**, 5–22 (in Russian).

Kharchenko, S. I., 1975: *Hydrology of the Irrigated Lands.* Leningrad: Gidrometeoizdat, 372 pp (in Russian).

Levchenko, G. P., V.V. Sumarokova and K.V. Tsytsenko, 1990: Research into irretrievable water consumption and return water from the irrigated lands of the USSR arid areas. In: *Tr. Vsesoyuzn. Gidrol. S'yezda, Volume 4.* Leningrad: Gidrometeoizdat, 511–18 (in Russian).

Nikolsky, G.V. and M. A. Fortunatov, 1959: Irrigation development and the fishery in the Aral Sea: Materials on productive forces of Uzbekistan. *Vyp. 10, Tashkent, Izd. An UzSSR*, (in Russian).

Shiklomanov, I. A., 1979: *Anthropogenic Changes of River Water Availability.* Leningrad: Gidrometeoizdat, 302 pp (in Russian).

Shults, V. A., 1965: *Rivers of Central Asia.* Leningrad: Gidrometeoizdat, 691 pp (in Russian).

Sumarokova, V.V. and G. M. Degtyarev, 1985: Crops water consumption, irrigation and return water flow in the Amudarya River basin. *Meteorologiya i Gidrologiya*, **11**, 93–102 (in Russian).

Sumarokova, V.V. and K.V. Tsytsenko, 1978: Decrease of river runoff in the Aral Sea basin. *Soviet Hydrology, Selected Papers,* **17**(4), 323–8.

Sumarokova, V.V. and K.V. Tsytsenko, 1991: Reorganization of water management in Central Asia is necessary! *Chelovek i Stikhiya* (in Russian).

Sumarokova, V.V., L. P. Babkina and V. E. Kriventsova, 1991: Landscape parameterization of the Amudarya lower reaches based on interpretation of air-space information. In: *Environmental Monitoring in the Aral Sea Basin*, ed. Yu. A. Izrael and Yu. A. Anochin. St. Petersburg, 200–8 (in Russian).

Tsinzerling, V.V., 1927: *Irrigation on the Amudarya.* Moscow: Izd. UVKh Sredn. Asii, 808 pp (in Russian).

Tsytsenko, K.V. and V.V. Sumarokova (eds.), 1990: *Hydrologic Principles of Irrigation Development in the Chu and Talas River Basins.* Leningrad: Gidrometeoizdat, 334 pp (in Russian).

Tsytsenko, K.V. and O.G. Vonsovskaya, 1985: Modern and long-term assessment of irretrievable water consumption and return water in the Syrdarya River basin. *Tr. GGI,* **278**, 3–17 (in Russian).

Unusov, G. R., 1974: Dynamics of the rivers' runoff within the Aral Sea basin in the context of an irrigation development. *Trydy GGI*, **221**, 128–159.

Voeikov, A. I., 1908: Irrigation of the Trans-Caspian lowland from the point of view of geography and climatology. *Izv. Russk. Geogr. Obschestva*, **44**(3).

Volftsun, I. B. and V.V. Sumarokova, 1985: Dynamics of anthropogenic and natural
  flow losses of the Amudarya and Syrdarya in a multi-year period.
  Meteorologiya i Gidrologiya, **2**, 98–194 (in Russian).
Volftsun, I. B., V.V. Sumarokova, and K.V. Tsytsenko, 1988*a*: Water resources of the
  Aral Sea basin: conditions and prospects for use. In: *Tr. Vsesoyuzn. Gidrol.
  S'yezda, Volume 2*. Leningrad: Gidrometeoizdat, 197–204 (in Russian).
Volftsun, I. B., V.V. Sumarokova and K.V. Tsytsenko, 1988*b*: On changing the structure
  of river flow consumption in the irrigation zone of the Amudarya and
  Syrdarya river basins. *Vodnye resursy*, **3**, 117–23 (in Russian).
Yunusov, G. P., 1974: River flow dynamics of the Aral Sea and Balkhash Lake as
  affected by irrigation development. *Tr. GGI*, **221**, 128–59 (in Russian).

10   Fish population as an ecosystem component and economic object in the Aral Sea basin

ILIYA ZHOLDASOVA

Encompassing an area over 2.3 million km², the Aral Sea basin was one of the largest economic regions of the former USSR. In the early 1960s, a program of extensive development of irrigated farming was launched using water from the Amudarya and Syrdarya, the largest Central Asian rivers flowing into the Aral Sea. By the mid-1970s and early 1980s, the intense use of the downstream flow resulted in the drying out of the riverbed in different seasons at a distance of about two hundred meters from the sea. This distance increased steadily at a rate of about 1 meter per year. The irrigated area in the mid-1980s had reached 6.8 million ha compared with 2.9 million ha in 1959. In the former USSR, 95% of the raw cotton and 40% of the rice was produced there. However, these Soviet attempts to conquer nature in the Aral region were accompanied by adverse environmental impacts. Steadily accumulating adverse impacts and their eventual interactions resulted in an environmental crisis in the early 1980s that evolved into an environmental disaster a few years later. As a result, the situation in the Aral Sea ranks among the largest human-induced environmental disasters in the twentieth century, in terms of geographic scope and degree of severity.

Water resource problems in the Priaralye region (the region around the Aral Sea) are a result of the diminution of the natural water supply to the Amudarya and Syrdarya deltas and their vegetation. The reduced water supply generated soil degradation, solonchak (saline soils) development on the newly exposed dry seabed, and large-scale salt and dust transport to adjacent areas, which have destabilized terrestrial ecosystems. One could argue, however, that aquatic ecosystems suffer the greatest losses.

The diversion of large volumes of water from the Amudarya and Syrdarya rivers has affected the runoff regime downstream, as shown in Table 10.1. In the low flow period (1975–1988), runoff was further reduced to 7 to 10 km³/year and, in 1975, for the first time in recent history the Syrdarya failed to reach the sea. The Amudarya failed to do so in 1982. Zero river runoff was reported far downstream for several seasons and years until the late 1980s. The greatest water deficit was reported in this part of the Aral region in the same period.

Ecological conditions were further aggravated by the implementation of government programs which used mineral fertilizers and pesticides in

| | **Table 10.1** | Annual runoff in the upstream and downstream of the Amudarya and Syrdarya (km³) (1959–1995) | | | |

	Amudarya		Syrdarya		Inflow in the
Years	Kerki	Kyzljar	Syrdarya	Kazalinsk	Aral Sea
1959	70.0	46.6	35.3	18.3	64.9
1960	62.7	43.0	35.8	21.1	64.1
1961	55.3	30.9	25.4	13.4	44.3
1962	52.5	27.6	25.2	5.8	33.4
1963	51.2	33.1	31.3	10.6	43.7
1964	62.8	38.3	34.5	15.0	53.3
1965	51.4	25.5	25.3	4.7	30.2
1966	69.6	33.1	36.1	9.6	42.7
1967	61.6	27.0	32.1	8.7	35.7
1968	62.6	28.0	32.2	7.3	35.3
1969	98.9	55.5	55.4	17.5	73.0
1970	61.0	28.0	37.0	9.8	37.8
1971	50.2	15.8	32.1	8.2	24.0
1972	52.1	13.2	32.1	7.0	20.2
1973	76.4	31.2	35.9	8.9	40.1
1974	41.7	6.3	22.2	1.9	8.2
1975	53.2	10.6	21.4	0.6	11.2
1976	56.8	11.1	23.8	0.6	11.7
1977	55.1	9.0	26.2	0.5	9.5
1978	65.8	21.3	31.9	0.8	22.1
1979	60.9	11.1	36.1	3.2	14.3
1980	61.9	8.6	30.3	2.5	11.1
1981	59.9	6.3	31.0	2.5	8.8
1982	47.6	0.5	24.9	1.7	2.2
1983	55.3	2.3	25.6	0.9	3.2
1984	62.9	8.0	26.3	0.6	8.6
1985	63.8	2.4	29.3	0.7	3.1
1986	40.8	0.4		0.5	0.9
1987	63.2	8.0	37.2	1.5	9.5
1988	72.0	16.3	39.0	6.9	23.2
1989	48.8	1.7	28.4[a]	5.3	6.0
1990	41.4[a]	6.1		1.6	7.7
1991		9.4		4.0	13.4
1992		21.2		4.6	33.4
1993		15.1		7.9[b]	26.7
1994		18.4		8.93	
1995		0.55[c]		2.32[c]	

Notes:
[a] No data available on the hydroregime observation stations.
[b] From 1993 on Karateren' hydroregime observation station.
[c] Runoff as of 1 July.

Fish population changes in the Aral Sea 205

Irrigation region	Volume of drainage flow (km³/year)	Average salinity (g/l)	Main recipients of drainage
Pyandzh	1.35	1.0	Kzylsu, Pyandzh
Vakhsh	2.67	1.8	Vakhsh
Kafirnigan	0.70	0.7	Kafirnigan
Surkhan-Sherabad	0.95	2.4	Surkhandarya, Amudarya
Turkmen	2.31	3.5	Amudarya
Tuyamuyun	4.71	4.0	Sarykamysh Lake
Takhiatash	2.35	4.1	Depressions
Karshi	1.22	7.7	Amudarya
Samarkand	0.75	1.0	Zaravshan
Navoi	0.49	2.3	Lakes
Bukhara	0.98	4.2	Lakes
Murgab	1.20	10.5	Depressions
Tedjen	0.44	14.2	Depressions
Total	20.12		

Table 10.2 Drainage flow of the Amudarya basin irrigation regions

Source: Chembarisov (1989).

agricultural production. Over time, this resulted in widespread pollution of the environment. Pesticides with high impacts on humans were used intensively (e.g., B-58, metaphos, cotoran, butyphos, hexachloran, Lindane, DDT). The result has been that almost all surface waters have become polluted as a result of drainage water discharge into the rivers and lakes from the irrigated agricultural areas. The volume of drainage water increased with the extension of irrigated farming and, by the end of the 1980s, more than 30 km³/year of these contaminated waters were accumulated in the Aral basin. Of this amount, almost 20 km³/year have been discharged back into the channels of the Amudarya and Syrdarya and their tributaries, and into irrigation canals. The remaining portion went into lakes and natural depressions along the side of the Aral Sea (Chembarisov, 1989) (Table 10.2).

The hydrochemical regimes of rivers and lakes were strongly affected by the increasing salinity levels (from 1 to 14.2 g/l) and by drainage water that had been polluted by pesticides, mineral fertilizers and soil salts. The hydrochemical map by Alekin (1950) classified most Central Asian rivers as hydrocarbonate with an average salt content ranging from 200 to 500 mg/l. The annual salt content of the Amudarya in the 1960s was 540 mg/l (Rogov, 1957). In the 1980s, the mean annual salt content downstream reached 600–1500 mg/l or more. The mean annual salt content at Nukus (215 km from the Amudarya estuary) measured 1525.5 mg/l in 1989 and 946.8 in 1990. The salt content in the Amudarya's lower reaches ranged from 900 to 4000 mg/l in 1989 and from 975 to 1870 mg/l in 1990 (*Annual Review of Surface Water Quality,* 1991). This increasing salt content was accompanied by excessive pollution of both rivers by petroleum products, phenols, heavy metals and organic compounds in addition to those mentioned earlier.

Recently, there has been a trend to reduce pesticide pollution of river water and lakes associated with the river. However, especially hazardous pesticides (e.g., hexachloran and DDT) are found regularly, although according to reports from Uzbekistan's Hydrometeorology Service, no DDT was found in Amudarya water in 1989–90. These substances exist in the rivers, and they have been found in the organs and tissue of fish living in the channel part of the Tuyamuyun Reservoir on the Amudarya (the water line Tuyamuyun-Nukus-Chimbai-Takhtakupyr begins here). This could explain the large-scale loss of silver carp (*Hypopthal michthys molitrix*) which took place in 1990–91. DDT and its metabolites have been found in the Amudarya in the amount of 0.16–5.0 mg/l; hexachlorocyclohexane, 0.015–0.616 mg/l; in the bottom sediments 5.35 to 100 mg/kg of DDT and 0.002–0.46 mg/kg of hexachlorocyclohexane were found (Zholdasova *et al.*, 1991*a*; Lyubimova *et al.*, 1992).

DDT and hexachlorocyclohexane were found everywhere In the lower reaches of the Syrdarya (Nilov *et al.*, 1994). In autumn 1993, polluted waters were found in the three river sections examined in the Kazalinsk region. Pollution of bottom sediments was also discovered. Near the city of Nukus, the pollution in 1989 and 1990 of the Amudarya by phenols was 4–10 times the maximum permissible concentration; oil products were 3 times maximum permissible concentration in 1989; copper compounds, about 10 times the maximum permissible concentration; and zinc, 5 times the maximum permissible concentration (1990).

Additional adverse impacts occurred when the lower downstream water supply triggered desertification processes and the drying up of lake/wetland complexes. The data of several researchers (e.g., Rogov *et al.*, 1968; Nikitin and Bondar, 1975) indicate that the surface area covered by water in the Amudarya delta was 80–100 000 ha and that floods brought 6.4 km³/year to the delta. With the decreased streamflow in the lower reaches, flooding ended in 1974 and water levels in lakes and bogs rapidly decreased. According to Tadjitdinov and Butov (1972), the flooded delta area had been reduced by twofold. By 1975, 25 large lakes and 62 minor ones had disappeared.

Between 1961 and the present, the Aral Sea has been transformed from a brackish-water basin to a saline water body. The present salt content of the sea is as high as 40‰ in some areas. Sea level has dropped over 16 m, the water volume has been reduced to about 38% of its pre-World War II level, and the sea's surface area has declined by more than half (31 000–32 000 km²). The coastline had retreated by more than 100 km. The largest freshwater bays of the Aral Sea have dried up (Muynak, Adzhibai, Abbas, Jyltyrbas); these were the major spawning areas of the Aral commercial fish species.

In 1987 the Aral Sea divided into the Large and Small Seas, with the Large Sea fed by the Amudarya and the Small Sea by the Syrdarya. With continued sea level drop since 1990, the chain of islands from the Muynak peninsula to

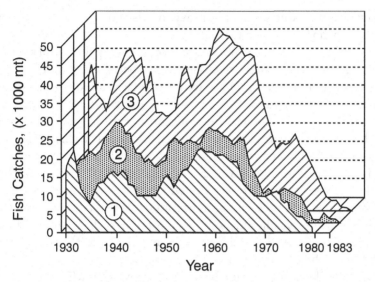

Figure 10.1 Dynamics of commercial fish catches in the Aral Sea. (1) North region; (2) South region; (3) total fish catch. Fish catch in the northern part of the Aral Sea was terminated in 1979, and in 1984 in the south.

Lazarev, Belinsgauzen, Komsomolsky and Vozrozhdenie islands, a distance of 150 km, linked up, thereby dividing the larger sea into the Eastern and Western basins.

A lowered river inflow to the Aral Sea, the increasing salt content, and the falling sea level have each contributed to adverse effects on the fish stocks. In 1980 commercial fishing was closed in the northern part of the Aral Sea, and in 1983 it was closed in the southern region. The southern fishing grounds accounted for 60% of the total catch and the northern area contributed the remaining 40% (Figure 10.1).

Until the mid-1960s the Aral Sea fish catch, valued for its high quality, accounted for up to 7% of the total catch of the former USSR's inland waters and for about 13% of its valuable fish species. The catch was composed of large fish: aral barbel, bream, aral carp, pike-asp, etc., which made up 80–85% of the total catch. As of 1960, the Aral Sea portion of the Uzbekistan fish catch was 98%, and 25% of Kazakstan's catch (Volodkin, 1964). By the mid-1980s, however, the native species of the Aral were extinct, except for a euryhaline (salt-resistant) species, the aral stickleback *Pungitus platigaster aralensis,* which was found in the freshwater inflow areas in the northern and southern Aral Sea (Lim and Ermakhanov, 1986; Zholdasova *et al.*, 1992a,b; Zholdasova, 1995).

Aral ichthyofauna was poor in species diversity. In terms of plant and animal biodiversity, the Aral Sea ranked last among the brackish seas of the former USSR (Karpevich, 1975). For example, it was inhabited by 20 species and subspecies of 7 fish families compared with 130 species of 19 families in the Caspian Sea (Nikolsky, 1938). The landings of about 12 commercial fish species, but primarily carp, amounted to 50 000 metric tons per year. Carp and perch species were the most numerous. Five other families were represented by a single species (Figure 10.2). The extinction of native species, because of an increase in salt content of the Aral Sea, was inevitable because the ichthyofauna comprised freshwater and brackish water species.

Confronted by disastrous changes in the characteristics of the Aral Sea during the past two decades, scientists and fishery experts were constantly searching for ways to preserve its importance for fisheries (Lim and Markova, 1981; Lim and Ermakhanov, 1986).

Present-day Aral ichthyofauna is represented by four acclimatized species. Two were introduced during planned acclimatization, and two were introduced accidentally. The acclimatization program was launched after 1927 and was aimed at enriching the local fauna and at increasing the commercial catch (Karpevich, 1975). Subsequently, it was dominated by introducing euryhaline ichthyofauna. Thirty hydrobiont (fish and invertebrate) species have been introduced into the Aral since the late 1920s (Andreyev, 1990).

Starred sturgeon (*Acipensez stellatus*) was among the first species to be introduced. The 1927–32 campaign to introduce this species failed, but attempts were repeated in 1948 and 1962 when a total of 12 million larvae and fry were introduced into the Amudarya and Syrdarya deltas. Simultaneously, almost 3 million bastard sturgeon were released. No details of their fate were reported in the scientific literature (Bykov, 1961; Osmanov, 1971). Until the late 1970s, however, adult sturgeon of up to 1 m in length and 4–6 kg in weight were often caught in the river deltas. The following species were also introduced: *Alosa caspia* (eggs and larvae) in 1929–32; baltic herring *Clupea harengus membras* (eggs) in 1954–59; two species of grey mullet, *Mugil auratus* and *M. saliens* (fingerlings) in 1954–56; and grass carp *Ctenopharyngodon idella* in 1960–61 (Karpevich, 1975).

Acclimatization programs often lacked a scientific basis and failed to produce the hoped-for results. Moreover, the rates of the Aral Sea level decline and of the salt-content increase surpassed all scientific projections. The acclimatization process was further complicated by accidental acclimatization. With the introduction of the grey mullet, for example, other species, such as the following, also appeared in the Aral Sea: silverside (*Atherina boyeri caspia*); pipefish (*Syngnatus nigrolinatus caspius*) and seven goby species (*Pomatoschistus caucasicus, Neogobius fluviatilis pallasi, N. kesslerigozlap, N. melanostomus, N. bathybis, N. syrman eurystomus* and *Proterorhinus marmoratus*; Baimov, 1964; Karpevich, 1975). Besides fish, a crab species (*Rhithropanopeus harrisi tridentata*) was accidentally imported from the Sea of Azov, as were shrimp (*Leander squilla*) from the Caspian Sea. Also, algae species (*Calanipeda aquateduclis, Nereis diversicolor*) and in the 1950s and 1960s, mollusk (*Abra ovata*) were introduced to the sea as fish fodder (Zholdasova *et al.*, 1991c, 1992a,b; 1995; Andreyev, 1990; Andreyev *et al.*, 1990).

Juveniles of the Russian sturgeon (*A. guldenstadti*) were released in 1978 but

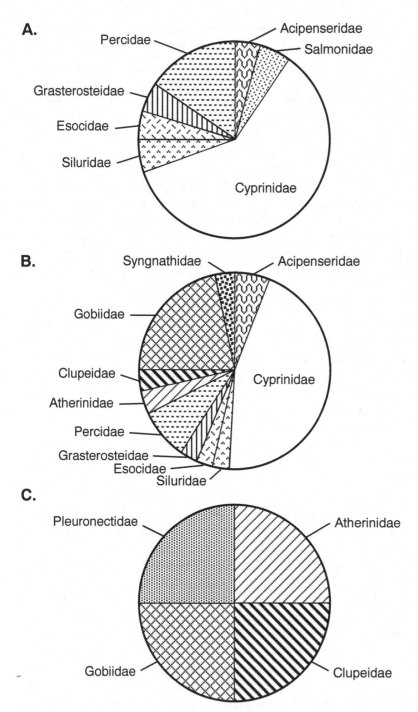

Figure 10.2 Anthropogenic transformation of Aral Sea ichthyofauna. A: Native ichthyofauna composition (prior to the 1930s). B: Ichthyofauna composition in 1950–70. C: Ichthyofauna composition in the 1990s.

failed to acclimatize. The last acclimatization project in the Aral Sea introduced glossa in 1979–87, when about 15 000 fingerlings of different ages were released in the sea. During a trial catch of glossa in 1991, about 112 metric tons were caught in the northwestern part of the sea (*Sovetskaya Karakalpakiya*, 1 February 1992).

A total of 18 fish species from 8 families were ultimately introduced in the Aral Sea (on purpose and accidentally). All of them except the bastard sturgeon (i.e., 95% of the total) are new to this sea. Of the nine introduced species only two acclimatized successfully (baltic herring and glossa) in contrast to the nine accidental species. They had been brought to the Aral in the mid-1950s, together with grey mullet from the southeastern Caspian Sea.

The eurybenthic species proliferated for a short period of time but then, because of harsh living conditions, gradually became extinct. Only the silverside and sand goby have survived until now. Since the early 1990s the Aral ichthyofauna has been represented by four acclimatized species: baltic herring, glossa, silverside, and sand goby. Under conditions of increasing salinity of the sea, the continued existence of these species in the Aral Sea is clearly in doubt. Their ability to survive is also related to the changing temperature, food supply, and other environmental factors. Two cases of large-scale baltic herring kills have been reported in the western basin of the large sea (Zholdasova, 1995; Zholdasova *et al.*, 1995).

Marked environmental changes have occurred as well in the ichthyofauna in the river and lake systems around the Aral Sea. Since the 1960s, 19 new species have appeared in the Syrdarya and the Amudarya (Aliev *et al.*, 1963; Mitrofanov *et al.*, 1986–1992; Zholdasova *et al.*, 1989, 1991a,b,c; 1995; Pavlovskaya and Zholdasova, 1991; Zholdasova and Pavlovskaya, 1995). This was accompanied by the rapid extinction of the native species. The extinction of river-fish species has been more serious than that of marine species. The Aral Sea was never inhabited by fish species living in the open sea or in benthic zones alone. Aral Sea fauna comprises fauna of the sea, Amudarya, and Syrdarya faunas. Most fish species (95%) were eurybenthic and euryhaline whose life cycles were directly or indirectly dependent on the rivers. Most of them spawned either in the rivers or in the estuaries and lakes interconnected by rivers. Other species pastured in the sea and rivers (e.g., sabrefish, pike perch). While in the sea, the goldfish and yazeh kept to freshwater estuary areas.

For reasons noted earlier, native marine species could survive in the lake and river systems. Yet, regulating the runoff of the Amudarya and the Syrdarya, erecting temporary earth dams and, since the 1970s, dams and reservoirs, river pollution from drainage, and water diversions for irrigation have dramatically affected their hydrological and hydrochemical regime. Such changes in the downstream rivers and lakes have resulted in the extinction of river species, as well as of the migrant and semi-migrant species.

The Red Books (listings of endangered species) of Kazakstan (1978),

Uzbekistan (1983), Kyrgystan (1985), Turkmenistan (1985), Tajikistan (1988), and of the former USSR (1988) included 8 Aral species: aral trout (*Salmo trutta aralensis*); 3 species of the *Pseudoscaphirhynchus* genus; bastard sturgeon (*Acipenser nudiventris*); pike-asp (*Aspiolucius esocinus*); aral barbel (*Barbus brachycephalus*); and Turkestan young sheat-fish (*Glyptostermur reticulatum*).

Extinction of the aral trout, a rare migrant species, is not related to the Aral Sea crisis, but was possibly the result of their small numbers. However, a few of them were caught annually in the Amudarya's lower reaches. The fish's length of up to 1 m and its weight of 10–13.5 kg made it a favorite catch of amateur fishermen. The aral trout spawned upstream reaching the Vaksh and Pyandj tributaries (Nikolsky, 1938: Maksunov, 1968). Zero catch has been reported in downstream Amudarya since the 1940s. Maksunov (1968) reported the existence of extremely large trout species but those fish were possibly migrant salmon species. Reports of resident trout species in the upstream Amudarya show optimistic prospects for restoring the salmon migrant species. To preserve and multiply the species it would be useful to stage an experiment on acclimatizing the Amudarya trout in the off-channels of the Tuyamuyun reservoir (i.e., Sultansanjar and Kaparas).

Three amudar shovelnose varieties are typical river species. This endemic species *Pseudoscaphirhynchus* was the first victim of the Aral sea crisis. The big and little shovelnose (*P. kaufmanni* and *P. hermanni*) are the Amudarya endemic species, and the *P. fedshenkoi* is the Syrdarya endemic species. They are close relatives of the American shovelnose (*Scaphirhynchus*) found in the Mississippi basin.

The shovelnose is a small-size species of no commercial value. The little amudar and the syrdar shovelnose are 21–27 cm long (Nikolsky, 1938; Tleuov and Sagitov, 1973). The big amudar shovelnose is the largest species, with a length of up to 58 cm (without a tail thread) and a weight of up to 760 g (Tleuov and Sagitov, 1973). Catches weighing up to 2 kg had been reported in the past (Berg, 1948–1949).

The little amudar shovelnose was a rare species found in the Amudarya midstream from Termez to the estuary. Only three cases of its catch had been reported from 1974 to 1982. The syrdar shovelnose inhabited the Syrdarya midstream from Balykchi to the estuary. The last reliable reports on its catches were in the Novo-Chily irrigation system and were dated 1952–53 (Dairbayev, 1958; Shilin, 1984; Dukravets, 1986). By now, both species are possibly extinct.

The big amudar shovelnose is also an endangered species. Compared with its relatives, it was more abundant and populated a larger area. It had been frequently found from the Pyandj and Vaksh to downstream Amudarya and the estuary (Nikolsky, 1938; Maksunov, 1968). It was also common in the piedmont and plains sections (Kerki-Chardjou-Ildjik) of the river and was commercially caught in the 1930s. The number of shovelnose in the downstream Amudarya was significant. The 1930s' brood made up 26% of the total fish progeny as of the mid 1960s (Tleuov *et al.*, 1967). It was often found in shallow irrigation ditches. As a river species, it was possibly salt-tolerant. One shovel-nose was reportedly caught in the pre-estuary area where water had a salt content of 8.5‰ (Gosteeva, 1953). Today the species is found only in the middle reach of the Amudarya. Studies carried out in 1989 and 1991 (after a 15-year interval) found the shovelnose in its traditional habitat from Kerki to Chardjou. It was not found then, however, in the Tuyamuyun reservoir (Zholdasova et al., 1991a). No reports of shovelnose catches in the downstream Amudarya have been made for the last 80 years.

Shovelnose inhabit the turbid flowing water. Common habitats are shallow waterways with sandy or sandy-pebble ground (Nikolsky, 1938; Tleuov and Sagitov, 1973). Between 1989 and 1991, the shovelnose was found in such an environment near shallow water (at a depth of 1–1.5 m) and near turbid backwater edges.

The maximum size of shovelnose reported in the literature (Berg, 1948–1949) is 75 cm and its maximum weight reached 2 kg. The mean fish weight in commercial catches (1965–66) was 241 g and the body length 37 cm. The length of the fish caught by the author in 1989 and 1991 ranged from 9.3 to 38 cm (23.6 cm on average) and weighed from 3.2 to 270 g (100.2 g on average). The range of age was 1–6 years. The largest was 69.4 cm long with the tail thread (33.5 cm without the tail thread) and weighed 250 g.

The age structure of the Amudarya shovelnose population is 8 groups shorter today, as compared with the 1960s (Zholdasova et al., 1990). At that time the fish age ranged from 1 to 14+ years with the 3–6 year olds dominant (Tleuov and Sagitov, 1973). The present-day fish population is primarily 1 to 6 years old with 3-year-olds (36.8%) and 4-year-olds (31.6%) dominating. The linear growth rate has slowed down, as compared with the 1960s, possibly because of a decrease in the amount of fish in the shovelnose diet (based on nutrition data analysis).

The shovelnose is a species with a marked predatory bias (Nikolsky, 1938; Berg, 1948–1949; Tleuov and Sagitov, 1973). Its diet in midstream Amudarya in the 1960s was 64.5% fish of five species; fingerlings of the aral barbel, pike-asp, bastard sturgeon, ostroluschka (*Capoetobrama kuschakewitschi*) and amudar stone loach (*Noemachelius oxianus*). Under present conditions, its diet is quite opportunistic including, apart from fish (36.6% by mass), the chiro-nomidae larvae of 15 species (30.5%). Fish species are represented by ostro-luschka and acclimatized goby (*Rhinogobius*).

Big amudar shovelnose were reported in the Ordabai duct in the Amudarya, during ichthyofauna studies in 1993. Juvenile species' length was 20 cm (without a tail thread) and they weighed about 50 g. A single shovelnose of similar size was reportedly caught near Nukus in the Kattyagar irrigation channel gate to the Amudarya. The shovelnose appearance in its historic habitat in downstream Amudarya is a result of recent (early 1990s) significant increase in volume and stable river flow into the downstream area; the average runoff volume of the Amudarya in the 1991–93 period exceeded 3.7

times the average volume of the previous decade. In addition, the runoff was turbid and discharged into the Tuyamuyun reservoir directly and not by way of off-stream storage reservoirs.

The big amudar shovelnose inhabits only Amudarya water that is flowing and turbid. There are no reports of it having been found in lakes. According to Nikolsky (1938), it dies quickly in still water. We can assume then that the shovelnose expanded its range from the midstream to the downstream Amudarya in the low-flow years of the 1970s and 1980s during increased transparency and a decrease in river flooding.

As noted earlier, numerous drainage collectors discharge their water into the Amudarya. There was a connection between shovelnose habitats and collector-water inflow into the river in 1989–90, possibly because of the organisms transported by collector water. In addition, some collectors cross lakes and lake systems, possibly resulting in a larger biodiversity of the Amudarya benthofauna in 1989 as compared with 1974. The species composition increased from 58 species of 10 systemic groups to 83 species of 17 systemic groups. Species variety of Chiromonidae, as well as may flies and mollusks, increased, and new species appeared: Mysidae and shrimp. Fifteen Chiromonidae species and forms were found in the food intake of the big amudar shovelnose in May and October 1989 (Zholdasova *et al.*, 1990).

Changes in river hydrobionts, including fish populations, have negatively affected the shovelnose population. Its age range was reduced by 8 groups because of its shorter life span, its extremely low reproduction level, and its slow linear growth. These were indications of the precarious state of this endemic species. Its extinction will mean the end of the endemic genus *Pseudoscaphirhynchus.* Two other representatives of this genus are already possibly extinct. All of these changes mean that urgent measures need to be taken to protect the sturgeon species as a whole.

Another endangered sturgeon species in the Aral Sea is the bastard sturgeon (*Acipenser nudiventris*). This large-size migrant species populates the Black Sea as well as the Caspian and Aral Seas, and it was introduced into Lake Balkhash in 1933–34. It was the only sturgeon species inhabiting the Aral Sea. Its length had reached 160 cm and its weight as high as 45 kg (Tleuov and Sagitov, 1973).

Prior to regulating the streamflow of the Amudarya and Syrdarya, the bastard sturgeon could be found throughout the Aral Sea. Its major spawning grounds were situated along the Syrdarya. The bastard sturgeon moved upstream more than 1800 km (Nikolsky, 1938; Dukravets, 1986). Its vast spawning area in the Amudarya stretched from Cape Kyzljar (103 km from the estuary) to Faizabadkal (over 1500 km from the estuary). Major spawning sites were located in the mainstream between Chardjou and Turtukul, accounting for 50–60% of the reproduction of the fingerlings (Tleuov and Sagitov, 1973). It also spawned successfully at Nukus (Tleuov, 1981). Its spawning age varied from 7 to 30 years with a predominance (i.e., 82%) of 12–21 year old species in the Amudarya.

The bastard sturgeon was more closely connected with the Aral sea than was the big amudar shovelnose. Upon spawning, it returned to the sea to pasture until the next season. It preyed primarily on mollusks (88%) *Hipanis minima, Dreissena polimorpha, and D. caspia*. Juvenile fish were rarely found in bastard sturgeon stomachs. With the appearance of the silverside and goby in the Aral as a result of acclimatization projects, the bastard sturgeon became a predator, usually in its second year. By the age of three, fish dominated its diet (about 60%) (Tleuov, 1981).

Although bastard sturgeon is a rare species of the *Acipenser* genus, it is a valuable commercial species. Its commercial catch in the Aral Sea in the 1928–35 period totalled 300–400 metric tons per year. With the development of the *Nitshia sturiones* and with large-scale fish kills in 1936–37, its number decreased, and a ban was imposed on bastard sturgeon catches in the Aral in 1940 (Tleuov and Sagitov, 1973; Dukravets, 1986). A trial catch to evaluate the fish population was carried out in the 1960s. The commercial catch in the 1960s to 1970s was about 700–9300 kg per year. The 1970 catch was the last statistically evaluated commercial catch.

In the following years, the bastard sturgeon population was destroyed because of its economic value and because it was an easy catch along its migration routes at dams and dikes. No bastard sturgeon have been reported in rivers downstream since the 1970s. No progeny were reported in the Amudarya midstream in 1989 and no bastard sturgeon have been found in the Amudarya estuary in the last 7–8 years. M. B. Mamedkuliev, a fishery inspector from Chardjou, reported two incidents of bastard sturgeon catches – a 4 kg fish caught in December 1990 in Ildjik and a 2 kg fish caught in March 1991 35 km above Chardjou. Such rare occurrences testify to the high possibility of species extinction of this Aral sea population.

The environmental setting of the Aral bastard sturgeon allows one to assume that, in the future, the Tuyamuyun reservoir could become a site for its reintroduction into the Amudarya. Since it is situated halfway along the bastard sturgeon's historical migratory routes and since it spawned in waterways above Tuyamuyun, its natural reproduction in the river is possible. Its spawning migration will be shortened. It is necessary, of course, to examine Tuyamuyun conditions as applied to the problem of reacclimatization of bastard sturgeon, along with a number of other Aral natives, generally the commercial species. Considering that bastard sturgeon is characterized by its late maturation (8–10 years or more), the fisheries research in Tuyamuyun should be pushed forward to the earliest possible time in order to begin the formation of a population.

The problem with sturgeon conservation in the Amudarya and Syrdarya should be considered as a common problem of the maintenence of biodiversity in the Aral Sea basin. This task will require the combined efforts of Uzbek, Turkmen, and Kazak practitioners and scientists involved with fish hatcheries. The following points are priority measures for the conservation and maintenance of the population under natural conditions:

- Inclusion of all three species of sturgeon of the *Pseudoscaphirhynchus* genus listed in the Red Book of the IUCN because of the critical reduction of their numbers and because of the adverse ecological situation in the Aral basin.
- Establishment of low-temperature genetic banks of fauna of the Aral Sea basin at the Institute of Bioecology (Karakalpak Division of the Academy of Sciences of the Republic of Uzbekistan) for the conservation and storage of genetic material of all endangered fauna.
- Inclusion of typical habitats of *Pseudoscaphirhynchus* on the list of protected environments, the prohibition of fishing in these areas, and their inclusion into the conservation zone of the reserves located along the Amudarya and Syrdarya (e.g., Aral-Paigambar, Kyzylkumsky, Badai-tugai).
- Investigation of the possibility for the reacclimatization of the Aral's bastard sturgeon in the middle reaches of the Amudarya using the Tuyamuyun Reservoir. The balkhash shoal of bastard sturgeon, duly formed by ways of acclimatization of the Aral's bastard sturgeon, should be used as an initial genetic fund (source) for colonization.

Pike-asp (*Aspiolicus esocinus*) is yet another rare endemic Aral species of commercial value. It is commonly found in the midstream sections of the Amudarya, Syrdarya, Vaksh, Pyandj, Kyzylsu, Chirchik and Naryn rivers, and also in the Farhad, Kairakkum and Toktogul reservoirs (Nikolsky, 1938; Maksunov, 1968; Sadykov, 1981).

Pike-asp is a rheophylic (flowing-water-loving) predator species which was a popular as well as common commercial species in the midstream of the Amudarya and Syrdarya in the 1930s. At that time, only fingerlings were reported in the Amudarya delta (Nikolsky, 1938). Present-day Amudarya and Syrdarya populations have markedly decreased in number. Its range has also decreased: no pike-asp have been found in the Amudarya below Chardjou (Zholdasova *et al.*, 1990). It had been found in the Tuyamuyun Reservoir (Sagitov, 1983). No pike-asp have been found in the commercial catch in Kazakstan's part of the Syrdarya.

In the 1930s pike-asp length was 30 to 40 cm and weighed about 1 kg on average (Nikolsky, 1938). Individual species of up to 5 kg were found in the Farhad Reservoir (Maksunov, 1968). The largest pike-asp size found in the Amudarya in the 1960s was 54 cm and weighed 1.2 kg (Urazbayev *et al.*, 1971; Sagitov, 1983). The author caught 10 pike-asps at the river section near the Karakum Canal intake at Chardjou (they were primarily juveniles of various lengths (19.9 to 44.2 cm long) and weights (70 to 860 g, average mg 431 g). Their age was 1+ years (1 fish) 3+ (6 fish) and 4+ (3 fish).

Like other predator species, pike-asp grow rapidly as follows: $l(1) = 11.7$ cm; $l(2) = 17.5$ cm; $l(3) = 30.2$ cm; $l(4) = 41.7$ cm. The linear growth of 3- to 4-year-old fish had improved, when compared with the 1930s and the 1960s because of the decreased competition from other predator species (e.g., shovelnose, bastard sturgeon).

Fish is the main diet component (75% by mass) of the pike-asp (Nikolsky, 1938; Urazbayev *et al.*, 1971). Its main prey, ostroluschka, is abundant. Pike-asp reach maturity at 6 to 7 years (a body length of 45–50 cm). They spawn in February-March at an air temperature of 5–10° C. Eggs are spawned in fast currents (Maksunov, 1968). Rates of natural reproduction have remained low, possibly because of its scarcity.

Catching pike-asp has been banned since 1988, but it is often caught in the nets of poachers seeking to catch relatively valuable species; they die quickly in the nets. To preserve pike-asp it is necessary to protect its habitats in the Amudarya and Syrdarya waterways and to populate those reservoirs which have favorable conditions for its reproduction and survival (Sadykov, 1981; Saifullayev, 1983). When introduced into the Toktogul Reservoir while it was being filled, pike-asp became somewhat abundant and in 1978 100 kg were caught.

Turkestan barbel (*Barbus brachycephalus*) is an extremely valuable commercial species that was at one time common in the Aral basin. It pastured in the Aral Sea and entered the Amudarya and Syrdarya for spawning, covering distances of 100 km or more. In 1930–31 the turkestan barbel was introduced in the Ili River near Kapchagai. Turkestan barbel is a large-size fish (with a length of 1 m or more and a weight of up to 20 kg). It makes up about 2–6% of the total commercial catch (Dukravets *et al.*, 1988).

Turkestan barbel males reach maturity at 6–7 years and females at 7–8 years. They enter the Amudarya from April–May until October–November (10–15 days later in the Syrdarya). Mass migrations of fish of the II-III maturity stage usually occurred in summertime; hibernating in winter, they spawned in springtime. Some fish, however, have entered the river in April–May with developed reproductive organs. It was classified into spring and winter species, depending on the migration period and reproductive behavior.

Turkestan barbel spawns in rivers and its semi-pelagic eggs develop in deep water. Its spawning period lasts 90 days. After spawning, the fish returns to the sea and fingerlings develop and reach the sea in their first year. Some juveniles stay in the river for 2–5 years. Its lowest spawning boundary on the Amudarya was at Nukus, and on the Syrdarya at Kyzyl-Orda.

Earth dam construction on the Amudarya in 1966 changed river runoff as did the regulation of runoff in 1974. The construction of the Takhiatash dam (1974), the Tuyamuyun dam (1984), and the Kazalin and Chardarin dams (1989) on the Syrdarya blocked turkestan barbel migration routes and its natural reproduction, therefore resulting in near-extinction of species. A rapid reduction of this major commercial species was reported in the mid-1970s. In 1976 a quota was introduced to preserve the species, but it proved ineffective. Since 1983, the turkestan barbel has been listed in Uzbekistan's Red Book of endangered species.

The species has survived in the Amudarya as a river variety spawning in the upstream piedmont river section (Zholdasova and Pavlovskaya, 1991). The Balkash-Ili population is estimated at about 3000. With the regulation of the downstream stretches of the Ili River, the turkestan barbel was reported to have increased in number as a result of cutting off its migration routes and of eliminating its competitive species – the marinka (Bashunov and Tsoi, 1983, 1986).

The number of turkestan barbel in the Amudarya increased slightly in the 1991–93 period. This increase was a possible result of the steady inflow of water during high water years in the Takiatash dam and regular discharges from it. Such a flow facilitated the accumulation of fingerlings in the Mezhdurechensk Reservoir bordering the Amudarya. Nevertheless, the prospects for the stabilization of turkestan barbel are hardly favorable in view of the Amudarya's unstable flow levels and the increase in water diversions in the mid- and upstream sections of the river. Protecting this valuable species and restoring its population are possible by cultivating it in various water reservoirs; ones with the best prospect for success are the Tuyamuyun, Sultansanjar, Kaparas and Koshbulak reservoirs.

In addition to the endangered species listed in the Red Books, there are several native species with low population numbers in need of protection as well. Among them are the aral white-eye (*Abzamis sapa aralensis*), aral shemaya (*Chalcalburnus chalcoides aralensis*), and turkestan ide (*Leuciscus idus*). Their situations have been aggravated by the deterioration of the regional water quality as well as quantity.

The decrease in species variety in the Aral Sea has been somewhat offset by the sea's fauna being continuously replenished since the 1930s by the government's introduction of new species. Since 1950, sixteen species have been introduced in the Aral (including those introduced by accident). Twelve of them became acclimatized and generated self-reproducing populations. Despite the adverse impacts caused by the introduction of some species, their presence seems to have sustained Aral fauna and contributed to the productivity of local fisheries.

Some decades ago, the grass carp species (about 8000 larvae) was introduced from China into the Syrdarya Delta (Samuha, 1961; Bykov, 1964). This Far Eastern species also entered the Amudarya from the Karakum Canal, where almost a half-million grass carp and silver carp juveniles had been released in 1960–61 (Aliev *et al.*, 1963). By the end of the 1960s, grass carp widely populated the Amudarya.

Carp and snakehead comprise 50% of the total catch of herbivorous species. However, introducing grass and silver carp into natural reservoirs in the Kazakstan part of the Aral Sea yielded zero commercial results (Dukravets, 1992; Mel'nikov, 1992). Snakehead catches in the Amudarya delta from 1968–83 were about 1.5 times greater than they were in the Syrdarya (Dukravets, 1992).

Acclimatization projects in the Aral basin dramatically affected local biota. They increased fish species variety and biodiversity and sustained productivity. Accidental acclimatization of various species (such as the goby and the sawbelly) seriously affected regional fauna. Highly important was the self-population of the snakehead. This fish took the place of native predators and restricted the population growth of some of the commercially favored species.

Alongside herbivorous species, numerous small undesired species appeared in the Priaralye water ecosystems (e.g., sawbelly, goby, and snakehead). The snakehead was introduced into Aral ecosystems together with its fodder. The snakehead/sawbelly/goby complex strongly affected the existing local ichthyofauna. Sawbelly and goby consumed the zooplankton and benthos and competed with young as well as mature native species. Low-value species (sawbelly, goldfish, goby) and high-value commercial species (carp, pike-perch and bream) made up about one-third of the snakehead's prey. Based on this information, the author evaluated the losses of carp, pike-perch and bream catch caused by the snakehead.

The catch of local predators, e.g., pike and pike-perch, has increased since 1990. Pike-asp and catfish catches have been relatively low. Predators made up about 48% of the total 1990–94 catch in the Karakalpak Republic. To a certain extent, these indices reflect a quantitative relationship in the ecosystem between predators and commercial species. Such a high number of predators in these waters is an important factor. Perhaps this explains low levels of fish productivity (within a range of 10–30 kg/ha) in the major part of the water bodies of the southern Aral Sea region, a region with relatively favorable ecological conditions and which is fed by river water. If one could reduce predators to an optimal level of 20–25%, fish productivity could be doubled. Thus, optimizing the structure of fish populations is one of the ways to increase the gross amount of fish resources in the natural water bodies of the Aral Sea region.

Another possibility is the creation of conditions for the natural reproduction of carp phytophile and other spring-spawning fish, because of the reliable water supply of lakes during the spring and summer. And the observation of young fish migration in 1994 in the lower reaches of the Amudarya River suggested an opportunity to replenish the fish population in lakes of the region at the expense of the young fish migrating into the river from their spawning grounds.

According to our calculations (Zholdasova and Pavlovskaya, 1995; Pavlovskaya and Zholdasova, 1991), within a two-month period (May–June), more than 100 billion larvae of 14 fish species were brought in with the water to the Takhiatash weir (i.e., a passive migration of young fish). Acclimatization was responsible for the largest population (72%), among which the commercial species of silver carp, grass carp, white bream, mottled carp, and of the native Eastern carp and pike perch dominated. Because of the use of the river flow in the spring and first half of summer, it is possible to form herds of commercial carp from year to year at the expense of the migrating young fish (but not at the expense of costly pond fish) in the largest fish hatchery water bodies of the southern Aral Sea region, i.e., in the

Mezhdurechensk and Dautkulsk reservoirs, in the Vostochny Karaten, Domalak, and Makpalkul lakes, and in the newly replenished Sarvas (Rybachiy) Bay.

A wise use of river runoff in May and June could also increase carp populations in the larger fishery areas such as the Mezhdurechensk and Dautkulsk water reservoirs. This is especially important, since the grass carp and the white, black and eastern bream do not spawn in lakes. They reproduce in rivers and then larvae and juveniles enter the lakes. Lakes in the downstream part of the Amudarya can provide enough fodder to support an abundance of herbivorous species.

Productive populations of native carp species cannot be created in the presence of abundant predators (snakehead, pike, pike-perch, pike-asp and catfish). The low number of carp species has resulted from a coincidence of carp and loach habitat with that of snakehead and pike, and of bream habitat with that of pike-perch and pike-asp. To increase carp numbers (and productivity), the predator stress should be reduced and the acclimatized carp species replenished.

Water passage in downstream Amudarya and Syrdarya for sanitary and fishery purposes should be planned, taking into account the interests of the fishery. Maximum water flow to the downstream areas should be planned for April–June and November–February in order to prevent winter fish-kill. Water levels in the lakes must also be sustained.

Fishing quotas should be coordinated with streamflow to avoid the recurrence of the situation that took place with Muynak fish farms. In 1995 water was cut off for a month, during the hatching period – water levels in the lakes dropped 2 m and, as a result, vast shallow water areas dried up. To prevent the fish-kill of the entire population, a sweeping net was used in June so that the fish stock could be replenished in 3–4 years. Clearly, rational fishery management is extremely difficult under water deficit conditions.

In 1993, the Kzyl-Orda Oblast (Kazakstan) in the northern Priaralye region was declared an ecological disaster area. However, the southern Priaralye region, suffering from similarly adverse conditions, had not yet reached crisis status. Karakalpak scientists have suggested that acknowledging a crisis status would be necessary in order to resolve downstream water supply issues. They have suggested, for example, that it would be necessary to set quotas for streamflow for sanitary and fisheries purposes on the order of 5–6 km³/year.

The present trend for fishery development in the Aral Sea basin is to create a network of lakes and ponds. In addition, in view of the impaired reproduction of native species, a network of fish farms should be created. The reservoirs fed by the Amudarya and Syrdarya have a tremendous potential for fishery development. To some extent, they can be used to compensate for fisheries losses that have resulted from the human-induced drying out of the Aral Sea.

References

Alekin, O.A., 1950: *Hydrochemical Types of USSR Rivers*. Trudy GGI. Leningrad: State Hydrochemical Institute, **25**, 79.

Aliev, D.S., *et al.*, 1963: Species composition of fish brought in with grass and silver carp from China. In: *Problems of the Use of Herbivorous Fish in Water Bodies of the USSR*. Ashkhabad: Turkmen Academy of Sciences, 178–80.

Andreyev, N.I., 1990: Zooplankton and zoobenthosis of the Aral Sea during initial period. Author's abstract, dissertation. Candidate M.S. (Biology). Moscow, 24 pp.

Annual Review of Surface Water Quality, 1991: Tashkent, Uzbekistan: Uzgidromet.

Baimov, U.A., 1964: With reference to the results of goby acclimatization in the Aral Sea. In: *Fish Reserves of the Aral Sea*. Tashkent: FAN Publishing House, 118–24.

Bashunov, V.S. and V.N. Tsoi, 1983: Assessment of population of spawning grass carp herd and the Aral Sea barbel in the Yli River. In: *Biological Fundamentals of Fisheries of Water Bodies of Central Asia and Kazakhstan*. Tashkent: FAN Publishing House, 168–9.

Bashunov, V.S. and V.N. Tsoi, 1986: Conditions for reproduction and population of spawning herd of the Aral Sea barbel in the lower reaches of the Yli River. In: *Rare Animals of Kazakhstan*. Almaty, Kazakhstan: Nauka, 196–8.

Berg, L.S., 1948–1949: Freshwater fish of the USSR and allied countries. Moscow-Leningrad: Izd-vo Soviet Academy of Sciences, **1-3**, 1382.

Bykov, L.S., 1961: Materials on acclimatization of fish and fodder invertebrates in the Aral Sea. *Proceedings of the Conference on Fisheries of the Republics of Central Asia and Kazakhstan*. Frunze: Kyrgyz Academy of Sciences, 45–50.

Bykov, L.A., 1964: New data on acclimatization of Baltic herring and other fish and their nutritive relations. In: *Fish Reserves of the Aral Sea*. Tashkent: FAN Publishing House, 61–70.

Chembarisov, E.I., 1989: Flow and mineralization of water of the large main drains of Central Asia. *Water Resources*, **1**, 49–63.

Dairbayev, M.M., 1958: Forming, composition, and distribution of ichthyofauna in water bodies of different types of the irrigation system of the Syrdarya. In: *Collection of Works on Ichthyology and Hydrobiology*. Almaty: Gylym, **2**, 286–99.

Dukravets, G.M., 1986: *Pseydoscaphirynchus fedtschenkoi* (Kessler): The Syrdarya River pseudo-shovelnose. In: V.P. Mitrofanov *et al.*, *Fish of Kazakhstan*, Vol. 1, Almaty: Gylym, 162–3.

Dukravets, G.M., 1992: *Ctenopharyngodon idella* (Valenciennes): Grass carp fish of Kazakhstan, Vol. 5. Almaty: Gylym, 126–59.

Dukravets, G.M. *et al.*, 1988: *Barbus brachycephalus* brachy-cephalus Kessler: The Aral Sea barbel. In: V.P. Mitrofanov *et al.*, *Fish of Kazakhstan*, Vol. 3, Almaty: Gylym, 24–39.

Gosteeva, M.N., 1953: The finding of shovelnose *Pseydoscaphirhynchus kaufmanni* (Bogd.) in brackish water. *Problems of Ichthyology*, **1**, 115–16.

Karpevich, A.F., 1975: *Theory and Practice of Acclimatization of Aquatic Organisms*. Moscow: Pischevaya Promyshlennost [Food Industry], 432 pp.

Lim, R.M. and Z. Ermakhanov, 1986: *Current Hydrobiological Regime of the Aral Sea and Prospects of the Use of its Basin Water Bodies under Conditions of the Sea's Regression*. Summary Reports of V Congress VGBO (Tol'yati, 15-19 September 1986). Part II. Moscow: Kuibyshev, 93–4.

Lim, R. M. and E. L. Markova, 1981: The results of colonization of sturgeon and plaice-glossa in the Aral Sea. *Rybnoe Khozyaistvo*, **9**, 25–6.

Lyubimova, C. K., I. M. Zholdasova and L. P. Pavlovskaya, 1992: About the loss of fish in the Amudarya River. In: *Collection: Biological Fundamentals of Fisheries of the Water Bodies of Central Asia and Kazakhstan.* Proceedings of XX Scientific Conference. Bibliographical List 'Scientific Works Deposited in KazNIINKI,' **1**, 3675-ka 92. Almaty, 61–7.

Maksunov, V. A., 1968: *Commercial Fish of Tadjikistan.* Dushanbe: Donish, 100 pp.

Mel'nikov, V. A., 1992: Subfamily *Hypophthalmichthyinae*: Thick-forehead-like. Mitrofanov *et al.*, *Fish of Kazakhstan*, Vol. 5, Almaty: Gylym, 198–230.

Mitrofanov, V. P., G. M. Dukravets, A. F. Sidorova *et al.*, 1986-1992: *Fish of Kazakhstan.* 5 Volumes. Almaty: Gylym.

Nikitin, A. M. and V. A. Bondar, 1975: On the dynamics of lakes of the Amudarya river delta. *Proceedings of KARNIGMI*, **25**(106), 79–90.

Nikolsky, G. V., 1938: *Fish of Tadjikistan.* Moscow-Leningrad: Izdvo Soviet Academy of Sciences, 228 pp.

Nilov, N. I., N. I. Bragin and N. Kh. Birzhanova, 1994: Organochlorine pesticides in aquatic ecosystems of the lower reaches of the Syrdarya River. In: *Collection of Medical, Social and Environmental Problems of the Aral Sea Region.* Proceedings of Research and Practice Conference, Part II. Almaty: Gylym, 56–7.

Osmanov, S. O., 1971: About the loss of *Aterinka* in the Aral Sea. *Journal of the Karalpak Division of the Academy of Sciences of the UzSSR*, **3**(5), 95–6.

Pavlovskaya, L. P. and Ilm. Zholdasova, 1991: Anthropogenic changes of fish populations of the Amudarya River (based on the data of migration of fish roe and larvae). *Problems of Ichthiology*, **31**(4), 585–95.

Red Book of the USSR, 1984: Vol. 1. Moscow: Lesnaya Promyshlennost, 390 pp.

Red Book of the Tadjik SSR, 1988: Dushanbe: Donish, 336 pp.

Red Book of the Turkmen SSR, 1985: Ashkabad, Turkmenistan: Ilym Publishing House, 414 pp.

Red Book of the Uzbek SSR, 1983: Vol. 1: Vertebrates. Tashkent: FAN Publishing House, 127 pp.

Rogov, M. M., 1957: *Hydrology of the Amudarya River Delta.* Leningrad: Gidrometeoizdat, 156 pp.

Rogov, M. M., S. S. Khodkin and S. K. Revina, 1968: *Hydrology of the Mouth Area of the Amudarya River.* Moscow: Gidrometeoizdat, 267 pp.

Sadykov, S. K., 1981: About the biology of pike-like *Aspius aspius* of the Toktogylskoe Reservoir. In: *Biological Fundamentals of Fisheries of the Water Bodies of Central Asia and Kazakhstan.* Summary Report of Scientific Conference. Balkhash: Ilym Publishing House, 146–8.

Sagitov, N. I., 1983: *Fish and Fodder Invertebrates of the Middle and Lower Reaches of the Amudarya River.* Tashkent: FAN, 115 pp.

Saifullayev, G., 1983: Distribtuion and place in the field of predatory fish in the water bodies of the lower reaches of the Zarafshan River. In: *Biological Fundamentals of Fisheries of the Water Bodies of Central Asia and Kazakhstan.* Proceedings of XVIII Scientific Conference (Tashkent, 27–29 September 1983). Tashkent: FAN Publishing House, 215–16.

Samuha, I., 1961: Acclimatization of grass carp in Kazakhstan. *Agriculture of Kazakhstan*, **7**, 83–4.

Shilin, N. I., 1984: Current state of rare and disappearing fish species. In: *Red Book of the USSR: Scientific Fundamentals of Conservation and Efficient Use of the Animal World.* 2nd Ed., Moscow: VNII of Nature Conservation and Reservation, 8–16.

Tadjitkinov, M. A. and K. N. Butov, 1972: *Vegetation of Present-Day Water Bodies of Karakalpakia.* Tashkent: FAN, 135 pp.

Tleuov, R. T., E. Adenbaev and L. N. Guseva, 1967: Migration and nutrition of young fish of the Aral Sea sturgeon. *Journal of the Karakalpak Division of the Academy of Sciences of the UzSSR,* **1**, 49–54.

Tleuov, R. T., 1981: *New Regime of the Aral Sea and its Influence on Ichthiofauna.* Tashkent: FAN, 190 pp.

Tleuov, R. T. and N. I. Sagitov, 1973: *Sturgeon Fish of the Amudarya River.* Tashkent: FAN, 190 pp.

Urazbayev, Zh., N. I. Sagitov and R. T. Tleuov, 1971: Some problems of biology of pike-like *Aspius aspius* of the middle reaches of the Amudarya River. *Journal of the Karakalpak Division of the Academy of Sciences of the UzSSR,* **1**, 83–6.

Voeikov, A. I., 1949: *Man's Impact on Nature.* Moscow: State Publications of Geographical Literature.

Volodkin, A. V., 1964: The state of and ways to increase fish reserves of the Aral Sea. In: *Fish Reserves of the Aral Sea.* Tashkent: FAN, 39–48.

Zholdasova, I. M., 1995: Limiting role of certain factors of the Aral Sea environment for Baltic herring *Clupea harangus membras L. Reports of the Academy of Sciences of the Republic of Uzbekistan,* **1**, 47–9.

Zholdasova, I. M. and L. P. Pavlovkskaya, 1991: Ecological fundamentals of preservation of genetic fund of rare and disappearing species of fish of the Amudarya River. Summary Report of the All-Union Conference, 'Ecological Problems of Living Nature Conservation.' Part II. Moscow, 90–1.

Zholdasova, I. M. and L. P. Pavlovskaya, 1995: About migration of fish posterity in the lower reaches of the Amudarya River. Proceedings of the Conference, 'Ecological Problems of the Amudarya River Region of Central Asia.' Bukhara, 83–5.

Zholdasova, I. M., L. N. Guseva and E. Adenbaev, 1989: Silver bream *Parabramis pekinensis* in water bodies of the lower reaches of the Amudarya River. *Problems of Ichthyology,* **29**(3), 475–82.

Zholdasova, I. M., L. P. Pavlovskaya, L. N Guseva and V. T. Utenbaeva, 1990: *State of Populations of Rare and Disappearing Species of Fish of the Amudarya River and Measures for their Conservation.* Information Report No. 483. Tashkent: FAN Publishing House, 12 pp.

Zholdsaova, I. M., L. P. Pavlovskaya and S. K. Lyubimova, 1991a: About the loss of fish in the Tuyamuyunskoe Reservoir. *Journal of the Karakalpak Division of the Academy of Sciences of the Republic of Uzbekistan,* **1**, 18–24.

Zholdasova, I. M., L. P. Pavlovskaya, S. Embergenov and S. Kazakhbaev, 1991b: reconstruction of hydrobionts community of the Aral Sea basin. Summary Report of the All-Union Conference, 'Ecological Problems of Living Nature Conservation.' Part II. Moscow, 198–9.

Zholdasova, I. M., L. P. Pavlovskaya, S. Embergenov, S. Kazakhbaev, L. N. Guseva and O. V. Babanazarova, 1991c: In: *Anthropogenic Transformation of Biotic Components of Aquatic Ecosystems of the Aral Sea Basin.* Summary Report of IV Congress of the All-Union Hydrobiological Community (Murmansk, 8–10 October 1991), Vol. 2. Murmansk: Polyarnaya Zvezda, 171–3.

Zholdasova, I. M., L. P. Pavlovskaya, S. Kazakhbaev and E. Adenbaev, 1992a: Present state of fauna of the Aral Sea. In: *Collection: Biological Fundamentals of Fisheries of the Water Bodies of Central Asia and Kazakhstan*. Proceedings of XX Scientific Conference. Bibliographical List, 'Scientific Works Deposited in KazNIINKI,' Issue No. 1, N 3675-Ka 92, Almaty, 32–8.

Zholdasova, I. M., L. P. Pavlovskaya, L. N. Guseva and E. Adenbaev, 1992*b*: Plaice glossa *Plathichthys flesus luscus* in the Aral Sea. *Journal of the Karakalpak Division of the Academy of Sciences of the Republic of Uzbekistan*, **3**, 21–30.

Zholdasova, I. M., E. Adenaev and L. N. Guseva, 1995: Analysis of the causes of loss of Baltic herring *Clupea harengus membras L.* in the Aral Sea. *The Uzbek Biological Journal*, **6**, 32–7.

11 Creeping environmental changes in the Karakum Canal's zone of impact

NIKOLAI S. ORLOVSKY

Turkmenistan occupies an area of 491 200 km². Eighty percent of the area is Karakum sandy desert, 17% is mountains, water area, and waste land. Only 3% of the area is occupied by natural and man-made oases. Turkmenistan has some fertile land and modest water resources located primarily in the eastern part of the country.

Prior to construction of the Karakum Canal, the 110–160 000 ha of irrigated lands were dependent on local runoff with the water supply being unstable and insufficient. In western Turkmenistan and the Pre-Kopetdag regions, water was insufficient not only for irrigation but for domestic use. Agriculture in southern Turkmenistan was unstable as water resources were frequently on the verge of collapse. The poor water supply in those regions restricted development of other industries, primarily oil, natural gas and chemicals.

To promote Turkmenistan's economic development, it was essential to supply water to its internal regions from the Amudarya, a river with an average annual runoff of 63.8 billion m³ near the town of Kerki. The problem was successfully resolved by the construction of the Karakum Canal. The canal enabled this Republic to irrigate lands and desert pastures, supply water to cities and industrial centers, create an agricultural infrastructure near the cities, establish recreational areas, develop inland fisheries, and create a navigable waterway. Karakum Canal construction was largely responsible for promoting the development of the republic's productive forces and contributed to its economic development. Agriculture in the Murgab, Tedjen, and Pre-Kopetdag oases was dramatically improved. Prior to the beginning of the Karakum Canal operation in 1958, the total irrigated area in Turkmenistan was about 166 000 ha. In 1992 more than 700 000 ha of new (additional) land was being irrigated. Water supply to the old irrigated areas increased from about 40% to 85%. It allowed the introduction of crop rotation and the creation of a stable basis not only for cotton growing but for cattle breeding, vegetable growing and horticulture whose water demands had not been fully satisfied in low-water years.

The Ashgabat-Erbent waterway was constructed to irrigate pastures situated more than 100 km from the canal. A 250-km waterway carried water from Kazandjik to Nebitdag, Kraznovodsk (now called Turkmenbashi) and

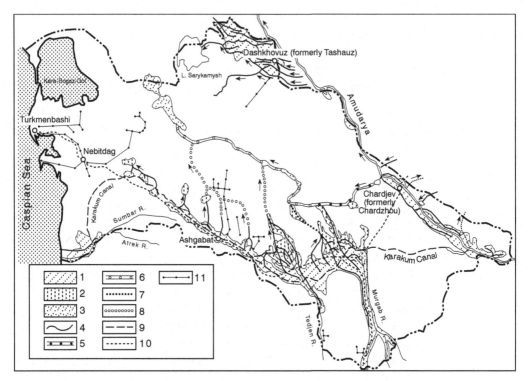

Figure 11.1 Schematic map of Turkmenistan water management (Kolodin, 1992): (1) irrigated lands; (2) flooded lands; (3) natural depression; (4) drainage collectors; (5) trans-Turkmenian collector (TTC); constructed pilot section; (6) TTC, designed section; (7) TTC branches operating Murgab collector; (8) TTC branches, designed section; (9) canals; (10) railways; (11) water mains.

Cheleken, solving the problem of water supply needs in these industrial centers. New canal waterways in this region of western Turkmenistan have recently been constructed in order to irrigate pastures and to supply water to settlements (Figure 11.1).

The canal flows from eastern to western Turkmenistan, a distance of over 1295 km, with 800 km of it navigable. The length of the canal's unlined bed of soil is 1150 km. According to the project, the last 140-km section of the canal must be lined to prevent seepage of water from the canal. The canal irrigates lands in eastern and southern Turkmenistan comprising one-third of the country's area, and its zone of influence is over 260 000 km² with a population of over 2.2 million.

In addition to the positive aspects of canal construction and land development it is important to note the various negative environmental changes. Identifying the latter was complicated by several factors: (a) specific features of the canal construction and operation and land development; and (b) the 'modified environmental situation emerges as a result of accumulating separate changes whose manifestations appear at different intervals in various physico-chemical regions' (Gerasimov, 1978, p.6). Such problems have been referred to as 'creeping environmental problems' – CEP (Glantz, 1994; see Glantz, this volume).

To comprehend CEPs in the context of the canal dynamics and varying environmental impacts, one should become familiar with the features of canal construction and the physio-geographical conditions of its impact zone.

Characterization of the Karakum Canal

Using the natural slope of the land's surface, which ranges from 250 m on the banks of the Amudarya to minus 25 m on the Caspian Sea coast, the canal runs from east to west across the southern Turkmenistan plains with different geological structures. The terrain is inclined westward at an average of 0.25 m/km. In addition, the terrain is depressed from the Kopetdag front ridges in the south to Badhyz and Karabil' mounds in the north.

The operational section of the canal crosses the southeastern Karakum desert and the Pre-Kopetdag piedmont plain bordering the central Karakum sandy desert to the north. The western section crosses the Danatin corridor between the Kopetdag mountains and the Minor Balkhan ridge and enters the western Turkmen plain. The terrain declines to the south towards the Atrek River and to the west of the mountains toward the Caspian Sea. This downward sloping pattern ensures the canal's natural course (gravity flow) along its length enabling water supply to irrigate distant lands.

The head of the Karakum Canal is on the left bank of the Amudarya near the village of Bosaga. Its diversion structure is 3 to 4 km from the river course. The water intake structure includes three canals for settling out silt from the river water. A navigable sluice is connected to the Amudarya by a separate channel. For a 360-km distance, the canal has a latitudinal direction. It then turns southwest and crosses the Murgab River. The Murgab-Tedjen section flows southwest, then turns to the west, crosses the Tedjen River at the 537-km point, and flows northwest across the Kopetdag piedmont plain. From Kazandjik, the canal flows to the south on the western Turkmen plain toward the Atrek River.

At a 30-km distance from its head, the canal lies in the bed of the Bosaga-Kerki Canal constructed in 1928. At the 105-km point it lies in the bed of the ancient Kelif Uzboi channel. At the 50–105 km section numerous settling lakes were constructed by banking; these were called Kelif lakes. Their surface area is about 70 km² at the maximum.

After crossing the Kelif Uzboi, the canal enters the southeastern Karakum (sandy-clay) plain that is gradually transformed to a sandy plain from the 160-km point. At the 180–300 km section, natural depressions were used to reduce the need for moving earth. When these depressions were filled with water, various lakes and flooded areas were created. Some lakes grew to 2000 m wide and up to 3000 m long. Gradually, they either silted up or were cut off by dams.

Below the 310-km point, the canal enters the sandy-clay plain of the ancient Murgab delta and at the 397-km point crosses it about 4 km above Merv. From there, up to the 420-km point, the canal crosses a cultivated zone in the Murgab oasis and sandy mounds with poor vegetation. At the 456-km point the canal is diverted to the Khauzkhan reservoir created in a natural depression. Water is supplied from the reservoir by a waterway over 60 km long to irrigate land in the Tedjen oasis.

The Karakum Canal rounds the Khauzkhan reservoir on the south and enters the Tedjen delta below the 474-km point. From the 601-km point it flows along the northern Kopetdag piedmont plain to Kazandjik (1100 km). The Karakum Canal was constructed in several stages.

STAGE I. THE AMUDARYA-MURGAB SECTION

The Murgab River basin was a natural frontier for Stage I. The canal length is 397 km of which 300 km crosses a sandy desert. The canal bed is not lined. The annual water consumption reached 130 m³/s and the annual runoff diverted from the river is about 3.5 billion m³.

Karakum Canal construction started in 1954 and in 1959 Amudarya water reached the Murgab oasis, irrigating 33 000 ha and improving the grave situation caused by limited water supply at this particular oasis. Prior to canal construction, from 80 000 to 100 000 ha were irrigated in the Murgab oasis, depending on the annual water content and crop pattern. Average perennial river runoff allowed for the irrigation of 82 000 ha. Taking into account the land irrigated by the canal, the total irrigated area in the Murgab basin upon completion of Stage I of the Karakum Canal was estimated at 170 000 ha. In 1975 the actual area of irrigated land was 212 000 ha, i.e., 42 000 ha above the designed capacity. In 1991 the irrigated area here was measured at 421 100 ha.

STAGE II. THE MURGAB-TEDJEN SECTION

With unused arable land estimated at more than 700 000 ha, up to 27 000 ha on average were irrigated in this section. Yet, in some years crops production failed because of zero runoff. The Murgab-Tedjen section was constructed in seven months and in the autumn of 1960 the Amudarya reached the Tedjen River.

Stage II involved the expansion of irrigated land in the Tedjen oasis of about 72 000 ha with the total irrigated area reaching 99 000 ha. This portion of the canal's length was 138 km of which 70 km crossed the sandy desert. Following the reconstruction of the Stage I canal, the annual consumption increased to 198 m³/s and water intake from the Amudarya reached 4.7 billion m³. The Khauzkhan reservoir with a 460 billion m³ capacity was constructed to use the free autumn–winter canal runoff that enabled a marked reduction in the canal's passage capacity above the reservoir. Construction of Stage II was finalized in 1966. Newly irrigated additional land in the Tedjen basin was estimated at 140 000 ha.

STAGE III. THE TEDJEN-GEOK-TEPE SECTION

A 258-km-long Tedjen-Ashgabat pilot canal was constructed to improve the water supply to the capital (Ashgabat) and to increase suburban agriculture. The maximum water consumption at the canal head near the Tedjen River was 13.7 m³/s, and 6 m³/s near Ashgabat. Two water reservoirs – Western (with a 48 million m³ capacity) and Eastern (with a 6.3 million m³ capacity) – were constructed near Ashgabat. This pilot canal construction was started in 1961 and in May 1962 the inhabitants of Ashgabat received Amudarya water.

Construction of Stage III started in 1966. After reconstruction of Stages I and II of the Karakum Canal and construction of the pilot canal, the head water consumption reached 317 m³/s. The annual water intake diverted from the river increased to 8.3 km³. The length of the canal became 873 km, including 44 km of the Ashgabat-Geok-Tepe section. The Khauzkhan reservoir capacity was increased to 875 million m³. The Kopetdag reservoir is currently (1995) under construction west of Geok-Tepe.

STAGE IV. THE GEOK-TEPE TO KAZANDJIK SECTION

This section (850–1100 km) was started in 1972. The canal reached Kazandjik in 1981. The Kopetdag reservoir capacity was increased to 500 million km³; the Amudarya intake reached 13.5 billion m³ per year.

Construction of the southwestern branch in order to irrigate Turkmenistan's dry tropics was launched in 1991 and irrigation put in operation in 1993. Construction of the large Zeid reservoir with a 3.5 billion m³ capacity was undertaken at the canal head. The maximum intake from the Amudarya is estimated at 580 m³/s.

The canal is designed to operate throughout the year because of the necessity to ensure year-round navigation, the cheapest and most reliable means of transportation and canal inspection given harsh desert conditions. It is also important to preserve the fish stock throughout the year. The canal water was used not only for irrigation but also for water supply for domestic and agricultural purposes.

Year-round canal operations are related to water passage in the autumn/winter season, when there is no agricultural demand for water. Water accumulates in reservoirs and is released in the summer. Water reservoirs with a total capacity of 4.8 billion m³ allow for the maximum use of canal water and for the irrigation of additional areas with no additional water intake required in the summer. Water reservoirs also serve to ameliorate canal level fluctuations thereby improving canal operations and navigation.

CEP in the canal zone

The environmental impacts of large-scale canals are approximately proportional to their size. The impacts are manifested during their construction because of related technological alterations to the landscape resulting in changes in topographic features. A canal also influences adjacent areas during filling (e.g., waterlogging, soil salinization, groundwater level changes) and afterwards by direct economic activity (e.g., the expansion of irrigated lands and the discharge of excess contaminated irrigation water). Even the most ardent champions of the canal underestimated the scope, rate and intensity of the land degradation that it brought about.

Karakum Canal construction and operation resulted in significant and occasionally radical changes in the landscape of adjacent areas. As the impacts were not always favorable, the canal's impacts became an object of public concern in the 1970s (a threshold of problem awareness).

There were and still are decision-makers who believe that the canal presents no danger to society. Although displaying an alleged concern for this issue in their statements, they have not been encouraged to modify their optimistic views by the fact that tens of thousands of hectares of arable land continue to be removed from agricultural use annually because of soil degradation. However, it is important to note that the public and agricultural workers sensed dangerous signs earlier than scientists.

The issues of environmental change in the canal's zone of impact were raised in the 1960s and 1970s in the works of researchers at the Institute of Geography of the USSR Academy of Sciences and at the Institute for Desert Research of the Turkmen Academy of Sciences (Niyazov, 1965a,b; Saparov, 1969, 1971; Saparov et al., 1970; Grave, 1974; Grave and Kostyuchenko, 1975). The scientific data were accumulated and the scientific awareness of the types and extent of creeping environmental problems increased in the 1970s and 1980s (Grave and Grave, 1981, 1983; Vostokova and Skaterschikov, 1980; Babayev and Dobrin, 1978; Zonn, 1981).

The canal's environmental impacts have been primarily (a) filtration, (b) water discharge at hydrosystems, (c) dam rupture, and (d) dredge discharge by suction dredges. Their direct consequences have been as follows: rising groundwater level, transformation of desert soils (serozems) to meadow-swamp and hydromorphic soils, increased salinization of soils and of groundwater, and the degradation of native desert vegetation and its replacement by primitive halophytic communities. Zoocomplexes radically changed their characteristics. Depending on local conditions, they are either markedly enriched by hydrophylic invertebrate and bird species, or impoverished because of the loss of native desert species and their replacement by several widely spread halophylic invertebrate species (Zaletaev, 1991).

Consider the example of the spatial and temporal aspects and of the CEP rate on Stage I section of the Karakum Canal. Features of the interactions between the canal and the environment are the most vivid and diverse in this segment of the canal because of the long term of impacts (i.e., 40 years), the dominance of sandy sediments and the lack of anti-filtration protection (e.g., the lack of a lining of the canal).

Field studies carried out in 1971–73 and 1980–85, as well as an analysis of mapping data for the area in 1961–67, allowed for the identification of features of the canal's impacts on the environment in general and on several of its components specifically, e.g., groundwater, vegetation, soil and geomorphological features (Grave, 1974; Grave and Kostyuchenko, 1975; Grave and Grave, 1981, 1983).

Hydrological and hydrogeological changes were the first indicators of the human-induced modification of natural ecosystems. The groundwater regime was adversely affected by water filtration through the canal's banks and its bottom. Filtration losses of canal water dominate, accounting for 70–99% of the total water losses in the canal. Particularly high filtration losses were reported at an early stage of construction: over 1500 l/s per 1 km in sandy ground, and about 200–300 l/s in sandy-clay ground (Balakaev, 1974). Filtration losses sharply decreased over time. According to aerial mapping in 1965, the distance between the most remote filtration lake and the canal was 1500 m. This indicates that in the first 7 years of operation the groundwater had moved at a rate of about 215 m per year (or 0.6 m per day). Eventually, the movement of groundwater slowed down sharply. According to aerial mapping in 1976, the groundwater had moved at a rate of about 73 m per year (or 0.2 m per day). Thus, in that 11-year period the rate decreased threefold (Babayev and Babayev, 1994).

The annual filtration loss of water from the canal (via its banks and bottom) at its section between the Murgab and the Amudarya is shown in Table 11.1.

The groundwater level increased markedly immediately on canal construction. From 1957 to 1960, the groundwater table at the 227-km and the 275-km points (located in the sandy desert) increased 25–28 m at a distance of 50 m from the canal. Within the boundary of the Kelif Uzboi (the 50–100 km section) with numerous filtration lakes, the groundwater level increased 10–12 m in 1967, and the canal's impact zone reached 25–30 km from the right bank and 5–7 km from the left bank.

Table 11.1 Filtration in the Murgab-Amudarya section of the Karakum Canal

Indicators	Natural areas						Total
	Pre-Amudarya	Kelif Uzboi	Obruchev Steppe	Pre-Murgab			
				Eastern	Western	Region total	
Mean annual filtration flow (m³/sec)	4.16	7.36	3.98	4.85	20.45	25.30	340.80
Mean annual filtration volume (× 10⁶ m³)	131.19	232.10	125.10	152.95	644.91	797.86	1286.66

Source: Compiled by Grave and Grave (1981).

| Canal section | Coastal shallow water complexes | Terrestrial complexes | | | Total area |
		Hydromorphic	Semi-hydromorphic	Transitory	
Pre-Amudarya barkhan (30–50 km)	1	19	11	5	36
Kelif solonchak (50–105 km)	48	180	79	33	340
Obruchev takyr (105–180 km)	2	17	20	4	43
Pre-Murgab sandy-mound (180–310 km)	11	51	38	14	114
Total	62	267	148	56	533

Source: Gerasimov (1978).

Typical of the Obruchev steppe (the 100–180 km section) is a relatively narrow canal impact zone not exceeding 6–8 km on the right bank. At an early stage of canal operation the groundwater level rose 16 m at a distance of 100 m from the canal (Niyazov, 1965a). Hence, variations in moisture content of the soils are related to distance from the canal and the period of time since the filling of the canal.

Hydrological and hydrogeological variations are closely related to the vegetative cover, which quickly responded to the newly emerging moisture conditions. As expected, a few moisture-loving species appeared in the first years of the canal's operation. Consequently, desert vegetation was replaced almost completely in the zone nearest to the canal, particularly in topographic depressions. Today, associations of moisture-loving plants (hydrophytes, phreatophytes, etc.) are common throughout the canal's impact zone (Grave and Grave, 1983). Areas of modified natural complexes are shown in Table 11.2.

Along with changes in water, the soils, too, become modified (but more slowly) and have affected the state of regional biota. More specifically, waterlogging and soil salinization reduced desert pastures for sheep breeding. In the late 1970s the breeding of sheep on the rangelands collapsed in that part of the Chardjou region within the canal's impact zone. Regional authorities requested experts to evaluate sheep fodder loss in the canal zone with the aim of compensating them with wild fodder or by allocating new pastures.

Figure 11.2 depicts spatial changes in manmade (i.e., irrigation) and natural complexes over a 10-year interval. The following conclusions can be drawn: Kelif Uzboi depressions that were initially used as 'settlers' or 'settling ponds' for the turbid Amudarya silted up and became a region of complexes of excessive moisture. The silting up of Kelif lakes reduced filtration processes; numerous filtration lakes, overgrown with hydrophytes (e.g., cane,

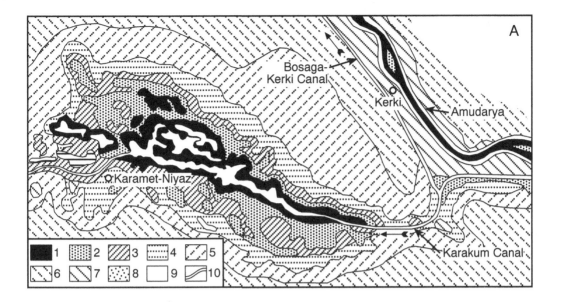

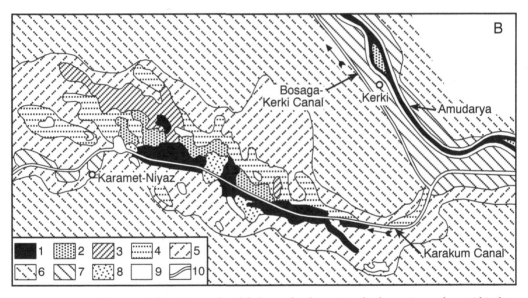

Figure 11.2 Schematic map of modified natural-technogenic and technogenic complexes within the Karakum Canal zone at the section from the Amudarya plain to the Karametz-Niyaz village: (a) early 1970s (Grave et al., 1987); (b) early 1980s (Grave *et al.*, 1987). *Modified natural-technogenic complexes:* (1) coastal shallow water with excessive moistening; (2) surface with high moistening; (3) surface with moderate moistening; (4) surface with weak moistening; (5) surface with weak episodic moistening. *Natural unmodified complexes:* (6) desert natural complexes; (7) irrigated lands; (8) shallow water and silted sites; (9) settling lakes and filtration lakes; (10) canals.

Table 11.3 Changes over time in the areas (km²) of clear water of filtration lakes and moisture-loving plants in Kelte-Beden (at the 286-km point)

Observation object	1963	1965	1972
Water area of filtration lakes	0.75	1.4	0.5
Area covered by moisture-loving plants:			
reed and cattail	0.85	1.5	1.0
camel's thorn and carelinia	1.58	2.0	2.4
Total	3.18	4.9	3.9

Source: Gerasimov (1978).

cattail), dried up and the hydrophytes were replaced by phreatophytes (e.g., carelinia, camel's thorn); reduced filtration losses promoted a lowering of groundwater level, affecting the occurrence of terrestrial complexes requiring different levels of moisture. Particularly noticeable were changes in moderately and weakly moistened areas, as well as in the total width of the canal's impact zone in the desert.

The above-mentioned vegetation complexes declined twofold in a decade. It is noteworthy that the process dynamics has an oscillating nature as characterized by moisture fluctuations (Table 11.3, Figure 11.3). Canal construction was followed by lake flooding, and lakes became surrounded by moisture-loving vegetation which was highly sensitive to moisture availability.

In 1963 the total area of filtration lakes was less than 0.75 km². In 1965 minor lakes united to form three larger ones. Hydrophytes that first appeared to the south of the canal proliferated and eventually outnumbered the phreatophytes. The total lake area almost doubled. In 1973 the filtration lakes were again reduced in number and size to 0.5 km². In 1983 the number of filtration lakes and the size of the lakes reportedly increased, accompanied by an increase in hydrophytes.

Fluctuations in moisture in the canal zone were possibly related to construction works aimed at increasing the canal's flow capacity, particularly in Stage III in the 1960s and in Stage IV in the 1970s. They were accompanied by an increase in water flow and filtration losses. The following conclusions can be drawn:

- New groups of shallow-coastal and terrestrial complexes were formed as a result of the canal's impact over a 15 year period.
- The zone of the above changes on the northern bank was 5–20 km and on the southern 1–8 km. In this zone the modified complexes are closely related to lithological and geomorphological conditions.
- The formation of new complexes starts with changes in soil moisture content which is related to vegetation changes.
- The increased wetness of saline areas has serious impacts resulting in the active salt transport in the soil and in the formation of bitter-salt lakes.

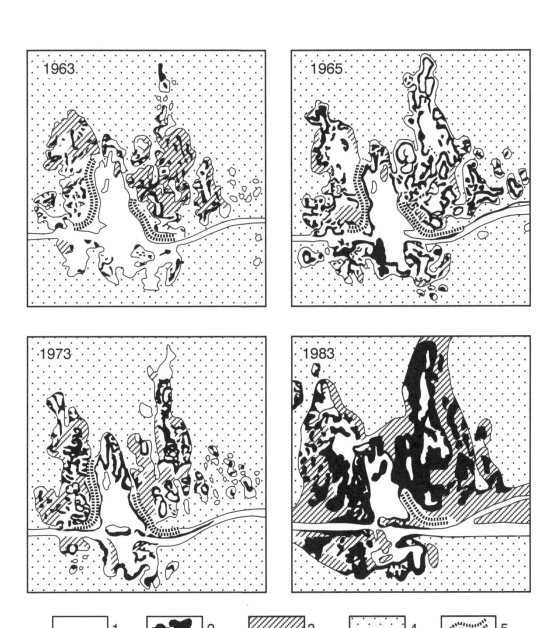

Figure 11.3 Changes in filtration and flowing water lakes and moisture-loving plants from 1963 to 1983 in Kelte-Beden Bay (1963–72 data compiled by Bogdanova and Kostyukovky; 1983 data compiled by Grave *et al.*, 1987). (1) water area; (2) cane and cattail; (3) camel's thorn and carelina; (4) desert xerophytes and psammophytes; (5) dams.

• The area encompassed by altered natural complexes may increase as the canal broadens and as water flow increases. This will result in a dynamic equilibrium because of the stabilization of the groundwater level and a balance between filtration on the one hand and groundwater outflow and transpiration on the other. Additional changes in the natural environment will be related to the increasing development of irrigation land.

CEPs in irrigation regions

Upon construction of Stage II of the Karakum Canal, irrigation in southern Turkmenistan developed at a high rate. In 1970 the canal's water irrigated 309 700 ha; in 1980 the area was 463 300 ha and in 1992, 700 200 ha.

Unwise water use and poor land management practices during the initial stages of irrigation, in addition to the absence of a collector and drainage network along with poor agricultural practices, have modified the hydrogeological and soil conditions. Adverse soil processes developed at a higher rate than had been anticipated in the pre-project assessments.

As a result of excessive moisture and secondary salinization, disastrous land degradation emerged in the Murgab (1962–70) and in the Tedjen (1970–90) oases and in the Khauzkhan area (1970–90) as well as in separate plots of land in the Kopetdag piedmont.

Consider creeping environmental changes that occur during irrigation, using the example of the Khauzkhan area (Figure 11.4). Located in the central part of the ancient Tedjen delta, it borders the Karakum Canal to the south, sandy-clay complexes along the Murgab-Tedjen watershed to the east, the railway to the south and the Tedjen River to the west. It stretches 60 km from east to west and 40 km from north to south.

The Khauzkhan irrigation project was launched in 1962 at sites on the plain with nonsaline soils of light mechanical composition. Earth ditches comprise 98% of the irrigation network; hence, the output does not exceed 50–60%. In the first year, about 5000 ha were irrigated with a water supply of 80 million m³. In 1972 the figures were 92 600 ha and 1.31 billion m³; in 1983 the figures were 95 450 ha and 1.18 billion m³ (Table 11.4). In different years the water supply reached 25 000 m³/ha. As compared with experience worldwide (8–10 000 cubic m/ha), this specific irrigation rate markedly exceeded biological consumption.

Such enormous water consumption has been the result of a low standard of technology for irrigated farming and irrigation networks, and of the absence of advanced irrigation hardware and techniques (e.g., sprinkler, drip, subsurface). An obsolete ditch irrigation technique has also led to significant filtration losses.

Prior to land development, the highest groundwater table (about 5 m below the surface) was reported along the Tedjen River, gradually lowering eastward to 20 m and below. An area of 111 900 ha (40.9%) was occupied by

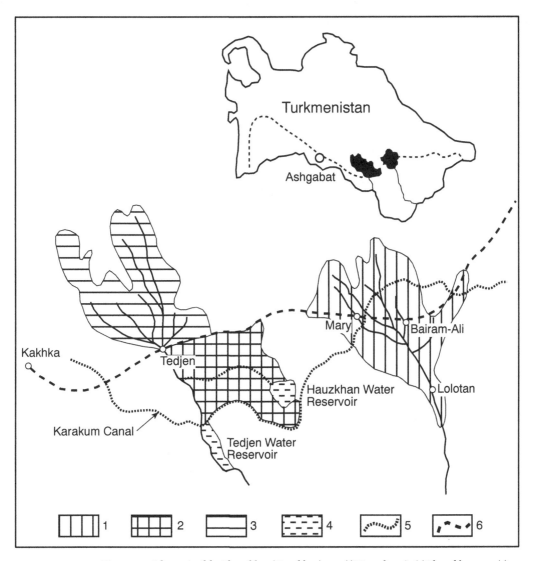

Figure 11.4 Schematic of the Khauzkhan (Hauzkhan) area. (1) Murgab oasis; (2) Khauzkhan area; (3) Tedjen oasis; (4) water; (5) canals; (6) railway.

profiles with a groundwater level of 10–15 m, and 46 700 ha (20.9%) by profiles with a groundwater level of 5–8 m. Five to eight years after launching into operation the Khauzkhan reservoir and Stage I of the Karakum Canal, the groundwater table rose rapidly by 10–11 m. In 1970–73 the groundwater reached a critical level of 0.7–1.5 m in the central part and 20 m on the periphery. The lands unfavorable in terms of reclamation are situated in the zone of intense irrigation with a groundwater level of 1–2 m (31 300 ha or 13.9%) and 2–3 m (52 900 ha or 23.6%). By 1973, the lands with a groundwater level of 3 m below the surface totalled 84 000 ha compared with the 30 000 ha that were

Table 11.4 Irrigated area and water intake in Khauzkhan area

Year	Irrigated area (× 1000 ha)		Water intake (× 10⁶ m³)	Average water supply (× 1000 m³/yr)
	Cotton	Total %	$(\times 10^6\,m^3)$	$(\times 1000\,m^3/yr)$
1965	5.8	11.45	96.37	8.41
1966	19.69	26.45	441.99	16.81
1967	29.83	42.33	862.73	20.38
1968	27.53	39.78	992.75	24.93
1969	35.77	49.78	1169.20	23.49
1970	59.66	76.05	1122.84	14.76
1971	62.86	79.40	1287.57	16.22
1972	62.62	80.43	1405.85	17.48
1973	64.87	86.32	1215.62	17.48
1974	69.52	92.60	1309.34	14.14
1983	75.08	95.45	1182.26	12.39

Source: Rejepbayev and Esenov (1987).

predicted by hydrogeologists for 1975. The average groundwater rise in the 1962–74 period was about 1 m per year. Groundwater, at a depth of 5 m in 1962, went to only 2–3 m below the surface after 1974 (Figure 11.5).

During a decade of intensive land development, the groundwater level reached 3 m on most of the area, having risen at a rate of 2–2.5 m per annum. On the periphery of the irrigated area, the groundwater rose at a rate of 1 m per annum in 1968–75. The mean annual groundwater rise in the area varied from 0.40 m to 2.0 m per annum, an unprecedented rate in the history of irrigation in Turkmenistan.

Intense irrigation for 10–15 years, under conditions of almost no drainage, contributed to the marked deterioration of the land. During the first phase of land development (5–8 years), the groundwater had not reached a critical level and irrigation was accompanied by soil desalinization (Table 11.5).

Nevertheless, the salt content in the groundwater remained high and its level rapidly increased, surpassing the estimates. It reached a critical level in 1968, when the first signs of secondary salinization appeared. Various experts were increasingly concerned with the land reclamation situation (e.g., the problem realization threshold was noted).

The second period (10–15 years after canal construction) was marked by the groundwater level reaching the critical level of 2–3 m. Since that period (1973–78), active salt accumulation started. About 80% of the area had been adversely affected by heavy salinization. Disregard for local environmental features accelerated the development of the crisis situation.

A comparison of salinization maps shows that irrigation pushed salts to the north of the central part, toward the nonirrigated peripheral lands. Salts

Table 11.5 Saline lands in the Khauzkhan area

Degree of salinity	Prior to irrigation 1958–1962		After 5–8 years of irrigation		After 10–15 years of irrigation	
	×1000 ha	‰	×1000 ha	‰	×1000 ha	‰
Nonsaline	15.1	6.73	47.9	21.36	36.1	16.08
Weakly saline	12.8	5.70	36.5	16.32	73.4	32.74
Medium saline	80.5	35.92	42.4	18.90	58.8	26.24
Strongly saline	45.0	20.04	34.9	15.53	27.1	12.09
Extremely saline-solonchak	45.2	20.19	40.8	18.21	21.0	9.35
Sands	25.6	11.42	21.7	9.68	7.6	3.50

Source: Rejepbaev and Esenov (1987).

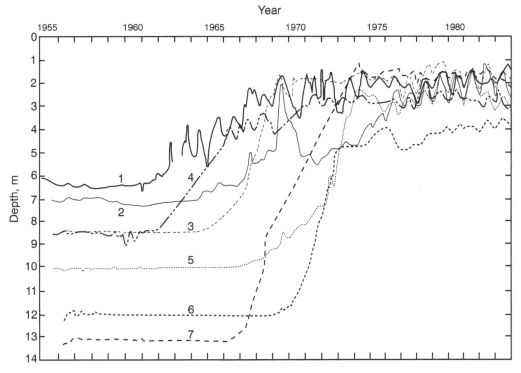

Figure 11.5 Groundwater dynamics of the Khauzkhan area. Number of the drill hole site: (1) 14; (2) 2A; (3) 112; (4) 24; (5) 3; (6) 115; (7) 113.

were less markedly redistributed in the area between the irrigated and the fallow lands.

The area of nonsaline and weakly saline soils increased in this period from 12.4 to 37.7% (Table 11.5). Under irrigation, the proportion of medium-level saline soils was reduced from 36% to 19%. Extremely saline soils remained in virgin plots and a strip bordering the main irrigation ditches. After 10 to 15 years of irrigation, the medium and heavily saline soils were converted into heavily to extremely saline. Thus, soil salinity in the area between the Karakum and Khauzkhan canals increased considerably.

Both the profile and redistribution of salts occurred during the short period of the expansion of irrigation. Salinization was increasing in both space and time, resulting in large measure from an uneven distribution of water supplies accompanied by rising groundwater with a high salt content. It became obvious by the mid-1970s that further expansion of irrigated lands without proper drainage would accelerate secondary salinization (the realization, i.e., threshold, of a 'need for action').

Construction of a drainage collector network (DCN) in the Khauzkhan area began in 1973 and was carried out slowly. Today, its length is 2817 km, of which 342 km is made up of interfarm collectors and 2476 of an interfarm drainage collector network. The length of covered interfarming drainage collector network is 1102 km. Specific DCN extension in the area averages 22.2 m/ha. Drainage runoff in 1991 was 621.3 million m³. The mean salt content in drainage water is 20.2 g/l.

The condition of the drainage collector network is extremely poor: almost half of the covered drains are not operational and the open ones are clogged. Hence, the drainage network did not have a serious impact on the water and salt regimes of the irrigated lands. The Khauzkhan area, once a major cropping zone in the Mary velajat (Turkmenistan administrative unit), symbolized unwise land management. The soil salinity of irrigated lands was above average on at least 68% of the farms.

Consider creeping environmental change under a growing discharge of collector drainage water (CDW) outside the irrigated area. Drainage sustains groundwater and soil productivity at an optimum level and is an integral part of land reclamation. However, CDW also causes soil degradation and environmental pollution. Drainage water is often discharged directly into rivers, large and small natural depressions and on the desert plains, thereby aggravating existing environmental problems.

Apart from a high salt content, collector drainage water also contains highly toxic compounds such as pesticides, phenols and heavy metals. Pesticide loads on irrigated land averages 9.37 kg/ha throughout Turkmenistan and reaches 12.7 kg/ha in the Kopetdag plains. In 1991, a total of 2.67 km³ of CDW was discharged in the southern Turkmenistan desert. This discharge resulted in the flooding and salinization of desert rangelands, and surface and groundwater pollution in pasture wells. The flooded area in the Tedjen-Murgab desert regions in 1985 is shown in Table 11.6. Thus, vegetation and botanical composition of fodder was dramatically modified in the irrigated areas.

A rapid increase in the areal extent of flooded desert pastures and plots of land bordering irrigated areas in the late 1970s increased public awareness (a problem awareness threshold) and prompted the undertaking of scientific

Table 11.6 **Spatial and temporal dynamics of biosystems in the Tedjen and Murgab deltas**

Biosystem	1976 (ha)	1985 (ha)
Flooded	3350	11 250
Dried	none	1375
Regularly flooded	80 975	122 650
Medium flooded	475 975	954 900
Strongly flooded	120 275	255 475

Source: Babaev and Babaev (1994).

studies of the environmental impacts of CDW. In the mid-1980s vast pasture areas were withdrawn from agricultural use because of flooding and excessive moisture in the soils and changes in the vegetative cover.

To prevent environmental pollution, in 1986 authorities decided to prepare a feasibility study on the neutralization and decontamination of CDW. A project to construct a Trans-Turkmenian collector was prepared in 1988. Initially, it was planned for the collector to reach the Caspian Sea. Following a ban by the Eastern Caspian water inspection on drainage water discharge in the Caspian Sea, the collector was channeled into two depressions with a total capacity of 130 km³ in northern Turkmenistan.

The length of the Trans-Turkmenian collector is 765 km. It crosses the Unguz depressions, a chain of solonchak hollows. The collector is made up of 29 natural hollows with a total surface area of more than 808 km² and capacity about 5 km³. CDW from irrigated lands in the Tedjen and Murgab oases will be channeled to the collector via the Tedjen Central and Murgab collectors.

The construction of the Trans-Turkmenian collector is aimed at preventing a continued degradation of irrigated lands. Yet the project will hardly improve the overall environmental situation in Turkmenistan, since pollution will not be eliminated but will only be transferred from one place to another. Construction technology is at a low inexpensive level; no filtration control is provided for. A flooded zone, bordering the main collector and ditches, has been estimated at 5–10 km. Flooded desert rangelands within the main collector zone of influence are estimated at 765 000 ha, reaching 1.5 million ha with all the ditches included.

Experts in different fields have criticized the project's economic and environmental aspects. Instead of improving the condition of the irrigation lands, the Trans-Turkmenian canal will adversely affect the desert environment. Scientists have proposed an alternative, environmentally friendly demineralization project (Kolodin, 1992). The scheme has ten demineralization sites to be created at large interfarm collectors and three large-scale chemical treatment installations. The project minimizes the adverse environmental impacts of drainage water but would require additional costs. In addition, such a large-scale demineralization program would require considerable international technical and economic assistance.

When discussing the Trans-Turkmenian collector project, scientists emphasized the need to decrease filtration losses by 5%–7%. Apart from improving land reclamation, this would add several hundred new hectares of arable lands. Yet the present land reclamation policy is based on the concept of inevitable filtration losses during irrigation. Drainage and the flushing of the soils are the most widely practiced techniques aimed at eliminating the adverse consequences of irrigation rather than their causes. In addition, engineering-drainage scientists have proposed biological and water-saving techniques.

The problem realization threshold occurred in the late 1980s when scientists accumulated experience in optimizing regional salt budgets during arid land reclamation (Kirsta, 1988, 1989, 1993, 1995; Rejepbayev and Ovsyannikov, 1974).

From 1956 to 1989 the Karakum Canal carried out 154 270 000 metric tons of salts. In the first years of the Amudarya diversion into the arid lands of Turkmenistan, salts were accumulated within the irrigation zone in virgin and abandoned lands with insufficient or no drainage.

The plains of Turkmenistan are subject to intense salinization irrespective of the salt accumulation site – the irrigated zone or the borders of the desert lands. Salts are accumulated primarily within the aeration zone in the upper horizons of the soil up to the surface. Salinization results not only from the annual salt transport by irrigation water but also from the movement of previously accumulated salts. Actually, two processes occur: (a) the gradual salinization of Turkmenistan plains because of salt transport by rivers flowing from the mountainous regions and (b) the redistribution in depth and in area of the 'ancient' (relict) salt stock because of a rising groundwater level as a result of irrigation. While the latter does not affect the overall salt budget in the plains, it can markedly modify it in different areas.

Salt transport is a natural process that occurred in the past and will continue in the future. In the past the river runoff containing salts accumulated in depressions forming solonchaks and lakes with high salt content in the arid zone. Today salts are distributed everywhere, as a result of the use of runoff from irrigation.

Salinization control measures are presently restricted to the cultivated areas. Irrigated soils are regularly flushed and drainage systems are being constructed to keep groundwater below the critical level (2–3 m). Excessive drainage water with salts is discharged to desert areas or back into rivers. Both flushing and drainage are aimed at solving a local problem – to prevent a further deterioration of the local condition of irrigated lands. However, these measures have negative environmental impacts, since salts that have accumulated in the deep soil horizons and in the groundwater are then transported to the surface. Hence, flushing and drainage should be accompanied by integrated measures aimed at the protection and rational use of water and land resources.

Such integrated measures should be based on an integrated management of the water and salt budgets in irrigated lands and areas adjacent to them. The amount of water for irrigation and for flushing should be sufficient for obtaining a maximum crop harvest with minimum water consumption. This should be the guiding principle in irrigated farming.

Integrated measures should include the following:

- reduce water losses in inter- and intra-farming irrigation networks through reconstruction and the use of water-resistant lining; replace open ditches by pipes and flumes;
- strictly account for the volume and salt content for irrigation (including flushing) and drainage water; monitor the groundwater level and salt content;
- bring irrigation rates in line with plants' biological needs for water; water should be sufficient to ensure the necessary amount of soil moisture for plants' normal growth;
- introduce water-saving irrigation technologies most suitable for local crops and soils (e.g., subsoil and drip irrigation, flexible hoses, sprinkler);
- strictly observe agrotechnology and irrigation rates for each crop, taking into account soils as well as annual and seasonal meteorological factors; coordinate flushing rates with crop, soil and salinity features;
- introduce crop rotation, taking into account soil and land reclamation features;
- put in practice a differentiated payment schedule for varying water uses;
- introduce and enforce payment for environmental damage caused by over-irrigation, improper drainage-water discharge, etc.;
- develop the automated management of irrigation and drainage networks.

Integrated measures always require considerable effort, time and investment. But, for this region there is no alternative. Their implementation will markedly reduce the salinity of the irrigated soils and improve land reclamation in the region. Such measures will also promote the rational use of water whose deficit is sure to grow from year to year. The above measures will reduce drainage runoff and the need for expansion of the collector drainage network, will decrease drainage water discharge onto the plains, and will impede the movement and redistribution of previously accumulated salts. A new concept of water management development in Turkmenistan must take such measures into account.

References

Babaev, A.G. and L.G. Dobrin, 1978: The dynamics of natural landscapes in Karakum Canal impact zone. *Problems of Desert Development*, **6**, 3–9.
Babayev, A.M. and A.A. Babayev, 1994: Aerospace monitoring of geosystems under irrigation. *Problems of Desert Development*, **1**, 21–8.

Balakaev, B. K., 1974: Some aspects of constructing large-scale irrigation canals in deserts. *Problems of Desert Development*, 4, 41–6.

Gerasimov, I. (ed.), 1978: *The Karakum Canal and the Alteration of Natural Environment in the Zone of its Influence*. Moscow: Nauka, 232 pp.

Glantz, M. H., 1994: Creeping environmental changes in the Aral Sea basin, *Problems of Desert Development*, 4-5, 51–64. (In Russian)

Grave, L. M., 1974: Karakum Canal interaction with the environment. *Izv. AN SSSR, Ser. geogr.*, 5.

Grave, L. M., M. K. Grave, V. V. Gorbachev and L. A. Krapkova, 1987: Remote sensing of dynamics of geotechnical Karakum Canal-desert interface. *Problems of Desert Development*, 1, 41–51.

Grave, L. M. and V. P. Kostyuchenko, 1975: Desert landscape changes in the Karakum Canal's impact zone. *Izv. AN SSSR, Ser. geogr.*, 5, 36–47.

Grave, M. K. and L. M. Grave, 1981: *Karakum Canal and Desert Environment*. Moscow: Znanie, 48 pp.

Grave, M. K. and L. M. Grave, 1983: Impacts of large-scale Central Asian canals on desert ecology. *Problems of Desert Development*, 4, 25–30.

Kirsta, B. T., 1988: Salt transport by the Karakum Canal into Turkmenistan plains. *Problems of Desert Development*, 1, 17–23.

Kirsta, B. T., 1989: The Aral Sea issue and the Karakum Canal. *Problems of Desert Development*, 5, 10–17.

Kirsta, B. T., 1993: The salinization of irrigated lands in Turkmenistan. *Problems of Desert Development*, 1, 16–21.

Kirsta, B. T., 1995: Problems of the Aral Sea basin using the example of Turkmenistan. *Problems of Desert Development*, 3, 11–13.

Kolodin, M. V., 1992: *Water Protection Projects in Arid Zones (Economic Aspect)*. Ashgabat: Ilym Publishing House, 232 pp.

Niyazov, O., 1965a: The groundwater regime in the Karakum Canal zone. *Izv. AN TSSR, Ser. phys.-math., chem. i geolog. nauk*, 2.

Niyazov, O., 1965b: Characteristics of the groundwater regime in the Karakum Canal impact zone in the Murgab delta. *Izv. AN TSSR, Ser. biol.*, 2.

Rejepbayev, K. and P. Esenov, 1987: *Irrigation-Induced Changes in Soil and Land Reclamation Conditions in the Khauzkhan Area*. Ashgabat: Ilym Publishing House, 268 pp.

Rejepbayev, K. and A. S. Ovsyannikov, 1974: Karakum Canal impact on soil salinization in the Murgab oasis. *Problems of Desert Development*, 5, 18–26.

Saparov, T. K., 1969: On methods for assessing filtration losses from canals. *Express Information*, Series 7, Issue 4.

Saparov, T. K., 1971: Dynamics of filtration losses in the Karakum Canal. In: *Irrigated Farming in Turkmenistan*. Ashgabat: Ilym Publishing House.

Saparov, T. K., V. A. Novitsky and N. A. Bakashev, 1970: Forecasting water losses in Stage I of the Karakum Canal. In: *Land Reclamation Studies in Turkmenistan*. Ashgabat: Ilym Publishing House.

Vostokova, E. A. and S. V. Skaterschikov, 1980: Use of satellite imagery for assessing Karakum Canal impact zone. *Problems of Desert Development*, 4, 39–44.

Zaletaev, V. S., 1991: Features of the present state and degradation of natural desert systems. *Problems of Desert Development*, 3-4, 56–64.

Zonn, I. S., 1981: Irrigation development of deserts and reclamation of irrigated lands. In: *Desertification Control in the USSR*. Moscow: Nauka, 70–86.

12 Environmental changes in the Uzbek part of the Aral Sea basin

ANATOLY N. KRUTOV

The environmental safety of a nation is determined by its environmental philosophy or a concise statement summarizing the set of official views and principles worked out by political leaders, a declaration of primary goals, trends, and ways a nation's activities can achieve an optimal interaction between society and nature.

To harmonize the interactions between society and nature it is essential to create conditions for environmental stability manifested as the ability of the ecosystems to sustain their structural and functional features under direct and indirect stress of anthropogenic impacts. Environmental problems that emerged in the Aral Sea basin are an integral part of the socio-economic policy pursued in the former USSR. They cannot be solved isolated from other acute issues facing society.

Creeping environmental phenomena in the Aral Sea basin have resulted from:

- orienting agriculture and industry toward maintaining the cotton independence of the former USSR at the cost of large-scale development of new (virgin) lands without improving the quality of those lands;
- widespread introduction of cotton and rice monocrops;
- large-scale non-dose-related 'chemicalization' of agriculture;
- use of water-wasting technologies and irrigation techniques resulting in enormous losses of irrigation water during its transport;
- irrational siting of production activities in locations with a deficit of water resources and without regard to air pollution.

These factors emerged under the then-dominant socio-economic and political conditions of the former Soviet Union, which can be characterized by the following factors:

- centralized command-and-control-type administration of the national economy based on state ownership of the means of production and natural resources with an absence of economic responsibility (i.e., penalties) for environmental damage;
- pattern of extensive economic growth aimed at the development of production forces at the cost of an effective use of natural resources;
- priority for the development of large-scale, highly concentrated production resulting in extremely high stress on local environments;
- extreme degree of monopoly in certain sectors that hampered either the closure or the revitalization of environmentally hazardous industries;

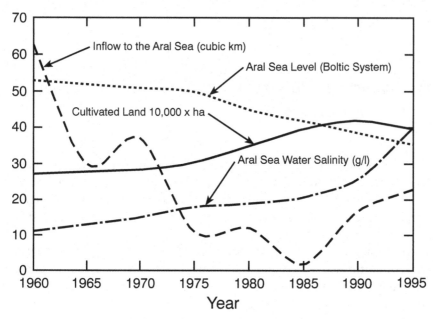

Figure 12.1 Environmental conditions of the Aral Sea, 1960–95.

- lack of direct competition with world producers resulting in the lack of available conservation technologies and management techniques;
- lack of public awareness of the qualitative and quantitative state of the nation's natural resources, leading to 'zero' impacts by the public on the decision-making processes related to the siting and construction of new (or reconstruction of existing) industries.

The peak of irrigated farming in the Aral Sea basin, particularly in Uzbekistan and Turkmenistan occurred in 1975–85. About one million hectares of new lands were developed in Uzbekistan in this period. By 1990 the area of irrigated agricultural lands had increased by 1.4 times compared with 1974 (Figure 12.1); the gross agricultural output increased by approximately the same amount.

Yet, an indicator such as the gross agricultural output per 100 ha of irrigated land was in 1990 at the level of 1975, and gross production per capita had decreased by about 19 times compared with 1980. New hard-to-reclaim saline lands were involved in agricultural turnover during a large-scale land development. Soil salinity of parts of the Republic of Karakalpakstan, Bukhara and the Golodnaya Steppe reached 90–95%. (Table 12.1).

Under conditions of extensive economic growth, the monocropping of cotton became a key environmental issue in Uzbekistan. Of cultivated land in Uzbekistan 73.4% was under cotton in 1986. Studies have shown that, if 50% of cultivated land is under cotton, rapid degradation of the soil is inevitable. A main indicator of soil fertility is humus and in the region the humus content of the soil had decreased by 30–40%. Soils with very low to low humus content

Table 12.1 Changes in soil salinity of agricultural lands in Uzbekistan

Land area assessed for soil salinity in 1982–87, and again in 1990–1992 (×1000 hectares)

Oblast	Surveyed land 1982–87	Surveyed land 1990–92	Nonsaline land 1982–87	Nonsaline land 1990–92	Saline land 1982–87	Saline land 1990–92	Degree of salinity — Slight 1982–87	Slight 1990–92	Moderate 1982–87	Moderate 1990–92	Heavy 1982–87	Heavy 1990–92	Type of change[a]
Karakalpakstan	455	443	24	17	430	426	180	167	179	184	71	75	1
Andijan	266	226	204	157	62	69	40	41	14	23	8	5	2
Bukhara	355	271	19	18	336	253	234	179	81	64	21	10	5
Djizak	269	249	94	128	175	121	74	78	79	29	22	14	5
Kashkadarya	461	422	178	215	283	207	172	135	81	41	29	31	5
Namangan	251	208	237	176	14	32	10	9	4	10	0	13	3
Samarkand[b]	333	303	284	248	49	55	43	44	6	11	0	1	4
Surkhandarya	284	263	186	149	98	114	59	76	32	29	7	9	1
Syrdarya	284	271	18	18	266	253	112	107	88	80	65	65	4
Tashkent	360	314	323	292	37	22	33	19	3	2	0	1	5
Fergana	323	287	276	226	48	61	34	19	11	32	8	10	5
Khorezm	239	212	65	75	174	136	126	86	33	43	15	7	5
Total in Republic	3879	3468	1908	1720	1971	1748	1118	961	611	547	242	241	
in percent	100	100	49	50	51	51	29	28	16	16	6	7	

Notes:

[a] Type of change between the 1982–87 and 1990–92 surveys: 1, slight deterioration; 2, deterioration; 3, serious deterioration; 4, stable; 5, improvement.
[b] Includes Navoi.

Source: State Committee for Nature Protection.

(0.4% and 1.0%) were estimated at about 40% of the total irrigated lands. The total low-productivity of plowed land amounted to about 0.5 million ha.

About 2.1 million ha of irrigated lands faced the risk of deflation of which 700 000 ha were at medium and high risk; 300 000 ha were subject to water erosion. Cotton monoculture dominated for several decades and striving for increases in gross production not only hampered crop rotation (a basic principle of soil protection in agriculture) but necessitated the large-scale application of fertilizer and about 80 different types of pesticides (Table 12.2). Enormous doses of fertilizers and an intense usage of the soil camouflaged the major losses of soil fertility, a fertility that had accumulated in the soil during thousands of years of evolution of natural systems in the region (Table 12.3). All this has resulted in the destruction of biological processes in soils, the degradation of natural regulatory mechanisms, and the transformation of the soil from a sophisticated ecosystem to a mere substrate transporting the applied mineral fertilizers to plant roots.

The established agricultural practices were the main cause of a difficult water management situation worsened by a cessation in the inflow of water to the sea. Waters of the Syrdarya and Amudarya were almost totally consumed on their way to the Aral Sea from the late 1970s (Figure 12.1). A high water consumption rate for irrigation, a low efficiency of the irrigation network, and a primitive use of irrigation exhausted regional water resources. Because of the low level of efficiency of water use, water resources were running out well in advance of their expected lifetime. Thus, the soils that could be continually irrigated turned out to be less than had been predicted.

Groundwater is an integral part of the republic's water resources. It is used as a reliable source of drinking water and for industry, rangeland, and irrigation. The total consumption of groundwater amounted to 52% of the estimated reserves (19 km³ per year).

The region's two major rivers were navigable down to their estuaries. They are polluted by effluents from cattle, municipal sewage networks, industries and by collector and drainage waters (Krutov, 1995). Water is polluted by heavy metals, fluorides, phenols, petroleum products and other toxins from ferrous and non-ferrous metallurgical, chemical, and petrochemical industries.

The hydrochemical and hydrobiological regimen of regional water sources has been strongly affected by the highly mineralized effluents from irrigation activities. A broad, hard-to-identify spectrum of chemicals has been supplied by municipal sewerage networks. For example, effluents from health care facilities with a high content of organic and biological pollutants have markedly increased in recent years. As another example, collector waters had been considered, from the standpoint of land reclamation, as a positive factor in enriching the fertility of the land (Rubinova, 1985). This view is no longer held today. Today these waters are discharged into rivers and water bodies and not only increase the water's salt content but also pollute them

Table 12.2 Detailed summary of pesticide use and application rates in oblasts of Uzbekistan

Oblasts	Year	Insecticides and fungicides			Herbicides			Defoliants			Sulfur compounds			Total of all pesticides		
		Total applied (Metric tons)	Area (×1000 ha)	Rate (kg/ha)	Total applied (Metric tons)	Area (×1000 ha)	Rate (kg/ha)	Total applied (Metric tons)	Area (×1000 ha)	Rate (kg/ha)	Total applied (Metric tons)	Area (×1000 ha)	Rate (kg/ha)	Total applied (Metric tons)	Area (×1000 ha)	Rate (kg/ha)
Karakalpakstan	1990	883	130	6.8	1085	68	16.0	2667	160	16.7	22	4	5.5	4657	362	12.9
	1993	128	87	1.5	618	45	13.7	2262	120	18.9	178	0	0.0	3008	252	11.9
Andijan	1990	573	154	3.7	134	14	9.6	1952	120	16.3	8041	170	47.3	10700	458	23.4
	1993	243	145	1.7	36	2	18.0	935	88	10.6	4048	432	9.4	5262	667	7.9
Bukhara (with Navoi)	1990	443	104	4.3	196	6	32.7	2782	167	16.7	673	15	44.8	4094	292	14.0
	1993	223	82	2.7	8	0	0.0	2089	119	17.6	506	8	63.2	2826	209	13.5
Djizak	1990	221	76	2.9	253	151	1.7	3971	231	17.2	5148	81	63.6	9593	539	17.8
	1993	127	70	1.8	81	12	6.8	4613	220	21.0	4621	66	71.5	9542	368	25.9
Kashkadarya	1990	826	84	9.8	151	35	4.3	7194	308	23.4	1711	46	37.2	9882	474	20.8
	1993	664	195	3.4	19	1	19.0	5574	232	24.0	826	18	45.9	7083	446	15.9
Namangan	1990	260	76	3.4	30	3	10.0	1606	104	15.4	1785	14	127.5	3681	197	18.7
	1993	261	94	2.8	24	2	12.0	1231	86	14.3	2587	52	49.8	4103	234	17.5
Samarkand	1990	811	142	5.7	77	24	3.2	4494	160	28.1	3043	59	51.6	8425	385	21.9
	1993	194	89	2.2	17	3	5.7	1286	91	14.1	2424	39	62.2	3921	222	17.7
Surkhandarya	1990	931	68	13.7	194	17	11.4	4460	190	23.5	1350	9	150.0	6935	284	24.4
	1993	633	108	5.9	128	14	9.1	3036	201	15.1	580	11	52.7	4377	334	13.1
Syrdarya	1990	285	61	4.7	414	111	3.7	3409	194	17.6	4248	74	57.4	8356	440	19.0
	1993	88	61	1.4	103	3	34.3	2993	144	20.8	2370	69	34.3	5554	377	14.7
Tashkent	1990	929	278	3.3	524	59	8.9	2995	166	18.0	5238	86	62.0	9776	589	16.6
	1993	347	226	1.5	132	14	9.4	2621	142	18.5	4541	106	42.8	7641	488	15.7
Fergana	1990	273	51	5.4	89	3	29.7	1785	113	15.8	5520	134	41.2	7667	301	25.5
	1993	205	64	3.2	14	2	7.0	1007	74	13.6	204	136	1.3	1430	296	4.8

Table 12.2 (cont.)

Oblasts	Year	Insecticides and fungicides			Herbicides			Defoliants			Sulfur compounds			Total of all pesticides		
		Total applied (Metric tons)	Area (×1000 ha)	Rate (kg/ha)	Total applied (Metric tons)	Area (×1000 ha)	Rate (kg/ha)	Total applied (Metric tons)	Area (×1000 ha)	Rate (kg/ha)	Total applied (Metric tons)	Area (×1000 ha)	Rate (kg/ha)	Total applied (Metric tons)	Area (×1000 ha)	Rate (kg/ha)
Khorezm	1990	254	98	2.6	10	4	2.5	1475	89	16.6	290	6	48.3	2029	197	10.3
	1993	88	46	1.9	40	1	40.0	463	36	12.9	91	1	91.0	682	84	8.1
Uzbekistan	1990	6687	1322	5.1	3157	497	6.4	38790	2003	19.4	37159	697	37.2	85793	4519	19.0
	1993	3202	1266	2.5	1222	99	12.3	28110	1553	18.1	24913	958	26.0	57447	3876	14.8

Source: State Committee for Nature Protection.

Table 12.3 Use of mineral fertilizers and pesticides in the oblasts of Uzbekistan

×1000 metric tons of active ingredients per year

Oblast	Total chemical fertilizers					Total pesticides					Percent reduction from 1989 to 1993	
	1989	1990	1991	1992	1993	1989	1990	1991	1992	1993	Chemical fertilizers	Pesticides
Karakalpakstan	137	121	111	96	99	5.6	4.7	3.4	3.0	3.2	27.7	42.9
Andijan	106	97	100	89	87	11.6	10.7	6.2	9.6	5.3	17.9	54.3
Bukhara[a]	113	113	111	111	82	4.3	4.1	2.7	3.4	2.8	27.4	34.9
Djizak	78	68	73	55	59	0.0	9.6	6.6	8.0	9.5	24.4	n/a
Kashkadarya	131	99	121	109	100	11.6	9.9	7.0	7.7	7.1	23.7	38.8
Namangan	87	94	90	83	70	3.6	3.7	3.6	3.4	4.1	19.5	−13.9
Samarkand	102	105	107	84	72	7.6	8.4	6.7	5.8	3.9	29.4	48.7
Surkhandarya	103	110	91	91	90	6.8	6.9	4.4	6.0	4.4	12.6	35.3
Syrdarya	162	83	84	65	59	14.3	8.4	5.0	5.4	5.6	63.6	60.8
Tashkent	105	83	97	83	72	9.0	9.8	7.4	6.4	7.6	31.4	15.6
Fergana	124	112	111	100	102	7.6	7.7	4.0	3.5	3.3	17.7	56.6
Khorezm	89	102	81	76	79	2.1	2.0	1.9	1.0	0.7	11.2	66.7
Uzbekistan	1338	1188	1177	1043	971	84.2	85.9	59.8	63.2	57.7	27.4	31.5

Note: [a]Includes data for Navoi oblast.

Source: State Committee for Nature Protection.

with pesticides and fertilizer residues (Krutov, 1995; Chembarisov, 1989). Organic compounds and toxic chemicals have been discharged into these water bodies. Of greatest concern is the environmental state of water bodies within large industrial areas and the collector and drainage network in the Priaralye region and in the Bukhara oasis.

Groundwater pollution in the Uzbek Republic could be classified as follows:

1. regional pollution – as a result of filtration losses from rivers and canals and the infiltration into the ground of contaminated irrigation water;
2. local pollution – at industrial and dumping sites, mineral fertilizers and pesticides depots, air fields for agriculture aviation, etc.

Large-scale irrigation farming with no attempts to maintain or improve land quality, growing and irreversible demand for irrigation water, and a series of low-flow dry years had reduced water flow into the Amudarya and Syrdarya deltas to 37.8 km³ in 1970 (Micklin, 1994). The rivers' inflow to the Aral Sea was almost terminated in the early 1980s. Not so long ago the Aral Sea was the fourth largest inland sea in the world and was economically important because of its fishing, hunting, transportation, and recreational value. The regime of the sea was directly related to Amudarya and Syrdarya inflow, the groundwater supply, precipitation and evaporation from the water surface. Under a regional water budget in equilibrium, the mean sea level was 53 m in the late 1950s and early 1960s, the volume of sea water was estimated at 1061 km³, its surface area was 67 000 km², and it had a maximum depth of 67 m. The Amudarya and Syrdarya inflow in that period was 56 km³, precipitation over the sea was 8 km³, the groundwater supply estimated at 1 km³ and evaporation estimated at 63 km³ (Figure 12.1).

Under conditions of a disrupted water balance, the sea level dropped by 16.3 m between 1961 and 1991. In the same period the water volume was reduced to a third, the water surface area was reduced by half, and the salt content of sea water increased from 9–10 g/l to 34–40 g/l.

The drying of the Aral Sea exposed an area of 32 000 km² of seabed; the coastline retreated over 80 km. The sea split into two parts, a Large and Small Sea, and it continues to dry up. It has also resulted in the shrinking of the tugai forest and bush vegetation, the drying of numerous lakes and their transformation into focal points of salt accumulation. Another major source of sand and salt aerosols emerged in the region. The exposure of over 3 million ha of dry seabed and parts of the deltas covered with fine earth and salt deposits led to the acceleration of wind erosion processes and the formation of new moving sand dune fields. As a result, the boundaries of the desert have expanded markedly.

The large-scale drying of the Aral Sea has also reduced its warming effect in winter and cooling effect in summer on the neighboring regions. It has intensified the arid and continental features of climate, particularly at a distance of 50–60 km from the sea's coast. The frost-free period has been reduced by 10–12 days and the total positive temperatures during the vegetation season increased to 120–170 °C. The Aral Sea contribution to regional moisture transport has been reduced from 5–8% to 2–4%. Increased temperature contrasts at the sea/dryland border generated a higher wind velocity and intensified the sea breeze phenomenon. This resulted in an increase in air turbidity and subsequently a change in the regional radiation budget.

The sea's natural regime allowed for the moistening of the area surrounding the Aral. Excessive condensed moisture near the sea and its islands favored the formation of freshwater lenses in sandy soils; various types of desert vegetation were found. However, a lowered groundwater table resulted in drifts of previously fixed sands of the newly exposed dried seabed with a disappearance of the freshwater lenses. The near-surface location of the groundwater with a high salt content (100–300 g/l), combined with the flat relief in the Priaralye, favored waterlogging and an acceleration of desertification processes.

Previously fertile, humus-rich, meadow-swamp soils in delta areas located within the zone of influence of irrigation have been transformed into low-productivity meadow-takyr and sandy-desert soils with much lower fertility.

The variety of mammal and bird species has been reduced by 50%. Dried areas rapidly became populated by rodents: vectors of dangerous diseases. Since 1983, the Aral Sea has lost its fishery, game, recreation, and transportation value.

The consequences of the above-mentioned factors in the context of the region's depressed social and economic development have contributed to the deterioration of the population's health. Preliminary studies in the Republic of Karakalpakstan and the Khorezm region have shown considerable adverse health conditions among the local population (see Elpiner, this volume). The field data correlate well with the conclusions conducted by various assessments of international organizations and independent experts (Kudat *et al.*, 1995).

In sum, the environmental phenomena accumulating in the Aral Sea region are unprecedented and require further investigation. As a result of the drying of the Aral Sea, a complex of environmental issues has emerged in the Priaralye region. That complex presents an intergovernmental challenge in terms of its origins and impacts of the economy and population health. The scope of this environmental disaster goes beyond the regional boundaries of Central Asia and has acquired a global concern.

Accelerated rates of land and water resources development have stimulated a dynamic (e.g., rapid) use of the Uzbek Republic's raw materials and its mineral and energy resources: Power and heat generation using local coal deposits using an open-pit mining technique, production of copper and lead by mining operations, production of mineral fertilizers, cement, ceramics and building materials by exploiting local quarries.

Over 100 million metric tons of waste, half of them toxic, have been generated in the Uzbek Republic annually (World Bank, 1995). They were partially used as secondary resources in the economy, while the bulk of it was accumulated at various storage sites with a total capacity of 2 billion metric tons. The highest volume of waste (1.3 billion metric tons) is generated by the mining and processing industries. The total volume of solid wastes has been estimated at 100 million metric tons. Wastes are not only dumped on lands suitable for farming but pollute soils, surface water and groundwater, and atmosphere.

The Uzbek Republic has numerous water- and power-consuming industries related to cotton production. The absence of closed air and water system technologies results in the large-scale emission to the atmosphere of pollutants, e.g., products of incomplete combustion of fuel, raw materials processing, and the like.

Scattered all over the republic are industrial projects using hydrocarbon fuel to generate heat and power for industrial and municipal purposes, such as enterprises processing sulphur- and nitrogen-containing raw materials. They discharge into the atmosphere many metric tons of sulphur, nitrogen and carbon oxides and particulates. Alongside the large-scale emissions of basic pollutants into the atmosphere, it has been estimated that local sites discharge about 150 specific hazardous chemicals.

The republic's fixed and mobile pollution sources have discharged about 4 million metric tons of hazardous chemicals annually of which about 50% was carbon monoxide, 15% hydrocarbons; 14% sulphur dioxide; 9% nitrogen oxide; 8% solid particles; and 4% specific high-toxic ingredients.

The locating of large-scale industrial sites in narrow mountain plains with a specific mountain–plain pattern of air circulation favored the exchange of discharged pollutants generating regional atmospheric pollution.

In Uzbekistan it is typical that the places exposed to high levels of pollution coincide with high-density populated areas. The same areas are usually characterized by the highest values of the atmospheric pollution index which take into account both pollutant concentration and toxicity.

National attempts at economic development and related environmental damage have negatively affected the republic's flora and fauna. In the recent past forests that had covered relatively large areas were reduced by an order of magnitude as a result of tugai forest-cutting and by river regulation. The uprooting of forests to clear potential agriculture lands accelerated water and wind erosion and secondary salinization processes in adjacent areas.

Over 2 million ha of forests were converted for agricultural activities. They were used as rangelands for several decades. Thus, these lands lost their economic value as forest. Forests were cut on the plains and on mountain slopes and the cleared sites were then turned into cultivated lands and pastures. Forest areas are shrinking and protection measures by the State Forest Fund have yielded zero results. Large-scale, uncontrolled procurement programs by various bodies have markedly curtailed the stock as well as the habitats of various wild species. The first edition of the Soviet Union's Red Book of endangered species contained 163 plant species compared with the listing of twice as many in its second edition. The destruction of mountain forests, floodplain soils and tugai forests, combined with cattle overgrazing on fields adjacent to water bodies, the growing mineralization and polluting of rivers, lakes and reservoirs, and the disastrous drying out of the Aral Sea have destroyed wildlife habitats and reduced species diversity by reducing the number of species and by slowing down their reproduction under natural conditions.

Typical of Uzbekistan fauna is their ancient history and sophisticated genetic ties, including various endemic species of Central Asia. The republic is inhabited by 650 vertebrate species, many which are endangered. Already the Turan tiger, red wolf, cheetah and striped hyena have become extinct. The Front Asian leopard, Ustyurt ram, markhor, great bustard, and sand viper, among others, are on the verge of extinction. Over 60 species are listed in the republic's Red Book (Uzbekistan Red Book, 1989).

The long-term orientation of agriculture and industry toward cotton growing with excessive chemicalization, the uneven distribution and high concentration of population in a limited area, limited and depleted water resources and their pollution by agricultural, municipal and industrial effluents, siting of settlements with high population density and environmentally hazardous industries in intermontane valleys, specific natural climatic features favoring pollutants accumulation in the environment – all this has led to the deterioration of ecosystems in the Aral Sea basin at the local level.

The Uzbekistan government attempted to control the progressing accumulation of environmental changes (e.g., creeping environmental change) by a series of decisions adopted in 1978–86, but they failed to improve the environmental situation (World Bank, 1996). An important landmark attempt to solve this complex multifaceted problem was an Integrated Scientific and Technological Programme of Environmental Protection for 1986–90 and up to 2000 elaborated in 1986 (UNDP, 1995).

Retrospective analyses have shown that most of the measures envisaged by the program for 1986–90 were either only partially implemented or not fulfilled at all. 'Perestroika' introduced its own corrections in this unfulfilled program. Nevertheless, the implementation of technological and administrative measures envisaged by the program yielded some positive results by, for example, controlling wastewater discharge into water bodies and by controlling emissions into the atmosphere.

The experience gained during the program's preparation was used in 1987 in the drafting of a detailed State Programme for Environmental Protection and Rational Use of Natural Resources for 1991–95 and up to the year 2000 (UNDP, 1995). Despite the previous program, it was relatively more environmentally biased with the planned measures taking into account environmental standards and economic benefits. However, a sectoral approach oriented toward state planning and the financing of environmental activities, under the conditions of the disintegration of the USSR, discouraged the adoption and implementation of the program.

Transition to a market economy was constrained by such negative phenomena as the disruption of economic ties with traditional partners in the former Soviet Union (World Bank, 1995), the non-payment of fines by various enterprises, and the growing inflation rates that resulted in a recession and low investments in the economy. Environmental protection funds allocated in a centralized manner were cut and prices for energy, materials, equipment, etc., were raised. This resulted in the short-term mitigation of anthropogenic loads of pollutants on the one hand, and abortive measures aimed at environment protection on the other. Target-oriented measures of the Uzbek government to ameliorate the crisis in the country's economy along with several high river flow years also contributed to temporarily lowering human-induced impacts.

Large-scale development of new lands in the Uzbek Republic ceased in 1991 for lack of funds. The commitment to cotton as a monocrop – a commitment that had lasted in the republic for several decades – appeared to have been abandoned. New trends appeared in agriculture aimed at environmentally friendly farming techniques based on sustaining a positive humus budget in the soil through crop rotations. The actual area under cotton production was estimated at 47.5% in 1993 compared with 56% in 1989.

In 1993 the total amount of pesticides was almost half and mineral fertilizers 30% less than in 1990 (in part due to the high cost of agricultural chemicals). Specific water consumption per hectare was lowered from 15 800 m^3 in 1985 to 12 600 m^3 in 1994. Gross agricultural output, however, did not decline but has remained stable in the last few years (World Bank, 1996).

Despite a considerable reduction in pesticide use, no marked changes in soil pollution by them was reported in the republic. As in past years, soils have remain polluted with residual DDT because of its high resistance to degradation (World Bank, 1997).

Of main concern are sites of local pollution near airfields of agriculture aviation, pesticide depots and isolated plots under agriculture crops and perennial plants.

Reduced application of pesticide and chemical fertilizers in agriculture, bringing the water content in the river back toward normal, and cutting the discharge of polluted effluents in open water bodies, has affected the pollution levels in a favorable way in the republic's main waterways. Favorable changes in pollution levels in the Amudarya and in the drainage collectors in the Bukhara region were reported in the mid-1990s.

Streams and bodies of water in the Zeravshan River basin up to Navoi stabilized at a 'moderate' level in 1993. Water quality markedly improved in the main water streams of the Fergana Valley with water pollution reaching a baseline level. Water quality improved in the Chirchik River, particularly in terms of reduced nitrogen pollutants. The HCH isomers in surface water bodies tended to decrease throughout the republic except in the Priaralye region. No DDT has been reported in water bodies in the last few years (Krutov, 1995).

Following the low-flow period of the 1980s, when flow from the Amudarya and Syrdarya did not reach the Aral Sea, the volume of water in these rivers had recently become closer to normal. Water supply in the deltas was 16.5 km^3 in 1991, 33.5 km^3 in 1992, and 30.6 km^3 in 1994. A growing water supply to the deltas was accompanied by a restructuring of the hydrological network, an appearance of lakes formed by flooding and new channels for river water to flow into the sea. The total area of lakes and floods in the Amudarya delta was more than 2000 km^2 in 1992, which was comparable to natural flood areas. Sites where the rate of desertification processes have slowed down have appeared. Neverthless, the 1993 sea level was 36.94 m.abs [above Baltic Sea level], the surface area was 33 300 km^2, the volume of water in the sea was 277 km^3, and the salt content was 40 g/l. As compared with 1960, the sea level had dropped by 16.06 m, the surface area by 32 700 km^2 and the volume by about 75%.

Economic recession and the lower use of fuels in Uzbekistan in 1991–94 have resulted in a 50% reduction in industrial pollutant discharge. The rate of the lowering of industrial discharges was so high that in 1991 the maximum allowable concentration levels in the republic were reported in two regions, and in 1994 in eight regions. A burning torch at the Minbulak oil deposit increased the density of aerosol precipitation by 1.5–2.5 times in 1992.

The Uzbek government adopted a special decree, withdrawing all licenses for the long-term use of forest lands (UNDP, 1995). A decision was made to create poplar plantations on low-productivity irrigated lands, a total area of 90 000 ha, Eldorado pine on 800 ha, and walnut on 500 ha.

New quotas and procedures have been established for procuring pharma-ceutical and decorative plants. The restriction of hunting increased pheasant numbers from 18 200 in 1991 to 65 000 in 1993. Commercial hunting of saiga antelopes has been banned since 1992–93 because of their already reduced number. A quota for muskrat was set in the Karakalpak Republic. For the first time in the past 25 years, the trapping of muskrat became illegal. Seven nature reserves have been created in recent years. The spatial extent of the Chatkal nature reserve, a UNESCO biosphere reserve, was increased.

Despite a short-term improvement in the environmental situation because of economic difficulties (e.g., no money for fertilizers) and several high-water years, the following long-term negative trends have been observed:

  • an environmental crisis because of the drying out of the Aral Sea;
  • an unsatisfactory state of land and water resources because of long-term

cotton monoculture and the excessive application of chemical fertilizers and pesticides;

- the adverse impacts of urban, agricultural and industrial development on the quality of land, water, air and biological resources and the atmosphere;
- poor collection, transportation and treatment of industrial and municipal wastes;
- the dumps at uranium mines that had been created in the 1940s without proper safeguards;
- the human-induced degradation of flora and fauna;
- the underdeveloped and unsatisfactory logistical setup and methodological framework for environmental (and pollution) monitoring;
- an inadequate legislative and regulatory framework; an absence of enforcement mechanisms for environmental protection and the rational use of nature.

Solutions to these problems were envisaged in the 1991 'Programme of Environmental Protection and Rational Use of Nature in the Republic of Uzbekistan to the Year 2005' (UNDP, 1995). Strategic trend projections of the need for environmental protection in 1991–95 as noted in the program have proven correct. However, newly emerged political and socio-economic realities in the republic have demanded that modifications of the program be made.

Large-scale agricultural, industrial and municipal development in Central Asia have resulted in enormous water consumption throughout the region. Such consumption has depleted various bodies of water, including the lower reaches of the Amudarya and Syrdarya. In addition, the continual discharge of highly mineralized collector and drainage water into the rivers has drastically modified the natural hydrochemical runoff.

A search for new patterns of economic development must take into account environmental aspects. From this viewpoint, water quality under arid conditions is a key factor for the stable functioning of ecosystems. Obviously, forecasts of changing qualitative indicators of surface water should be based on the analysis of trends and assessments of the actual state of water resources. Results of studying the dynamics of mean annual concentrations of pollutants, using the example of the Amudarya, allow for the evaluation of major trends of variations in the analyzed parameters.

Anthropogenic impacts in the Amudarya basin are expressed by growing volumes of water diversions and the discharge of polluted collector and drainage waters directly back into the river. Natural geographic and hydrographic features of a region play an important role. The hydrochemical composition of drainage and collector waters discharged into the Amudarya and its impacts on the river's water quality have been studied in detail. However, the changes in indicators taking into account industrial, agricultural and municipal wastewater are often completely unknown.

In the context of specific features of the Amudarya, the issue of changes in the salt content cannot be ignored. Such changes are closely related to the removal of river water and the discharge of drainage and collector waters into the river.

The correlation between water expenditure and salt content is well-known for practically all the sections of the Amudarya. It is worth noting that, in the Termez-Kerki section, effluents have had negligible impacts on Amudarya runoff because of the relatively low volume of water withdrawals and salt content. On the other hand, the impacts at the Kerki-Tuyamuyun section are most strongly expressed. About 3 km³ of drainage water with a salt content ranging from 2.0 to 12.0 g/l has been discharged annually into the river. Chemical and biological oxygen demand (COD and BOD) are indicators closely related to the organic compound content and they allow for an integrated assessment of water quality. Industrial and municipal wastewater are the main sources of organic compounds. Their content increases as one progresses from upstream to midstream locations; in the Tuyamuyun reservoir the self-purification processes with a lower organic content in the Kipchak section of the river have been noted. Polluted water discharge below this point increases the oxygen demand and, at the Nukus section of the river and below, self-purification processes play an important role in decreasing the COD and BOD.

The following comments related to BOD and COD:

1. It is impossible to establish direct relationships between these indicators. For instance COD in the river sections at Termez, Tuyamuyun, Kipchak, and Kzyljar tended to decrease in 1984–94. Lower BOD values were reported at the river sections at the Tuyamuyun Canyon and Nukus.

2. The level of pollution by poorly oxidized organic compounds in the river section at Nukus is unstable (fluctuates). Analyzing the available data has yielded no results.

3. The BOD value in the river sections at Termez and Kipchak is characterized by a stable pollution level. The minimum content of easily oxidized organic compounds in the river section at Kzyljar was reported in 1987. It rapidly increased in the following years.

In the upper watershed area, phenol concentrations were above the maximum allowable concentration for the municipal and domestic water supply. Mean annual concentrations of phenols were similar in the river section from Termez to Tuyamuyun. At the river section at Kipchak, their relative content increased by more than 27%. An analysis of available data of wastewater being discharged into the river allowed us to conclude that they were a source of phenols.

Petroleum products are another indicator of municipal and domestic water pollution above maximum allowable concentrations along the Amudarya. Mean annual concentrations decrease at the Termez-Kerki section of the river and increase at the lower section from Chardjou to the Tuyamuyun Canyon.

Analysis of the industrial water supply in the Republic of Uzbekistan and particularly of the content of petroleum byproducts in effluents failed to detect the source(s) of pollution.

A positive impact of the Tuyamuyun reservoir is the lowered concentration of petroleum byproducts in the downstream section of the river. Nevertheless, pollution levels will likely remain high over the long term.

In conclusion, it is worth noting that analyses of the available data on water diversions for irrigation obtained from calculations aimed at identifying pollution trends enabled us to conclude that there is an absence of a direct relationship between water content in a given year and pollutant concentrations in a long-term context. Against the background of lower water levels, environmental protection measures, and particularly water management projects, have improved water quality. For instance, mean annual concentrations of organic pollutants decreased, particularly phenols and petroleum byproducts. Yet, a change for the better has not occurred and Amudarya water is still polluted by various compounds. This water cannot be used for municipal and drinking water supply without special treatment.

The availability of a database and of software enables us to carry out timely analyses of historical and current data and to make forecasts for all rivers in the Republic of Uzbekistan.

References

Chembarisov, E. I., 1989: *Hydrochemical Characteristics of River and Drainage Water in Central Asia.* Moscow, 212 pp.

Krutov, A. N., 1995: The present conditions of surface water quality in the Amudarya River Basin. SANIGMI, **134**, 18–24. Tashkent. (In Russian.)

Kudat, A., A. Sholdasov, A. Ilkhamov, N. Ozmen and J. Bernstein, 1995: *Needs Assessment for the Proposed Uzbekistan Water Supply, Sanitation and Health Project.* Washington, DC: World Bank.

Micklin, P. P., 1994: *The Aral Sea Problem: Civil Engineering.* Proceedings of the Institution of Civil Engineers, 114–21.

Rubinova, F. E., 1985: *Changes in the Amudarya River Flow under the Influence of Water Management Activities in the Basin.* Moscow: SANII. 115 pp.

UNDP (United Nations Development Programme), 1995: *International Conference on Sustainable Development of the Aral Sea Basin: Final Report.* Nukus, Uzbekistan: UNDP.

Uzbekistan Red Book, 1989: Tashkent, Uzbekistan.

World Bank, 1995: *Annual Report.* Washington, DC: World Bank.

World Bank, 1996: *The Aral Sea Basin Program 3.1.A: Water Quality Assessment and Management.* Tashkent, Uzbekistan. Washington, DC: World Bank.

World Bank, 1997: *The Aral Sea Basin Program 3.2: Preparation Study of the Uzbekistan Drainage Project.* Tashkent, Uzbekistan. Washington, DC: World Bank.

13 Creeping changes in biological communities in the Aral Sea

NIKOLAI V. ALADIN

In recent years the environmental crisis of the Aral Sea basin has attracted worldwide interest. International, governmental, and public organizations, as well as different agencies of the United Nations and the World Bank have tried to comprehend the reasons for the initiation of this crisis and to overcome its adverse effects. The adverse effects of the crisis on socio-economic processes in the region and on the health of the local population have been reviewed in the scientific literature (e.g., Micklin, 1991; Glantz *et al.*, 1993; Elpiner, this volume) and in popular science publications (Ellis, 1990). Meanwhile, when purely biological problems are considered in such publications, they have either been considered only superficially or have not been considered at all. The notion of creeping environmental changes *per se* in the Aral Sea proper has been considered only slightly in the scientific literature. This is even more true for aquatic communities because the studies of such communities are more labor-intensive compared with studies of terrestrial ecosystems.

This paper attempts to fill the gap in this area, based on field observations over the past 16 years.[1]

In spite of insufficient detailed information on many of the ecosystems of the Aral Sea, it is possible to state with a great degree of certainty that creeping environmental changes began to appear in the mid-twentieth century. These changes were not initially connected with human-induced desiccation of the Aral Sea but were caused by large-scale acclimatization processes. For example, beginning in 1927, 18 fish species were introduced into the Aral Sea, of which 15 survived and inhabited the sea for some period of time (Karpevich, 1975). Stellate sturgeon (*Acipenser stellatus*) was introduced into

---

1. Aladin, 1982, 1983*a,b*, 1989*a,b,c*, 1990, 1991*a,b,c*, 1994*a,b,c*; Aladin and Andreyev, 1984: Aladin and Boomer, 1993, 1994: Aladin and Kotov, 1989; Aladin *et al.*, 1992*a,b*, 1994*a*; Aladin and Plotnikov, 1993, 1995*a,b*; Aladin and Filippov, 1993; Aladin and Eliseev, 1991; Aladin and Potts, 1992; Aladin and Williams, 1992, 1993). This chapter also relies on the available published information of the past 172 years beginning with the expedition of Lieutenant A. I. Butakov in 1848–49 (Andreyev and Andreyeva, 1981; Andreyev, 1981, 1989; Andreyev *et al.*, 1992*a,b*; Beklemishev, 1922; Bening, 1935; Berg, 1908; Blinov, 1956; Bortnik, 1990; Butakov, 1853; Dobrynin *et al.*, 1990; Dobrynin and Koroleva, 1991; Dogiel and Lutta, 1937; Elmuratov, 1988; Filippov, 1995; Karpevich, 1953, 1958, 1975; Khusainova, 1958; Kortunova, 1975; Kosarev, 1975; Lim and Markova, 1981; Lukonina, 1960; Meisner, 1908; Okolodkov, 1989; Orlova and Rusakova, 1995; Osmanov, 1962; Osmanov *et al.*, 1976; Pichkily, 1970, 1981; Plotnikov, 1991, 1993*a,b,c*, 1995; Plotnikov *et al.*, 1991; Rusakova, 1995; Trusov, 1947; Yablonskaya *et al.*, 1973; Zenkovich, 1963; Zernov, 1903).

the Aral Sea from the Caspian Sea during 1927–34 and from the lower reaches of the Ural River during 1948–63. Since 1958, it was recorded in catches. During 1929–32, attempts were made to introduce the Caspian shad, but it did not survive. In 1954–59, the Baltic herring (*Clupea harengus membras*) was introduced and, since 1957, it was recorded in catches. In 1954–56, two species of mullet were introduced from the Caspian Sea (*Mugil auratus, M. saliens*). In 1960–61, the grass carp (*Ctenopharingodon idella*) and silver carp (*Hypophthalmus idella*) were introduced from water bodies of China to the Aral Sea. Beginning in 1963, they were recorded in catches and even became important commercially.

In the attempt to introduce the mullet from the Caspian Sea to the Aral Sea six species of goby, friar (*Atherina mochon*) and pipefish (*Syngnatus nigrolineatus*) were introduced by accident; they became successfully acclimatized. It should be mentioned that the friar and three species of goby (*Pomatoschistus [Bubyr] caucasicus, Neogobius fluviatilis and N. melanostomus*) were widely spread in the Aral Sea by 1958–59. At the end of the 1970s, as a result of acclimatization, flatfish (*Platichthys flesus luscus*) were observed in the Aral Sea (Lim and Markova, 1981). The delta areas and the Syrdarya and Amudarya rivers proper were inhabited by *Mylopharyngodon piceus* and snakehead (*Ophiocephalus argus*).

In the process of acclimatization of the stellate sturgeon, two types of parasites were introduced from the Caspian Sea to the Aral Sea: a branchial parasite of the monogenetic fluke (*Nitzschia sturionis*) and a representative of Coelenterata (*Polypodium hydriforme*) living in the sturgeon roe (Trusov, 1947; Osmanov, 1962: Osmanov *et al.*, 1976). In 1934 the *Nitzschia sturionis* caused heavy epizootia among the Aral spinefish (an indigenous species of sturgeon) and massive deaths of the new host (Dogiel and Lutta, 1937). Amur parasites were brought from the Far East, together with the grass carp, *Mylopharyngodon piceus*, silver carp and snakehead. The *Bothriocephalus gowkongensis*, which was carried by the above species, adversely affected the indigenous Aral carp (Osmanov *et al.*, 1976).

In addition to the fish and attendant parasites, beginning in 1954, benthic and planktonic moving invertebrates were introduced into the Aral Sea. From 1954 to 1956, two species of shrimp were brought from the Caspian Sea; beginning in 1954, only shrimp were brought from the Caspian Sea; beginning in 1954, only one species of the two was found, i.e., *Palaemon elegans*. Originally, the representatives of this species of shrimp were found mainly in coastal waters. Later on, however, they were found in the middle of the sea. From 1958 to 1960, four species of opossum shrimp were introduced into the Aral Sea from the lower reaches of the Don River; starting in 1961, only three species were found: *Paramysis lacustris, P. intermedia, P. ullskyi.* From 1960 to 1963, the worm (*Nereis diversicolor*) and clam (*Syndosmya segmentum*) were introduced to the Aral Sea from the Berdyansk brackish lagoons of the Sea of Azov. The former was found in 1963, while the latter was found in 1967. At present, both species are widespread throughout the Aral Sea and are absent only in its deepest part.

In 1976 crab (*Rhithropanopeus harrisii tridentata*) was recorded as being in the Aral Sea. Evidently, while in the larval stage, it was accidentally introduced to the Aral Sea from the Caspian Sea or the Sea of Azov, along with the species targeted for acclimatization. In 1987–88 an attempt was made to acclimatize the Mediterranean mussel (*Mytilus galloprovincialis*) to the Aral Sea. However, that attempt was not successful, because by that time all of the hard rocks suitable for mussel settlement were located in the zone of desiccation (i.e., the recently exposed, dried seabed).

As mentioned earlier, planktonic invertebrates were introduced into the Aral Sea. Vegetable- and detritus-consuming Copepoda (*Calanipeda aquaedulcis*) were introduced into the Aral Sea from the Kuban brackish lagoons during 1965–66 and from the Taganrog Bay of the Sea of Azov in 1970. By 1972, representatives of this species were spread all over the sea and were dominant in the zooplankton. New planktonic larvae of benthos forms were found in the zooplankton, as a result of the acclimatization of new representatives of benthic invertebrates.

In making a retrospective review of creeping environmental changes related to the introduction of fish and invertebrates into the Aral Sea, it is important to note that these changes were both positive and negative. The most adverse effects were connected with fish acclimatization. The introduction of new representatives of the ichthyofauna and, primarily, such plankton-eating fish species as the Baltic herring, friar, goby fry, etc., contributed about tenfold to the reduction of the average biomass of the summer zooplankton. According to Kortunova (1975), from 1959 to 1968 the summer zooplankton biomass did not exceed 15 mg/m³; however, in the early 1950s it was usually about 150 mg/m³ (Lukonina, 1960).

New plankton-eating fish species consumed practically all coarse representatives of the zooplankton. *Arctodiaptomus salinus, Moina mongolica, Ceriodaphia reticulata, Cercopagis pengoi aralensis* were eaten up and fine forms, particularly Rotifera, became prevalent in zooplankton. Since 1973, the indigenous *Arctodiaptomus salinus* has not been observed in the Aral Sea, because it had been consumed. The same thing happened with *Moina mongolica* after 1975. This was not the only reason for the extinction of the two representatives of zooplankton. They disappeared quickly because they came into a competitive relationship for their food supply with the introduced *Calanipeda aquaedulcis*. However, in our opinion, the main reason they disappeared was related to their having been consumed by the above species of new plankton-consuming fish.

Similar negative processes occurred in the benthic communities. The introduction of benthos-consuming fish species and the competition between the indigenous species of the benthic fauna and the new acclimatized benthic species resulted in a threefold reduction of the summer benthic biomass, which in 1966–67 dropped to 9 g/m³ (Yablonskaya *et al.*, 1973). In previous years the biomass was usually within the range of 23–30 g/m³ (Karpevich, 1975). New benthos-consuming fish soon ate up the freshwater shrimp (*Dikerogammarus aralensis*); starting in 1964, its biomass sharply decreased, and from 1973 on it was not recorded in the sea. As in the case of the planktonic organisms, the competition of the *Dikerogammarus aralensis* with the newly introduced benthic species contributed to its extinction.

The introduction of the above-mentioned pathogenic parasites with new fish species had drastic adverse effects. The occasional introduction of six species of goby, friar and pipefish during acclimatization of the mullet can also be considered negative. These small non-commercial fish propagated very quickly in the Aral Sea and noticeably affected the nutritional base of the commercial fish. Today, we can state that in spite of the significant increase in species diversity in the ichthyofauna during acclimatization activities, the population of commercial fish changed only slightly; and the increase of the annual catch was insignificant (Karpevich, 1975).

The only positive effect of fish acclimatization might be associated with the introduction of flatfish into the sea. As of 1995, it was the only object of exploitation of the commercial fishery, whose fishing can theoretically be undertaken under the existing polyhaline conditions. From the late 1980s until the early 1990s such experimental fishing was successfully performed under the supervision of the Aral branch of the Kazakh Fishery Research Institute. In 1991 experimental large-scale fishing of flatfish yielded 112.2 metric tons (Zholdasova *et al.*, 1992). However, it was stopped because of its labor-intensive demands.

Contrary to the results of fish acclimatization, the introduction of moving invertebrates resulted in more positive effects, particularly, with respect to the *Calanipeda aquaedulcis*, *Syndosmya segmentum*, *Nereis diversicolor*, and *Palaemon elegans*. These introduced species became widespread under the polyhaline conditions of the mid-1990s. Their high tolerance for saline conditions and their ability of quick mass propagation made them the dominant species. Today these invertebrates are the main nutritional base for the surviving fish in the Aral Sea.

To summarize the information on creeping environmental changes in the Aral Sea related to acclimatization activities, one may conclude that these changes were more negative than positive, although there were some positive effects. Unfortunately, until 1954, only fish had been introduced into the Aral Sea, and all acclimatization activities failed to take into account recommendations of the scientific community. As a result, new parasites intruded into the sea and the sea's ecosystem was affected by the increased food load on the zooplankton and zoobenthos. Therefore, the epizootia of the Aral spinefish and the complete extermination of some planktonic and benthic invertebrates by the introduced fish were observed in the Aral Sea.

The environmental changes associated with these attempts at acclimatization were followed by creeping environmental changes related to the desiccation and salinization of the sea. The processes of sea level drop and of salinity increase began in 1960. However, the noticeable signs of their adverse effects were revealed only in the late 1960s and early 1970s. Fish roe, larvae, and fry were the first to be affected, and somewhat later the processes of ichthyofauna reproduction were disturbed (Karpevich, 1975). The point was that in the shallow-water spawning areas the salinity increased more rapidly than in the open sea areas and, by 1965–67, it exceeded 11–14 g/l. This level of salinity drastically affected the development of roe, larvae, and fry of fish of freshwater origin. According to Karpevich (1975), in the late 1960s the situation was aggravated in the spawning areas of the river's anadromous fish. By that time, in addition to the salinity increase, the area of spawning pools had decreased by 5–8 times because of the drop in sea level. The reduction of spawning areas was also caused by the construction of dams in the lower reaches of the Syrdarya and Amudarya. One of the dams (near the town of Kyzyl-Orda) cut off 90% of the spawning area of the Aral spineback and 80% of the spawning area of the barbel (Karpevich, 1975). The sum of the adverse effects of the acclimatization activities, construction of dams, salinization increases, and sea-level drop resulted in the fact that in the space of only one decade (1960–70) the total annual fish yield decreased by two-thirds from 44 000 to 17 500 metric tons.

Beginning in 1971, when the average water salinity in the open sea exceeded 12 g/l, the first signs of the adverse effects of a salinity increase on adult fish became apparent. The rate of growth of many fish species was slowed down; the death rate sharply increased and the population number decreased; numerous morphological aberrations and a general 'exhaustion' were observed (Karpevich, 1975). By the mid-1970s, when the average salinity was over 14 g/l, the natural reproduction processes of the majority of fish species were almost completely disturbed. By the early 1980s, when salinity exceeded 18–20 g/l, the Aral Sea lost its fishery importance and organized fishing activities ceased. By the mid-1980s all commercial fish in the Aral Sea, except for the flatfish, were completely lost. However, some old fish specimens were found in the deltas of the Syrdarya and Amudarya. A total of 20 indigenous fish species and the 15 introduced fish species were lost.

Among the most valuable of the extinct fish species were the following: the Aral spineback (*Acipenser mudiventris*), Aral salmon (*Salmo trutta aralensis*), *Chalcalburnus chalcoides aralensis*, bream (*Abramis brama orientalis*), carp (*Cyprinus carpio*), and roach (*Rutilus rutilus aralensis*). At present only five fish species, instead of 40, are observed in the Aral Sea, and four of those were introduced: the flatfish (*Platichthys flesus luscus*), Baltic herring (*Clupea harengus membrans*), friar (*Atherina mochon caspia*), and the monkey goby (*Neogobius fluviatilus pallasi*). The one indigenous species was the stickleback (*Pungitius platygaster aralensis*).

The invertebrates, both planktonic and benthic, were affected by the salinity increase and sea-level drop somewhat later than fish species. Below, creeping changes in each of the two ecological groups will be considered in detail.

The first signs of the adverse effects of salinization on zooplankton date back to 1971, when average salinity in the sea rose above 12 g/l. Until that time, all changes which occurred with the inhabitants of the aquatic environment were associated with the impact of the acclimatization activities. By the mid-1970s, when the average salinity exceeded 14 g/l, practically all freshwater and most brackish-water planktonic organisms were lost. However, euryhaline species of zooplankton survived. At that time, the following species were widespread: Rotifera (*Synchaeta vorax, S. gyrina, S. cecilia*); two species of Cladocera (*Podonevadne camptonyx, E. anonyx*); and two species of Copepoda (*Calanipeda aquaedulcis, Halicyclops rotundipes aralensis*).

In those years, some other zooplanktonic organisms were sometimes observed in separate parts of the sea. They included three species of Rotifera (*Brachionus plicatilis, B. calyoiflorus, Notholca squamula*) and one species of Copepoda (*Acanthocyclops bisetosus*) (Andreyev, 1989). Between the mid-1970s and the early 1980s, the total number of species of zooplankton that became extinct numbered about 20. Beginning in 1969, in spite of the reduction in the number of species, the average summer biomass of zooplankton gradually increased from 22 mg/m³ to 123 mg/m³ in 1981. On the one hand, this increase can be explained by the successful acclimatization of *Calanipeda aquaedulcis* and, on the other hand, by the reduction of plankton-eating fish species.

From the mid-1970s, a relative stability was observed in the species composition of zooplankton. Simultaneously, the average summer biomass increased, because the number of zooplankton-eating fish had dropped to a minimum. By the mid-1980s, when the average seawater salinity exceeded 22–23 g/l, signs of a new crisis in the zooplankton became apparent. In those years the number of Rotifera, *Podonevadne camptonyx* and *E. anonyx* sharply decreased. By the early 1990s, out of 56 species only 11 species remained in the plankton of the Aral Sea. They involved five species of Rotifera (*Synchaeta vorax, S. cecilia, Brachinous plicatilis, Notholca squamula, N. acuminata*), one Cladocera (*Podonevadne camptonyx*), five Copepoda (*Calanipeda aquaedulcis, Mesocyclops leuckarti, Schizopera aralensis, Malectinosoma abrau*, and *Nitocra hibernica*) (Plotnikov, 1995). However, it should be noted that, beginning in 1987, only *Calanipeda aquaedulcis* and larvae of benthic clams *Syndosmya segmentum* and *Cerastoderma isthmicum* dominated in the plankton. Thus, 10 species of the zooplankton out of the surviving 11 were seldom found.

Creeping changes in the benthic communities were roughly similar to the changes in the plankton described above. The first signs of the adverse effects of salinization on the zoobenthos of the Aral Sea dates back to 1971 (as in the case with zooplankton), when the average salinity level exceeded 12 g/l. Until that time (also as in the case with the plankton) all changes in benthic inhabitants were associated with the result of the acclimatization activities. Oligochaete and insect larvae were the first that could not survive. Then, mollusks of the genus *Dreissena* and *Hypanis* disappeared nearly completely. Only brackish-water and euryhaline species could endure salinity levels above 12–14 g/l. They involved such indigenous species as the clam (*Cerastoderma isthmicum*), Gasteropoda of the genera *Caspiohydrobia* and *Theodoxus*, Ostracoda (*Cyprideis torosa, Amnicytere cymbula, Tyrrenocythere amnicola donetz-ciensis, Limnocythere (Galolimnocythere) aralensis*) and the introduced species: clam (*Syndosmya segmentum*), polychaete worm (*Neries diversicolor*), shrimp (*Palaemon elegans*), and crab (*Rhithropanopeus harrisii tridentata*).

Of the 159 species of benthic inhabitants, which were noted in the *Atlas of Invertebrates of the Aral Sea* (1974) in the mid-1970s, only about 50 species survived (Andreyev, 1989; Filippov, 1995). Unfortunately, today it is impossible to give a more accurate figure, because only representative samples of the macrozoobenthos of the Aral Sea were regularly recorded; information on smaller benthic organisms was scarce and incomplete. As in the case with the zooplankton, from 1969 the average summer biomass of zoobenthos of the Aral Sea gradually increased from 11.7 g/m^3 to 184 g/m^3 in 1980 (Andreyev, 1989). The reasons for a more than tenfold increase of the biomass of zoobenthos were the same for the zooplankton, i.e., it was a result of the successful acclimatization activities and the loss of benthos-eating fish species.

By the mid-1980s, when the average water salinity was more than 22–24 g/l, the zoobenthos showed signs of a new crisis similar to that of the zooplankton. In those years many brackish-water species were lost and only euryhaline invertebrates coming from the Mediterranean-Atlantic Basin or continental brackish water bodies of the arid zone survived. By that time, only 15 species of benthic dwellers remained in the Aral Sea: two species of clam (*Syndosmya segmentum, Cerastoderma isthmicum*), nine species of gasteropod of the genus *Caspiohydrobia*, one species of polychaete (*Nereis diversicolor*), one species of shrimp (*Palarmon elegans*), one species of crab (*Rhithropanopeus harrisii tridentata*) and one species of ostracod *Cyprideis torosa* (Filippov, 1995). Because of the reduction in species diversity and the increase of benthos consumption by the propagated flatfish, by the late 1980s and early 1990s, the average summer biomass of the benthos dropped nearly twofold to 92.3–125.9 g/m^3 (Filippov, 1995), as compared with the period from the late 1970s to the early 1980s.

Unfortunately, the impacts of salinization and the drop in sea level on the phytoplankton and phytobenthos were not studied in the period of the anthropogenically induced desiccation of the Aral Sea. On the basis of individual works such as Pichkily (1970) and Elmuratov (1977), it is impossible to judge with certainty the creeping environmental changes in the sea's plant communities. We may only anticipate, in agreement with Kosarev (1975), that an intense substitution of freshwater and brackish-water plants by sea-water and hyperhaline plants occurred in the phytoplankton and phytobenthos.

Bear in mind that the change in the balance of biogenic components must have a great impact on plants. Thus, during desiccation of the Aral Sea, the phytoplankton and phytobenthos were, evidently, affected by the impact of the following four factors: (a) salinization, (b) sea level drop, (c) acclimatization activities, and (d) changes in the balance of biogenic components.

By the mid-1970s, when the average salinity level was above 14 g/l, the biomass and the number of phytoplankton decreased threefold, and diatomaceous algae dominated to a greater extent (over 60%) than earlier (Kosarev, 1975). In view of the lack of direct observational data, we may only suggest that the majority of freshwater and of slightly brackish-water plankton algae was lost, and that euryhaline species of phytoplankton reached their mass development. According to Okolodkov (1989), the marine euryhaline forms have made up the greater part of the dominant species of phytoplankton ever since the late 1970s. By 1980, the complex of freshwater and brackish-water species in the middle of the sea decreased to 23% (Pichkily, 1981). In 1985 this index was as low as 9.4% in the southern bays of the Aral Sea (Elmuratov, 1988). By that time, many species of blue-green and green algae disappeared from the phytoplankton composition. Thus, from an ecological viewpoint, the phytoplankton of the Aral Sea became more homogeneous, mainly composed of diatomaceous and dinophytic species of algae typical for brackish and sea water (Okolodkov, 1989; Dobrynin *et al.*, 1990; Dobrynin and Koroleva, 1991; Rusakova, 1995).

It is difficult to point out the degree of reduction of species diversity in phytoplankton as a result of salinization and water level drop. The point is that different authors have reported different numbers of phytoplankton species. According to the data published by Khusainova (1958), there were 67 phytoplankton species in the Aral Sea, and according to Zenkovich (1963), there were only 39 species. The greatest number, including 306 species, varieties, and forms of phytoplankton, was given by Pichkily (1970) for the mid-sea, estuaries of the Syrdarya and Amudarya, as well as for the desalinized southern bays. Thus, the initial number of species of planktonic algae in the Aral Sea prior to its anthropogenically induced salinization and water level drop remains unknown. The values of average summer biomass of phytoplankton prior to the start of the Sea's desiccation can be given with a greater degree of certainty. In the mid-1960s, the average value for the water body in summer months varied from 0.5 to 2.6 g/m³.

In the late 1980s, when salinity exceeded 28 g/l, there was an apparent relative stability of the Aral phytoplankton. This view was supported by data obtained in the area of the Island of Barsakelmes in 1989–90 and in Butakov Bay in 1991. In the former case, 52 species of phytoplanktonic algae were found at the salinity level of 28–29 g/l (Dobrynin *et al.*, 1990); in the latter case, 50 species were found at higher salinity values of 35–36 g/l (Dobrynin and Koroleva, 1991). In the same years and in the same areas, the average summer biomass of the phytoplankton varied from 0.22 to 6.83 g/m³ (Dobrynin *et al.*,

1990; Dobrynin and Koroleva, 1991; Orlova and Rusakova, 1995; Rusakova, 1995).

The phytobenthos was, evidently, more affected by the Aral Sea level drop than by the increase in salinity. Already in the late 1970s, the rapid and complete drying up of the Sea's shallow-water areas and the disappearance of the branched deltas of the Syrdarya and Amudarya resulted in the widespread loss of flowering plants. The famous fields of reed (*Phragmites australis*) of the Aral Sea were most affected. In a relatively short period of time, no more than 10–15 years, they became completely extinct.

By the mid-1980s, only one species of flowering plants (*Zostera noltei*) was still widespread in the sea. The drying up of large seabed areas caused the loss not only of flowering plants but of many species of algae, particularly *Chara*. According to Dobrokhotova (1971), the water milfoil and *Vaucheria*, plants most sensitive to salinization, were the first to disappear. The *Chara* existed until the salinity exceeded 25–26 g/l. Thus, the species diversity of phytobenthos decreased more then fourfold. At present, only 7 species are found, instead of the 32 species mentioned by Zenkovich (1963) for this sea. They include two species of higher aquatic plants (*Ruppia cirrhosa, Zostera noltei*), three species of green filament algae (*Chaetomorpha linum, Cladophora glomerata, Cladophora fracta*), two species of green microphytic algae (*Rhizoclonium hieroglyphicum, Enteromorpha intestinalis*) (Zhakova, 1995). However, the amounts of the last two species was insignificant.

Unfortunately, it is practically impossible to trace the dynamics of changes of the phytobenthos biomass that were linked to the increase in salinity and the drop in sea level, because such studies were not carried out over time. However, as is known, prior to the human-induced desiccation, the total weight of bottom plants was a little more than 0.5 kg/m². As for the biomass, the *Chara* and green algae of the *Vaucheria* type ranked first and made up 75% and 13%, respectively, of the total biomass of bottom plants. The share of phytobenthos in the total biomass of Aral Sea plants reached 90% (Karpevich, 1975). Thus, prior to the recent desiccation, the high degree of water transparency and the small average depth of the sea were responsible for the fact that the bulk of organic matter was created not by the phytoplankton but by the phytobenthos. This made the ecosystem of this water body quite different from other aquatic ecosystems. As for the present values of the phytobenthos biomass, they are available only for one northern bay and vary from 0.5 to 3.7 kg/m² (Zhakova, 1995). Unfortunately, these data cannot be taken as a basis for an assessment of the situation in the sea as a whole. However, we may assume with a great degree of certainty that at present (1995) the average summer biomass of the phytobenthos in the Aral Sea has decreased significantly.

In assessing creeping environmental changes in the Aral Sea, brief mention should be made of the trophic condition of this water body. In the opinion of Karpevich (1975), prior to the beginning of the human-induced desiccation of the Aral Sea, the trophic condition was characterized by low productivity. This was caused by the specific structure of the first trophic level (producers), which manifested itself in the primary development of the macrophytobenthos and semi-submerged weeds in the shallow-water areas of the sea and in the deltas of the Syrdarya and Amudarya. As mentioned earlier, the share of the phytoplankton in the total biomass of plants in the sea was less than 10%, while that of the phytobenthos was over 90%. In the early 1960s, the average biomass of higher aquatic plants and macroweeds in the sea was about 100 g/m^2, and the biomass of the phytoplankton was between 0.5–2.6 g/m^3. Such a structure of the phytocommunity of the sea and, hence, the low rate of primary production and the cycle of organic matter were related to the regime of inputs of biogenes. Suspended nutrients settled mainly in the delta areas of the Syrdarya and Amudarya and, therefore, were available for the phytobenthos, whereas the amount of dissolved biogenes available for the phytoplankton was insignificant (Karpevich, 1975). According to Novozhilova (1973), prior to the beginning of the recent desiccation, the Aral Sea could be characterized as an oligotrophic or slightly meso-trophic water body with the Syrdarya and Amudarya deltas being the most productive mesotrophic areas.

The development of irrigation farming contributed to the increase in the input of biogenes into the sea through the river flow and, primarily, to the increase of phosphates from fertilizers washed off the fields (Alekin and Lyakhin, 1984). From the late 1980s, the content of mineral phosphorus in water became relatively stable and did not exceed 14–30 µg/l (Sviridova, 1990; Tsytsarin, 1990). However, these values were more than tenfold higher than similar indices prior to the beginning of the drop in water level and increase in salinity. According to Blinov (1956) and Novozhilova (1973), in those years, the concentration of mineral dissolved phosphorus in the water layer was 1–4.2 µg/l, and in the near-bottom layers it dropped to zero with considerable quantities of silicic acid and carbonates.

The conservation data of the late 1980s to the early 1990s (Dobrynin *et al.*, 1990; Dobrynin and Koroleva, 1991; Orlova, 1993, 1995) reveal an increase in the primary production rate of the phytoplanktonic community and the sub-stitution of the trophic condition in the northern water area of the Aral Sea by the α– β mesotrophic one. Because of the lack of information, it is difficult to assess the situation in the southern part of the Aral Sea. However, the water level drop and the transport of biogenes by way of dust storms allow us to ascertain a growth of trophicity in this part of the Aral Sea.

Although data from continuous observations of variations in trophic conditions in the Aral Sea are not available, we shall try to reproduce in brief its possible dynamics. Most likely, from the early 1960s to the mid-1970s, the trophic condition of the Aral Sea started changing from oligotrophic to mes-otrophic, as a result of the increased input of mineral fertilizers washed away from irrigation fields and transported to the sea. However, in the late 1970s and early 1980s, when the water flow to the sea of the Amudarya and Syrdarya practically ceased because of intense development of irrigation farming in the region, the input of biogenic elements with the streamflow drastically decreased and the water body soon became oligotrophic. In the early and mid-1980s the trophicity of the Aral Sea was, evidently, at a minimum for the entire period of observation in the twentieth century. This happened not only as a result of a reduction in the input of biogenes with river flow but also as a result of restructuring in the phytocommunity. As mentioned above, in those years, many species of freshwater and brackish-water plants disappeared and were substituted by sea-water and hyperhaline forms. Extremely high water transparency (up to 15–20 m according to Secchi's disk) provided the indirect evidence of a high degree of Aral Sea oligotrophicity (Aladin and Kotov, 1989).

In the late 1980s, river flow to the Aral Sea increased slightly, resulting in an increase of the biogene input. At the same time, the number of days with dust storms in the area adjoining the Aral Sea increased, resulting in more biogenes being transported to the sea by wind action. According to Subbotina (1990), the number of days in 1995 with dust storms and wind drifts in the area of the town of Aralsk reached 110. This means that the eolian factor has become more important. Noting that the species composition was evidently at a new level by that time, the sea again showed a tendency toward the transformation from an oligotrophic to a mesotrophic condition. This tendency was particularly evident in the northern water area of the Aral Sea, as shown in the data cited above.

In conclusion, mention should be made of one more phenomenon which greatly affected the Aral Sea ecosystem, the division of the whole Sea's water area into two independent parts in 1989–90. In those years two lakes were formed as a result of the drop in sea level: the Small Aral Sea in the north and the Large Aral Sea in the south. From that moment on, the increase of salinity and decrease in sea level were observed only in the Large Aral Sea. Meanwhile, in the Small Aral Sea, there was a rise and a stabilization of the water level as well as a decrease in salinity. This resulted from the fact that evaporation in the Small Aral Sea was less than the total input, which involved a combination of Syrdarya water, precipitation, and groundwater flow. The rise in the water level in the northern Small Sea resulted in the flow of excess water to the southern Large Aral Sea. As the southern part continued to dry, the water level in the northern sea increased; the hydraulic gradient of the channel between them increased and caused the turbulent flow from the Small to the Large Aral Sea. By the spring of 1992, the flow through the channel from north to south amounted to 100 m³/sec. In the summer of 1992, the channel was dammed in order to hold water in the Small Aral Sea. This cofferdam lasted for about one year, and was then partially destroyed by a spring flood in 1993.

For the short period of the cofferdam's existence, the water level in the Small Aral Sea increased by more than 1 m, contributing to the partial filling of previously dried-up bays. For example, a part of the gently sloping depression of the Large Saryshaganat Bay was filled with water, and the same thing happened with some other unnamed bays. It is also worth noting that the elimination of the flow from the north to the south prevented the Small Aral Sea from disintegration and had other positive effects as well (Aladin *et al.*, 1994*b*; Aladin and Plotnikov, 1995*a*).

After the cofferdam failure, the flow rate from the Small Aral Sea to the Large Sea was less than 100 m³/s, as it had been before the cofferdam was constructed. Most likely, the reduction in flow rate took place for several reasons. First, after the cofferdam's construction, the surrounding area was subject to heavy wind erosion. The intense drifting of sand because of wind action was observed to the south of the cofferdam. Actually, the entire area between the coast of the Large Aral Sea and its southern edge was covered by barkhans, whose heights had reached 2–3 m. These dunes noticeably strengthened the cofferdam from its southern side and after its breach hampered the formation of the flow to the Large Aral Sea. Second, the water area of the Small Aral Sea near the cofferdam was extremely shallow. As a result of wave action, numerous shoals were formed parallel to the cofferdam, thereby strengthening it from the northern side.

As mentioned earlier, in spite of the natural protection of the cofferdam from both of its sides, it was partially destroyed by floods. However, even in its partially collapsed state, the cofferdam hindered (and still hinders) the former high volume of water flow to the Large Aral Sea. In 1993 it was this partially collapsed barrier that contributed to the rise of the water level by more than 1 m and its stabilization in 1994–95 at an elevation of +40 m (possibly even +41 m), according to an absolute scale. Today, the Small Aral Sea level is at least 2–3 m higher than that of the Large Aral Sea.

Thus, at present the destinies of two seas, which had formed in 1989–90 at the site of the larger Aral Sea, are different. Because of the flow of the Small Aral Sea, its salinity will decrease from year to year. The level of the Small Sea will be stable or very slowly drop only as a result of the scouring of the existing partially collapsed cofferdam. In the case of its reconstruction or the construction of a new dam, the Small Aral Sea level may rise. As for the Large Aral Sea, its salinity will increase and its level will drop.

Since the present hydrological conditions of the Small and Large Aral seas are different, it is most likely that in the future the creeping environmental changes in these two water bodies will be also different. All negative trends of the Large Aral Sea, which were described above, will remain and its ecosystem will turn into the ecosystem typical of hyperhaline water bodies in arid zones. As for the Small Aral Sea, the development of its ecosystem will go on as if in reverse of recent trends and will represent an ecosystem typical for slightly saline lakes. Theoretically, it can even be assumed that in the case of a stable flow of the Syrdarya of about 5–7 km³/year, a complete desalinization of the Small Aral Sea will be possible within several decades.

While mentioning possible positive trends in the development of the Small Aral Sea, it should be noted with regret that the sea will go on accumulating considerable amounts of chemical pollutants carried into it by Syrdarya flow. In this connection, it would be desirable to minimize the use of chemical fertilizers, pesticides, herbicides, and defoliants on irrigated areas adjoining the sea. A reduction of pollutants would allow for more rapid and effective conservation and the subsequent rehabilitation of aquatic communities of the Small Aral Sea.

The creeping environmental changes in the Aral Sea described in this chapter have been summarized in chronological order in Table 13.1. In analyzing this table, one may conclude that this water body has gone through at least the following three critical periods. The fourth critical period may begin in the near future.

- The peak of the first crisis was observed in the mid-1960s and was associated with the acclimatization activities.
- The second crisis was observed in the mid-1970s, when the average sea-water salinity rose to more than 11–14 g/l and when aquatic organisms of freshwater origin were lost.
- The third critical period occurred in the middle and late 1980s, when average sea-water salinity exceeded 22–28 g/l. In those years many aquatic organisms of brackish-water origin were lost.

The beginning of the fourth period will be noticeable when the sea-water salinity is more than 36–42 g/l (see e.g., Aladin, 1990). The peak of the crisis will occur at a salinity level of 50–55 g/l. At such a high salinization level the existing Aral flora and fauna of marine origin will not survive. Single forms of hyperhaline invertebrates, algae, and bacteria will remain in the Aral Sea. It should be emphasized that the fourth critical period brings a threat only to the ecosystem of the Large Aral Sea and will not affect the ecosystem of the Small Aral Sea.

In assessing the ecosystem of the Aral Sea as a whole, it would be reasonable to suppose that, because of the history of this water body, its biota has gained the capacity to adapt to and to withstand both periods of salinization and sea level drop as well as periods of desalinization and sea level rise. During increasing salinization conditions, freshwater aquatic organisms endure an unfavorable period in deltaic water bodies and branches of the rivers entering the Aral Sea. During decreasing salinization conditions, brackish-water, sea-water and hyperhaline inhabitants find refuge for survival in shallow-water salinized bays.

It has clearly been proven (Kvasov, 1976; Rubanov *et al.*, 1987; Shnitnikov, 1969, 1983; Aladin and Plotnikov, 1995*b*) that the Aral Sea, as a continental water body, periodically dried up and became saline, or was filled to overflowing and became desalinized. In the past these fluctuations were caused by natural climatic variations. However, beginning in ancient times, the water level and salinity of this inland body of water were affected by human

Table 13.1 Chronology of creeping environmental changes in the Aral Sea

Period	Abiotic factors		Action	Response
	Elevation (m)	Salinity (g/l)		
1927–1934	+53–53.4	8–10	Introduction of stellate sturgeon	
1927–1934	+53–53.4	8–10	Accidental introduction with the stellate sturgeon of 2 parasites (*Nitzschia* and *Polypodium*)	
1929–1932	+53–53.4	8–10	Introduction of Caspian shad	
1934	+53–53.4	8–10		Mass death of indigenous sturgeon caused by parasites accidentally introduced with stellate sturgeon
1954–1956	+53–53.4	8–10	Introduction of 2 shrimp species and 2 mullet species	
1954–1956	+53–53.4	8–10	Accidental introduction of 6 species of goby, friar and pipefish	
1954–1959	+53–53.4	8–10	Introduction of Baltic herring	1 species of shrimp was revealed; Baltic herring was recorded in catches
1957	+53–53.4	8–10		Stellate sturgeon was recorded in catches
1958	+53–53.4	8–10		Wide-scale propagation of friar and 3 goby species
1958–1959	+53–53.4	8–10		
1958–1960	+53–53.4	8–10	Introduction of 4 opossum shrimp species	Consumption of coarse representatives of zooplankton
1959–1968	+53–51.5	9.8–11.4		Tenfold decrease of summer zooplankton biomass
1959–1967	+53–51.5	9.8–10.9		Wide-scale propagation of fine representatives of zooplankton
1960–1961	+53–53.3	9.8–10	Introduction of grass carp; introduction of silver carp	
1960–1963	+53–52.7	9.8–10.4	Introduction of *Nereis diversicolor*; introduction of *Syndosmya segmentum*	

Period				
1960–1963	+53–52.7	9.8–10.4		3 species of opossum shrimp were recorded
1961	+53.3	10		Transition of parasitic cestoid from Far Eastern fish to Aral carps
1963–1979	+52.7–46.7	10.4–15.8	Occasional introduction of Amur parasites brought with the fish from the Far East	
1963	+52.7	10.4		Grass carp was recorded in catches; silver carp was recorded in catches; *Nereis diversicolor* was revealed
1964–1973	+52.5–50.3	10.7–12.1		Eating-up and complete extinction of Aral fresh-water shrimp
1965–1967	+52.5–51.7	10.7–10.9	Beginning of adverse impact of salinity growth on roe, larvae, and fry	
1965–1970	+52.5–51.6	10.7–11.5	Drying up of spawning areas of fluvial anadromous fish; introduction of *Calanipeda aquaedulcis*	
1965–1975	+52.5–49.4	10.7–13.7		Loss of such aquatic plants as water milfoil and *Vaucheria* sensitive to salinization
1966–1967	+52.1–51.7	10.9		Threefold reduction of summer zoobenthos biomass
1967	+51.7	10.9		*Syndosmya segmentum* was recorded
1967–1970	+51.7–51.6	10.9–11.5		Mass loss of flowering plants
1969–1981	+53.3–45.5	11.1–17.6		Gradual 10–15-fold increase of average summer biomass of zoo benthos; gradual 5–6-fold increase of average summer biomass of zooplankton
1970	+51.6	11.5		Threefold reduction of fish yields
1970–1975	+51.6–49.4	11.5–13.7		Complete loss of reed fields
1971	+51.3	11.5	Beginning of adverse impact of salinity on zooplankton; beginning of adverse impact of salinity growth on zoobenthos	
1972	+50.8	12		Wide-scale development of *Calanipeda aquaedulcis*
1974–1976	+50.1–48.5	12.8–14.1		Disturbance of natural reproduction processes due to salinity growth; wide-scale development of euryhaline species of zooplankton; loss of fresh-water species of zoobenthos:

Table 13.1 (cont.)

Period	Abiotic factors		Action	Response
	Elevation (m)	Salinity (g/l)		
			Introduction of flatfish, snakehead and *Mylopharyngodon piceus*	worms, insect larvae, mollusks *Dreissena* and *Hypanis*; wide-scale development of euryhaline species of zooplankton; threefold reduction of phytoplankton number and biomass; loss of much fresh-water planktonic algae
1975–1979	+49.4–46.8	13.9–15.9		
1975–1981	+49.4–45.5	13.9–17.6		Loss of about 20 species of zooplanktonic organisms; loss of about 100 species of zoobenthos
1975–1985	+49.4–42.2	13.9–22.5		Relative stability of species composition of zooplankton
1976	+48.5	14.1		Accidentally introduced crab was recorded
1977–1985	+47.9–42.2	14.5–22.5		Marine euryhaline planktonic algae dominated
1979–1981	+46.8–45.5	15.9–17.6		Lack of fry
1981	+45.5	17.6		Loss of commercial fishery importance, cessation of fishing
1984–1986	+43.1–41.4	20.2–24.5		Loss of all commercial fish species, except for flatfish; loss of many brackish-water species of zooplankton because of growing salinity; loss of many brackish-water species of zoobenthos because of growing salinity; only marine *Zostera* was widely spread; complete loss of *Chara*
1985–1989	+42.2–39.0	22.4–28.7		Only 5 fish species survived: flatfish, Baltic herring, friar, monkey goby, and indigenous stickleback
1987–1988	+40.1–39.9	26.5–27.8	Introduction of mussel	
1987–1994	+40.1–37.5[a]	26.5–39.4[a]		Relative stability of species composition of phytoplankton numbering about 50 species

1989–1990	+39.0–38.6	28.7–30.6	2–2.5-fold increase of average summer biomass of phytoplankton
1989–1991	+39.0–38.0[a]	28.7–35.0[a]	1.5–2-fold reduction of average summer biomass of benthos as a result of its consumption by flatfish
1990–1993	+38.6–37.9[a]	28.7–37.4[a]	Fivefold reduction of number of zooplankton species; only 11 species of zooplankton survived: 5 species of rotifera, 5 species of *Calanipeda* and 1 *Podonevadne camptonyx*; tenfold reduction of number of zoobenthos species; only 15 species of zoobenthos survived: 2 species of clams, 9 species of gasteropod, 1 *Nereis diversicolor*, 1 *Palaemon elegans*, 1 *Rhithropanopeus harrisii tridentata* and 1 *Cyprideis torosa*
1990–1994	+38.6–37.5[a]	28.7–39.4[a]	More than fourfold reduction of phytobenthos species diversity; average summer biomass of phytobenthos decreased several times

Note: [a]Data on the Large Aral Sea only.

activities. Evidently, many of the present-day environmental changes in the Aral Sea Basin are not new to the region. Similar changes and their impacts had arisen earlier and repeatedly. The only thing that makes the creeping environmental changes in the Aral Sea in the twentieth century different from similar changes in the earlier centuries are the consequences of large-scale acclimatization activities, heavy chemical pollution and very high rates of water withdrawal from the Syrdarya and Amudarya to meet the demands of the irrigation farming sector.

References

Aladin, N.V., 1982: Saliferous adaptation and osmo-regulating capacity of the *Podonevadne camptonyx* 2: Species from the Caspian and Aral seas. *Zoological Journal*, **61**(4), 507–14.

Aladin, N.V., 1983*a*: Saliferous adaptation and osmo-regulating capacity of the Ostracoda from the Caspian and Aral seas. *Zoological Journal*, **62**(1), 51–7.

Aladin, N.V., 1983*b*: About the displacement of the critical salinity barrier in the Caspian and Aral seas on crustacea. *Zoological Journal*, **62**(5), 689–94.

Aladin, N.V., 1989*a*: The critical character of biological activity in Caspian Sea water with a 7–11‰ salinity level and of the Aral Sea water with a 8–13‰ salinity level. In: *Biology of the Salinity Hyperhaline Water: L. Proceed. Zoologic. Inst.*, **196**, 12–21.

Aladin, N.V., 1989*b*: Zooplankton and zoobenthos of the coastal water of the Barsakelmes Island (Aral Sea). In: *Hydrobiological Problems of the Aral Sea: L. Proceed. Zoologic. Inst.*, **199**, 110–13.

Aladin, N.V., 1989*c*: Morphology of the osmo-regulating organs in the Cladocera with special reference to the Cladocera from the Aral Sea. 2nd Symposium on Cladocera, *Tatranska Lomnica*, 5.

Aladin, N.V., 1990: General characteristics of the Aral Sea hydrobionts from the viewpoint of osmo-regulation physiology. In: *The Contemporary State of the Aral Sea Under Conditions of Progressing Salinization: L. Proceed. Zoologic. Inst.*, **223**, 5–18.

Aladin, N.V., 1991*a*: Study of the influence of water salinity on hydrobionts in the separating bays of the Aral Sea. In: *The Modern State of the Separating Bays of the Aral Sea: L. Proceed. Zoologic, Inst.*, **237**, 4–13.

Aladin, N.V., 1991*b*: Thanatocoenoses of separating bays and gulfs of the Aral Sea. In: *The Modern State of the Separating Bays of the Aral Sea: L. Proceed. Zoologic. Inst.*, **237**, 60–4.

Aladin, N.V., 1991*c*: Large salt lakes of the USSR and their modern status. *Proceedings of the VI Congress VGBO, Part I*. Murmansk, 32–3.

Aladin, N.V., 1994*a*: Changes of biodiversity in the Aral Sea during the Holocene. NATO–Meeting: Tashkent, 31–42.

Aladin, N.V., 1994*b*: Adaptation and changes of biocenoses in the Aral Sea from historic times to the present: changes of the hydrogeological record in the Aral Sea basin. Aral Sea project seminar, UNESCO: Tashkent, 7–10.

Aladin, N.V., 1994*c*: The conservation ecology of the Podonidae from the Caspian and Aral Seas. *Hydrobiologia*.

Aladin, N.V. and N.I. Andreyev, 1984: Salinity of the Aral Sea impact on changes in fauna structure of the *Podonevadne camptonyx*. *Hydrobiological Journal*, **20**(3), 23–8.

AAladin, N.V. and I.A. Boomer, 1993: Lithopogical and micropaleontological analysis of the bottom sediments with the preliminary determination on the paleosalinity of the Aral Sea. In: *Ecological Crisis in the Aral Sea: L. Proceed. Zoologic Inst.*, **250**, 6–20.

Aladin, N.V. and I. Boomer, 1994: Paleolimnology of the Aral Sea over the past three million years. *Abstracts of the Vi International Symposium on Saline Lakes*: Beijing, **5**, 125.

Aladin, N.V. and D.O. Eliseev, 1991: The Aral Sea, USSR: A case study of a wetland conservation Issue. *Proceedings of IIth Int. Symp. Wetland and Waterfowl Conservation in South and West Asia*: Karachi, 11–12.

Aladin, N.V. and A.A. Filippov, 1993: On the viability of eggs of *Artemia salina* and *Moina mongolica* from bottom sediments from dried-up bays of the Aral Sea. In: *Ecological Crisis in the Aral Sea: L. Proceed. Zoologic Inst.*, **250**, 114–20.

Aladin, N.V. and S.V. Kotov, 1989: Natural status of the Aral Sea ecosystems and its change due to anthropogenic impacts. In: *Hydrobiological Problems of the Aral Sea: L. Proceed. Zoologic. Inst.*, **199**, 4–24.

Aladin, N.V. and I.S. Plotnikov, 1993: Large saline lakes of the former USSR: a review. *Hydrobiologia*, **267**, 1–12.

Aladin, N.V. and I.S. Plotnikov, 1995a: On the problem of the possible conservation and rehabilitation of the Small Aral Sea. In: *Biological and Environmental Problems of the Aral Sea and Aral Region: L. Proceed. Zoologic. Inst.*, **262**(I), 3–16.

Aladin, N.V. and I.S. Plotnikov, 1995b: Changes in Aral Sea level: paleolimnological and archeological evidence. In: *Biological and Environmental Problems of the Aral Sea and Aral Region: L. Proceed. Zoologic Inst.*, **262**(I).

Aladin, N.V. and W.T.W. Potts, 1992: Changes in the Aral Sea ecosystem during the period 1960–90. *Hydrobiologia*, **237**, 67–79.

Aladin, N.V. and W.D. Williams, 1992: Recent limnological and biodiversity changes in the Aral Sea and their conservation significance. *XXV SIL Int. Congress*: Barcelona, 655 pp.

Aladin, N.V. and W.D. Williams, 1993: Recent changes in the biota of the Aral Sea, Central Asia. *Verh. Internat. Verein. Limnol.*, **25**, 790–2.

Aladin, N.V., I.S. Plotnikov, and A.A. Filippov, 1992a: Changes of the Aral Sea ecosystem as a result of the anthropogenic impact. *Hydrobiological Journal*, **28**(2), 3–11.

Aladin, N.V., I.S. Plotnikov, and A.A. Filippov, 1992b: Hydrobiological environmetrics for the Aral Sea in XXth Century. *9th Int. Conference on Statistical Methods for the Environmental Sciences*: Espoo, 113.

Aladin, N.V., M.I. Orlova, A.A. Filippov, I.S. Plotnikov, A.O. Smurov, D.D. Piryulin, O.M. Rusakova, and L.V. Zhakova, 1994a: Changes in the biota of the Aral Sea. In: *Ecology of the Aral Sea*. Springer-Verlag.

Aladin, N.V., I.S. Plotnikov, and W.T.W. Potts, 1994b: Aral Sea desiccation and possible ways for the rehabilitation and conservation of its North part. *Int. J. Environmetrics*.

Alekin, O.A. and Yu. I. Lyakhin, 1984: Ocean chemistry. L., 345.

Andreyev, N.I., 1981: Some data about influence of water salinity on Aral Sea invertebrates fauna. In: *Biological Basis of the Fish Economy of the Central Asia and Kazakhstan Water Bodies*. Frunze: Ilim, 219–20.

Andreyev, N. I., 1989: Aral Sea zooplankton under initial period of its salinization. *L. Proceed. Zoologic Inst.* AS USSR, 199, 26–52.

Andreyev, N. I. and S. I. Andreyeva, 1981: Some regularity changes of the Aral Sea invertebrate fauna. In: *IV Congress of the VGBO. Part I.* Kiev: Naukova Dumka, 50–1.

Andreyev, N. I., S. I. Andreyeva, A. A. Filippov, and N. V. Aladin, 1992*a*: The fauna of the Aral Sea in 1989: Part 1. The benthos. *Int. J. Salt Lake Res.,* 1, 103–10.

Andreyev, N. I., I. S. Plotnikov, and N. V. Aladin, 1992*b*: The fauna of the Aral Sea in 1989: Part 2. The zooplankton. *Int. J. Salt Lake Res.,* 1, 111–16.

Beklemishev, V. N., 1922: New data about Aral Sea fauna. *Russian Hydrobiological Journal,* 1(9–10), 276–89.

Bening, A. L., 1935: Hydrological materials for the composition of the Aral Sea fishery map. *Proceeding of the VNIRO Aral Division,* 4, 138–95.

Berg, L. S., 1908: Aral Sea: Scientific results of the Aral expedition. *Izvestiya Turkestanskogo otd. Imper. Rossiyskogo Geograf. Obsch.* St. Petersburg, 5(9), 508.

Blinov, L. K., 1956: Hydrochemistry of the Aral Sea. *L.,* 252.

Bortnik, V. N., 1990: Water balance. In: *Hydrometeorology and Hydrochemistry of USSR Seas: V. VII, Aral Sea. L. Hydrometeoizdat,* 34–42.

Butakov, A. I., 1853: Information on the expedition sent to describe the Aral Sea in 1848. *Vestnik Rossiyskogo geograficheskogo obschestva,* Part VII, Section VII, 1–9.

Dogiel, V. A. and A. S. Lutta, 1937: Destruction of the Aral spinefish in 1936. *Rybnoe khoziaystvo,* 12, 26–7.

Dobrynin, E. G. and N. G. Koroleva, 1991: Production and microbiological processes in Butakov Bay (the Aral Sea). *L. Proceed. Zoologic. Inst. of the Academy of Sciences of the USSR,* 237, 49–59.

Dobrynin, E. G., N. G. Koroleva, and T. M. Burkova, 1990: Estimation of the ecological state of the Aral Sea near Barsakelmes Island. *L. Proceed. Zoologic. Inst. of the Academy of Sciences of the USSR,* 223, 31–5.

Ellis, W. S., 1990: The Aral: A Soviet sea lies dying. *National Geographic,* Feb., 73–92.

Elmuratov, A. E., 1977: Composition and distribution phytoplankton in the South Aral Sea under conditions of changing regime. In: *Structure of Aquatic Communities of the Amudarya Downstream,* Tashkent: FAN Publishing House, 25–34.

Filippov, A. A., 1995: Macrozoobenthos of the inshore zone of the north Aral Sea under modern polyhaline conditions: quantity, biomass and spatial distribution. *L. Proceed. Zoologic. Inst., Russian Academy of Sciences,* 262.

Glantz, M. H., A. R. Rubinstein, and I. S. Zonn, 1993: Tragedy in the Aral Sea basin: Looking back to plan ahead? *Global Environmental Change,* 3(2), 174–98.

Karpevich, A. F., 1953: Relation of the northern Caspian Sea clam and the Aral Sea clam to changes of salinity. *Avtoreferat dissertatsii doktora biologich: Nauk, M.,* 20.

Karpevich, A. F., 1958: Survival, reproduction and respiration of *Mesomysis kowalevskyi* (Paramysis lacustris kowalevskyi Czern.) in the salinized water bodies of the USSR. *Zoological Journal,* 37(8), 1121–35.

Karpevich, A. F., 1975: Theory and practice of the acclimatization of aquatic organisms. *M. Pischevaya promyshlennost',* 432.

Khusainova, N. Z., 1958: *Biological Features of Some Mass Benthic Invertebrates of the Aral Sea.* Alma-Ata: Kazakh University, 116 pp.

Kortunova, T. A., 1975: On changes in zooplankton of the Aral Sea in 1959–68. *Zoological Journal,* 54(5), 657–69.

Kosarev, A. I., 1975: *The Hydrology of the Caspian and Aral Seas.* Leningrad: MGU, 272 pp.

Kvasov, D. D., 1976: The reasons for the cessation of flow in the Uzboi and the Aral Sea problem. *Problems of Desert Development,* **6,** 24–9.

Lim, R. M. and E. L. Markova, 1981: Results of the introduction of sturgeon and flatfish into the Aral Sea. *Rybnoe khoziaystvo,* **9,** 25–6.

Lukonina, N. K., 1960: Zooplankton of the Aral Sea. *Proceedings VNIRO,* **43**(1), 177–97.

Meisner, V. I., 1908: Microscopic representatives of the Aral Sea water fauna and the inflowing rivers related to the conditions of their distribution. *Izvestiya Turkestanskogo otd. Rossiysk. geograf. obsch.,* **4**(8), 1–102.

Micklin, P., 1991: *The Water Management Crisis in Soviet Central Asia.* The Carl Beck Papers in Russian and East European Studies, No. 905. University of Pittsburgh Center for Russian and East European Studies. 120 pp.

Novozhilova, M. N., 1973: Microbiology of the Aral Sea. Alma-Ata, 160.

Okolodkov, Yu, B., 1989: Phytoplankton of the Barsakelmes Island coastal waters (Aral Sea). *L. Proceed. Zoologic, Inst.,* **199,** 103–9.

Orlova, M. I., 1993: Materials for general valuation of production and destruction processes in the coastal zone of the northern part of the Aral Sea: Part 1. The results of field observations and experiments in 1992. *Proceed. Zoologic, Inst., Russian Academy of Sciences,* **250,** 21–37.

Orlova, M. I., 1995: Materials for the general evaluation of the production and destruction processes in the coastal zone of the northern part of the Aral Sea: Part 2. About some properties of functioning of the ecosystem in the district of Syrdarya's delta and in shallow areas of the sea's bays. *Proceed. Zoologic, Inst., Russian Academy of Sciences,* **262,** 47–64.

Orlova, M. I. and O. M. Rusakova, 1995: Structural and functional characteristics of the phytoplanktonic community in the district of Tastubec coge in September 1993. *L. Proceed. Zoologic, Inst., Russian Academy of Sciences,* **262,** 208–30.

Osmanov, S. O., 1962: Parasitic fauna of fish which were acclimatized in the Aral Sea. *Vestnik Karakalpakskogo filiala AN UzSSR,* No. 2, 18–22.

Osmanov, S. O., E. A., Arystanov, K. K. Ubaydullaev, O. Yu. Yusupov, and A. T. Teremuratov, 1976: *Questions of Aral Sea Parasitology.* Tashkent: FAN, 200 pp.

Pichkily, L. O., 1970: Composition and dynamic of phytoplankton of the Aral Sea. *Autoreferat candidat. dissert. L.,* 49 pp.

Pichkily, L. O., 1981: *Phytoplankton of the Aral Sea under Anthropogenic Impacts (1957–1980).* Kiev, 142–4.

Plotnikov, I. S., 1991: Zooplankton of Butakov Bay (the Aral Sea) in September 1990. *L. Proceed. Zoologic. Inst.,* **237,** 34–9.

Plotnikov, I. S., 1993*a*: Aral Sea zooplankton in May 1991 (an example of the coastal zone of the Barsakelmes Island). *L. Proceed. Zoologic. Inst.,* **250,** 38–41.

Plotnikov, I. S., 1993*b*: Aral Sea zooplankton (Small Aral Sea and Butakov Bay) in September 1991. *L. Proceed. Zoologic. Inst.,* **250,** 42–4.

Plotnikov, I. S., 1993*c*: Aral Sea zooplankton in 1992. *L. Proceed. Zoologic. Inst.,* **250,** 46–51.

Plotnikov, I. S., 1995: Zooplankton of the coastal waters of the northern part of the Aral Sea under modern polyhaline conditions. *Avtoreferat candid. dissert.* St. Petersburg, 23 pp.

Plotnikov, I. S., N. V. Aladin, and A. A. Filippov, 1991: Past and present fauna of the Aral Sea fauna. *Zoological Journal,* **70**(4), 5–15.

Rubanov, I.V., D. P. Ishnijarov, M. A. Baskakova, and P. A. Chistyakov, 1987: *Geology of the Aral Sea*. Tashkent: FAN Publishing House, 248 pp.

Rusakova, O.M., 1995: Concise characteristic of the Aral Sea phytoplankton composition in the spring and autumn 1992. *L. Proceed. Zoologic. Inst.*, **262**, 195–207.

Shnitnikov, A.V., 1969: Century-long variability of the components of humidity. *L.*, 245.

Shnitnikov, A.V., 1983: The Aral Sea in the Holocene and the natural trend of its changes. In: *Paleogeography of the Caspian and Aral Seas in the Cainozoic: Part II.*, 106–18.

Sviridova, I.V., 1990: Regime of biogenic matter. In: *Hydrology and Hydrochemistry of the USSR Seas: Aral Sea L.*, **7**, 114–40.

Subbotina, O. I., 1990: Brief climate characteristic of the Aral Sea. In: *Hydrology and Hydrochemistry of the USSR Seas: Aral Sea L.*, **7**, 12–28.

Trusov, V. Z., 1947: Biological and experimental measures on the reproduction of *Acipenser nudiventris*. *Proceedings of the Laboratory of Fishery Basic Principles*, **1**, 186–200.

Tsytsarin, A.G., 1990: The modern state of the elements of the hydrological regime of the Aral Sea. *Proceed. GOIN Academy of Sciences of the USSR*, **62**, 72–92.

Yablonskaya, E.A., T. A. Kortunova, and G. B. Gavrilov, 1973: Long-term changes of benthos. *World Ocean Classification: Food Fish of the USSR's Southern Seas and Its Use*, **80**(3), 147–58.

Zenkovich, L.A., 1963: *Biology of the USSR Seas*. Moscow: Nauka, 739 pp.

Zernov, S.A., 1903: About zooplankton of the Aral Sea collected by L. S. Berg in 1900. *Izvestiya turkestanskogo otdel. Rossiyskogo geogr. obsch. Issue 3: Scientific results of the Aral expedition.*, 1–42.

Zhakova, L.V., 1995: Notes about the structure, distribution and biomass of the community of higher aquatic plants and green filament algae from the Bolshoy Sary-Chaganak Bay of the Aral Sea. *L. Proceed. Zoologic. Inst. of the Russian Academy of Sciences*, **262**, 47–64.

Zholdasova, I.M., L. P. Pavlovskaya, L. N. Guseva, and E. Adenbaev, 1992: *Platichthys flesus luscus* in the Aral Sea. *Vestnik Karakalpakskogo otdel. Academy of Sciences of the Uzbekistan Republic*, **3**, 23–30.

INDEX

Abbas, 207
Academy of Sciences
  Turkmen, 230
  USSR, 143, 151, 230
  UzSSR, 162
Adzhibai, 58, 80, 207
aerosols, 84
Agenda 21, 17
air
  and pesticides, 142–3
  temperature, changes in, 90–2, 95
Akbashli, 72, 77
Akdar'inskaya plain, 83
Akdarya River, 72, 111
Akhchadarya plain, 34
Akpetkinski archipelago, 80, 107, 124
Aksai-Kuvandarya, 72
Akshatau Lake, 77
algae, 209, 268–9, 273
Almaty, 144
aluminum, 139
amudar stone loach, 213
Amudarya
  -Bukharsky canal, 31, 166
  -Murgab section, Karakum Canal, 228
amur parasites, 262
Andizhan reservoir, 166
anemia, 130, 133, 135, 144, 146
antelope, saiga, 257
aral
  barbel, 212–13
  carp, 208, 262
  salmon, 265
  shemaya, 218
  spineback, 265
  spinefish, 262, 264
  stickleback, 208
  trout, 212
  white-eye, 218
Aral Sea
  basin
and CEPs (creeping environmental problems),
      182–7
    description of, 191
    history of, 204
    irrigation, 167, 191
    regions of, 29–30
  crisis periods, 273
  Governmental Commission, 180
  ichthyofauna, 210

level
  description of, 11
  history of, 48–60
  regime, 164
  region, map, 1
description of, 8–16, 21–2, 30–1, 47
division of, 271–2
importance of, 21–2
Aral-Paigambar reserve, 216
Aralsk, 13, 55, 60, 94, 216, 271
Armenia, 158
Arnasai
  depression, 33
  lake, 31, 33, 37
aryks, 137
Ashgabat, 229
  -Erbent waterway, 225
Asian donkey, 105–6
atmospheric
  circulation epochs, 89–90
  pollution index, 254
  protection, 149
Atrek river, 227
Ayakkum station, 124
Azerbaijan, 158
Azov, Sea of, 209, 262–3

bacteria, 273
  and water quality, 139–41
Badai-Tugai reserve, 106, 121, 216
Badhyz, 227
Baidal's classification, 92–3
Baikal, lake, 149
Balkash-Ili, 217
Balkhan ridge, 227
Balkhash, lake, 37, 214
baltic herring, 209, 211, 262–3, 265
Baltic Sea, 149, 257
balykchi, 212
barbel, 208, 265
  aral, 212–13
  turkestan, 217–18
barium, 139
barkhans, 72, 83, 272
Barsakelmes island, 49, 268
bastard sturgeon, 211–14
  history of, 215–16
bay
  Butakov, 268
  Dzhiltyrbas, 35, 58

coelenterata, 262
cofferdams, 199, 271–2
Cold War, 1
collapse of fishery, 265
collector drainage water, 240–1, 243
Committee on Environment Protection,
    Karakalpak, 121
Communist Party, 161, 175
copepoda, 263
copper, 144, 207
cotoran, 206
cotton, 146, 179, 238, 246–8
  and air temperature, 91
  and changes in planting patterns, 192
  and chemicalization, 255–6
  and military use, 161
  and post-Soviet period, 181
  production, 9, 14, 157–8, 180
    Brezhnev period, 165–73
    Khrushchev period, 161–5
    Lenin period, 158
    Stalin period, 159–61
Council
  of Ministers of USSR, 60, 149, 151, 161, 169,
    174, 181
  of People's Commissars, 158–9
CPSU (Communist Party of Soviet Union),
  164–5
  Central Committee, 161, 165, 169–70, 174, 181
crab, 209, 268, 267
Crimea, 158
Crimean hemorrhagic fever, 130
crisis
  awareness, Karakum Canal, 16, 231
  periods of the Aral Sea, 273
crop areas, Central Asia, 145
crop rotation, 256
  failure, 170
cyclones, 88

dams, 192, 199
Dashkovuz, 23, 140 see also Tashauz
DDT, 141–2, 206–7, 256–7
deer, 105–6
defoliants, 273
delta
  ecosystems, decline in, 12
  plains, desertification of, 69–70
demineralization project, 241–2
Dengizkol' depression, 31
desert, 191
  ecosystems, 105
  expansion, 252
  moss, 124
desertification, 34, 66–84, 146, 199, 253
  and remote sensing, 83–4
  and waterlogging conditions, 75
  causes of, 36
  description of, 66–7
  stages of, 72–3

diatomaceous algae, 268
dibromchlormethane, 139
diphtheria, 131
diseases, 130–3
  human, increase in, 13
djeiran, 124
Domalak lake, 220
Don river, 262
donkey, Asian, 105–6
drainage collector network, 240
drinking water
  laws to improve, 148–9
  pollution, 135
dust storms, 94, 136, 143, 146, 204, 271
  impacts of, 10–11
dysentery, 143
Dzanadarya, 34
Dzhiltyrbas bay, 35, 58

Earth Summit, 17
eastern
  bream, 220
  carp, 219
echynococcosis, 132
ecological changes, periods of, 28–9
economics and fish populations, 204–20
ecosystem
  changes along main coast, 122–4
  classifications of, 102
  deterioration, 255
  stability, 104–6
egret, 121
eluvial soil, 73–4
enteric bacillus, 137
environmental
  changes in Uzbekistan, 245–60
  conditions, 146
  impacts of Karakum Canal, 230–1
  issues in Priaralye region, 253
epizootic diseases, 110
Eskidaryalyk channel, 77
esophageal cancer, 132
euryhaline species, 266–7

failure of water supply to sea, 172
Faizabadkal, 214
Far East, 262
Farhad reservoir, 216
Farkhadskoye, 162
Fergana (also Ferghana) Valley, 30–1, 33, 257
Ferghanasky oasis, 31
fertilizer and pesticide use, 12, 15, 141, 171, 204,
  248, 256, 270, 273
  in Uzbekistan, 251–4
fish, 261–73
  and economics, 204–20
  and parasites, 264
  and pesticides, 142–3
  and salinity, 265–9
  catch, 110–11, 200

pneumonia, 133
poliomyelitis, 131
politics and CEPs, 157–81, 245–6
pollution, 206–7, 254–6
    bacterial, 137
    chemical, 273, 278
    index, atmospheric, 254
post-Soviet period and cotton, 181
postage stamp, Russian, 2
pre-Kopetdag region, 225, 227
precipitation, trends in, 90–2, 136
Priaralye region, 204, 252, 257
    description of, 100–4
    ecosystems, 100–25
    environmental issues in, 253
    vegetation in, 101–2
protein-vitamin deficiency, 152
psammophytes, 33, 80, 83, 122, 235
public health, 128–54
    thresholds in, 151–4
Pyandj
    river, 216
    tributary, 212

Raushan, 35
Red Books of endangered species, 105, 211–12,
    216–18, 255
red wolf, 255
reed vegetation, 75, 117–18, 199, 269
remote sensing, 36, 44, 83–4, 199
Research
    Center of Regional Problems of Nutrition,
        144
    Institute of Epidemiology, 143
reserve
    Aral-Paigambar, 216
    Badai-Tugai, 106, 121, 216
    Chatkal, 257
    Kyzylkumsky, 216
    Nurumtubek, 116, 121
    Sudochinsky, 121
reservoirs, construction of, 192
resolutions on land improvement, 176–7
rice, 158, 165–7, 171, 179, 192
Richter Scale, 108
rickets, 130, 146
Rio de Janeiro, 17
river
    diversion, Siberian, 169, 174, 180, 200
    fish species, extinction of, 211
    flow curves, annual, 197
    flow, changes in, 191–201
    pollution, 206–7, 255
    runoff, annual, 205
    water, 34, 192
rodents, 110, 253
roe, 262, 265
Roosevelt, President, 160
Rotifera, 263, 266
Russian

Academy of Sciences, 106
    see also Academy of Sciences, USSR
Civil War, 195
sturgeon, 209
-American Commercial Treaty, 157
Rybat, 80

sabrefish, 211
saiga antelope, 257
salinity, 13, 68, 194, 253
    alteration in, 47–64, 240
    and fish, 265–9
    and natural climate variations, 273
    causes of, 50–1
    effects on zooplankton, 265–6
    of lands in Uzbekistan, 247
salinization
    of soil, 37, 103, 122, 136, 232, 239
    processes in Turkmenistan, 242
salmon, 212, 265
salmonellosis, 131
salt
    accumulation, 238
    budget, average annual values, 54, 57
    content, changes in, 258
    formations, 80
    removal from seabed, 143
    sedimentation, 51, 59
    storms, 84, 146, 204
    transport, 242
saltworts, 68, 74, 77, 80
Samanbai, 198
Samarkand, 159, 171
    oasis, 31
sand viper, 255
SANIGMI, 84
Sanitary-Epidemiological Service, Kyzyl-Orda
    region, 152
Sariyazinskoye, 162
Sarvas bay, 220
Sarykamysh
    depression, 26, 35, 51
    lake, 27, 35, 37, 48, 121, 194
Saryshaganak bay, 49, 272
sawbelly, 219
saxaul, black, 35
sea level (Aral)
    changes, description of, 53–6, 157
    decline, 10–11, 265
Sea of Azov, 209, 262–3
Secchi's disk, 271
sedimentation, salt, 59
sewerage
    networks, 248
    system, inadequacy of, 136–7, 174
shad, Caspian, 262
Shagyrlyk, 77
Shaumyan concept, 160
sheat-fish, 212
sheep breeding, 232